Food Processing Operations Modeling

FOOD SCIENCE AND TECHNOLOGY

A Series of Monographs, Textbooks, and Reference Books

1. Flavor Research: Principles and Techniques, *R. Teranishi, I. Hornstein, P. Issenberg, and E. L. Wick*
2. Principles of Enzymology for the Food Sciences, *John R. Whitaker*
3. Low-Temperature Preservation of Foods and Living Matter, *Owen R. Fennema, William D. Powrie, and Elmer H. Marth*
4. Principles of Food Science
   Part I: Food Chemistry, *edited by Owen R. Fennema*
   Part II: Physical Methods of Food Preservation, *Marcus Karel, Owen R. Fennema, and Daryl B. Lund*
5. Food Emulsions, *edited by Stig E. Friberg*
6. Nutritional and Safety Aspects of Food Processing, *edited by Steven R. Tannenbaum*
7. Flavor Research: Recent Advances, *edited by R. Teranishi, Robert A. Flath, and Hiroshi Sugisawa*
8. Computer-Aided Techniques in Food Technology, *edited by Israel Saguy*
9. Handbook of Tropical Foods, *edited by Harvey T. Chan*
10. Antimicrobials in Foods, *edited by Alfred Larry Branen and P. Michael Davidson*

11. Food Constituents and Food Residues: Their Chromatographic Determination, *edited by James F. Lawrence*
12. Aspartame: Physiology and Biochemistry, *edited by Lewis D. Stegink and L. J. Filer, Jr.*
13. Handbook of Vitamins: Nutritional, Biochemical, and Clinical Aspects, *edited by Lawrence J. Machlin*
14. Starch Conversion Technology, *edited by G. M. A. van Beynum and J. A. Roels*
15. Food Chemistry: Second Edition, Revised and Expanded, *edited by Owen R. Fennema*
16. Sensory Evaluation of Food: Statistical Methods and Procedures, *Michael O'Mahony*
17. Alternative Sweeteners, *edited by Lyn O'Brien Nabors and Robert C. Gelardi*
18. Citrus Fruits and Their Products: Analysis and Technology, *S. V. Ting and Russell L. Rouseff*
19. Engineering Properties of Foods, *edited by M. A. Rao and S. S. H. Rizvi*
20. Umami: A Basic Taste, *edited by Yojiro Kawamura and Morley R. Kare*
21. Food Biotechnology, *edited by Dietrich Knorr*
22. Food Texture: Instrumental and Sensory Measurement, *edited by Howard R. Moskowitz*
23. Seafoods and Fish Oils in Human Health and Disease, *John E. Kinsella*
24. Postharvest Physiology of Vegetables, *edited by J. Weichmann*
25. Handbook of Dietary Fiber: An Applied Approach, *Mark L. Dreher*
26. Food Toxicology, Parts A and B, *Jose M. Concon*
27. Modern Carbohydrate Chemistry, *Roger W. Binkley*
28. Trace Minerals in Foods, *edited by Kenneth T. Smith*
29. Protein Quality and the Effects of Processing, *edited by R. Dixon Phillips and John W. Finley*
30. Adulteration of Fruit Juice Beverages, *edited by Steven Nagy, John A. Attaway, and Martha E. Rhodes*
31. Foodborne Bacterial Pathogens, *edited by Michael P. Doyle*
32. Legumes: Chemistry, Technology, and Human Nutrition, *edited by Ruth H. Matthews*
33. Industrialization of Indigenous Fermented Foods, *edited by Keith H. Steinkraus*
34. International Food Regulation Handbook: Policy • Science • Law, *edited by Roger D. Middlekauff and Philippe Shubik*
35. Food Additives, *edited by A. Larry Branen, P. Michael Davidson, and Seppo Salminen*
36. Safety of Irradiated Foods, *J. F. Diehl*
37. Omega-3 Fatty Acids in Health and Disease, *edited by Robert S. Lees and Marcus Karel*
38. Food Emulsions: Second Edition, Revised and Expanded, *edited by Kåre Larsson and Stig E. Friberg*
39. Seafood: Effects of Technology on Nutrition, *George M. Pigott and Barbee W. Tucker*
40. Handbook of Vitamins: Second Edition, Revised and Expanded, *edited by Lawrence J. Machlin*

41. Handbook of Cereal Science and Technology, *Klaus J. Lorenz and Karel Kulp*
42. Food Processing Operations and Scale-Up, *Kenneth J. Valentas, Leon Levine, and J. Peter Clark*
43. Fish Quality Control by Computer Vision, *edited by L. F. Pau and R. Olafsson*
44. Volatile Compounds in Foods and Beverages, *edited by Henk Maarse*
45. Instrumental Methods for Quality Assurance in Foods, *edited by Daniel Y. C. Fung and Richard F. Matthews*
46. *Listeria*, Listeriosis, and Food Safety, *Elliot T. Ryser and Elmer H. Marth*
47. Acesulfame-K, *edited by D. G. Mayer and F. H. Kemper*
48. Alternative Sweeteners: Second Edition, Revised and Expanded, *edited by Lyn O'Brien Nabors and Robert C. Gelardi*
49. Food Extrusion Science and Technology, *edited by Jozef L. Kokini, Chi-Tang Ho, and Mukund V. Karwe*
50. Surimi Technology, *edited by Tyre C. Lanier and Chong M. Lee*
51. Handbook of Food Engineering, *edited by Dennis R. Heldman and Daryl B. Lund*
52. Food Analysis by HPLC, *edited by Leo M. L. Nollet*
53. Fatty Acids in Foods and Their Health Implications, *edited by Ching Kuang Chow*
54. *Clostridium botulinum*: Ecology and Control in Foods, *edited by Andreas H. W. Hauschild and Karen L. Dodds*
55. Cereals in Breadmaking: A Molecular Colloidal Approach, *Ann-Charlotte Eliasson and Kåre Larsson*
56. Low-Calorie Foods Handbook, *edited by Aaron M. Altschul*
57. Antimicrobials in Foods: Second Edition, Revised and Expanded, *edited by P. Michael Davidson and Alfred Larry Branen*
58. Lactic Acid Bacteria, *edited by Seppo Salminen and Atte von Wright*
59. Rice Science and Technology, *edited by Wayne E. Marshall and James I. Wadsworth*
60. Food Biosensor Analysis, *edited by Gabriele Wagner and George G. Guilbault*
61. Principles of Enzymology for the Food Sciences: Second Edition, *John R. Whitaker*
62. Carbohydrate Polyesters as Fat Substitutes, *edited by Casimir C. Akoh and Barry G. Swanson*
63. Engineering Properties of Foods: Second Edition, Revised and Expanded, *edited by M. A. Rao and S. S. H. Rizvi*
64. Handbook of Brewing, *edited by William A. Hardwick*
65. Analyzing Food for Nutrition Labeling and Hazardous Contaminants, *edited by Ike J. Jeon and William G. Ikins*
66. Ingredient Interactions: Effects on Food Quality, *edited by Anilkumar G. Gaonkar*
67. Food Polysaccharides and Their Applications, *edited by Alistair M. Stephen*
68. Safety of Irradiated Foods: Second Edition, Revised and Expanded, *J. F. Diehl*
69. Nutrition Labeling Handbook, *edited by Ralph Shapiro*

70. Handbook of Fruit Science and Technology: Production, Composition, Storage, and Processing, *edited by D. K. Salunkhe and S. S. Kadam*
71. Food Antioxidants: Technological, Toxicological, and Health Perspectives, *edited by D. L. Madhavi, S. S. Deshpande, and D. K. Salunkhe*
72. Freezing Effects on Food Quality, *edited by Lester E. Jeremiah*
73. Handbook of Indigenous Fermented Foods: Second Edition, Revised and Expanded, *edited by Keith H. Steinkraus*
74. Carbohydrates in Food, *edited by Ann-Charlotte Eliasson*
75. Baked Goods Freshness: Technology, Evaluation, and Inhibition of Staling, *edited by Ronald E. Hebeda and Henry F. Zobel*
76. Food Chemistry: Third Edition, *edited by Owen R. Fennema*
77. Handbook of Food Analysis: Volumes 1 and 2, *edited by Leo M. L. Nollet*
78. Computerized Control Systems in the Food Industry, *edited by Gauri S. Mittal*
79. Techniques for Analyzing Food Aroma, *edited by Ray Marsili*
80. Food Proteins and Their Applications, *edited by Srinivasan Damodaran and Alain Paraf*
81. Food Emulsions: Third Edition, Revised and Expanded, *edited by Stig E. Friberg and Kåre Larsson*
82. Nonthermal Preservation of Foods, *Gustavo V. Barbosa-Cánovas, Usha R. Pothakamury, Enrique Palou, and Barry G. Swanson*
83. Milk and Dairy Product Technology, *Edgar Spreer*
84. Applied Dairy Microbiology, *edited by Elmer H. Marth and James L. Steele*
85. Lactic Acid Bacteria: Microbiology and Functional Aspects: Second Edition, Revised and Expanded, *edited by Seppo Salminen and Atte von Wright*
86. Handbook of Vegetable Science and Technology: Production, Composition, Storage, and Processing, *edited by D. K. Salunkhe and S. S. Kadam*
87. Polysaccharide Association Structures in Food, *edited by Reginald H. Walter*
88. Food Lipids: Chemistry, Nutrition, and Biotechnology, *edited by Casimir C. Akoh and David B. Min*
89. Spice Science and Technology, *Kenji Hirasa and Mitsuo Takemasa*
90. Dairy Technology: Principles of Milk Properties and Processes, *P. Walstra, T. J. Geurts, A. Noomen, A. Jellema, and M. A. J. S. van Boekel*
91. Coloring of Food, Drugs, and Cosmetics, *Gisbert Otterstätter*
92. *Listeria*, Listeriosis, and Food Safety: Second Edition, Revised and Expanded, *edited by Elliot T. Ryser and Elmer H. Marth*
93. Complex Carbohydrates in Foods, *edited by Susan Sungsoo Cho, Leon Prosky, and Mark Dreher*
94. Handbook of Food Preservation, *edited by M. Shafiur Rahman*
95. International Food Safety Handbook: Science, International Regulation, and Control, *edited by Kees van der Heijden, Maged Younes, Lawrence Fishbein, and Sanford Miller*
96. Fatty Acids in Foods and Their Health Implications: Second Edition, Revised and Expanded, *edited by Ching Kuang Chow*

97. Seafood Enzymes: Utilization and Influence on Postharvest Seafood Quality, *edited by Norman F. Haard and Benjamin K. Simpson*
98. Safe Handling of Foods, edited by Jeffrey M. Farber and Ewen C. D. Todd
99. Handbook of Cereal Science and Technology: Second Edition, Revised and Expanded, *edited by Karel Kulp and Joseph G. Ponte, Jr.*
100. Food Analysis by HPLC: Second Edition, Revised and Expanded, *edited by Leo M. L. Nollet*
101. Surimi and Surimi Seafood, *edited by Jae W. Park*
102. Drug Residues in Foods: Pharmacology, Food Safety, and Analysis, *Nickos A. Botsoglou and Dimitrios J. Fletouris*
103. Seafood and Freshwater Toxins: Pharmacology, Physiology, and Detection, *edited by Luis M. Botana*
104. Handbook of Nutrition and Diet, *Babasaheb B. Desai*
105. Nondestructive Food Evaluation: Techniques to Analyze Properties and Quality, *edited by Sundaram Gunasekaran*
106. Green Tea: Health Benefits and Applications, *Yukihiko Hara*
107. Food Processing Operations Modeling: Design and Analysis, *edited by Joseph Irudayaraj*

Additional Volumes in Preparation

Handbook of Microwave Technology for Food Applications, *edited by Ashim K. Datta and Ramaswamy C. Anantheswaran*

Applied Dairy Microbiology: Second Edition, Revised and Expanded, *edited by Elmer H. Marth and James L. Steele*

Alternative Sweeteners: Third Edition, Revised and Expanded, *edited by Lyn O'Brien Nabors*

Control of Foodborne Microorganisms, *edited by Vijay K. Juneja and John N. Sofos*

Handbook of Dietary Fiber, *edited by Susan Sungsoo Cho and Mark Dreher*

Food Additives: Second Edition, Revised and Expanded, *edited by Alfred Larry Branen, P. Michael Davidson, Seppo Salminen, and John H. Thorngate*

Food Processing Operations Modeling

Design and Analysis

edited by

Joseph Irudayaraj

The Pennsylvania State University
University Park, Pennsylvania

MARCEL DEKKER

MARCEL DEKKER, INC. NEW YORK • BASEL

ISBN: 0-8247-0488-6

This book is printed on acid-free paper.

Headquarters
Marcel Dekker, Inc.
270 Madison Avenue, New York, NY 10016
tel: 212-696-9000; fax: 212-685-4540

Eastern Hemisphere Distribution
Marcel Dekker AG
Hutgasse 4, Postfach 812, CH-4001 Basel, Switzerland
tel: 41-61-261-8482; fax: 41-61-261-8896

World Wide Web
http://www.dekker.com

The publisher offers discounts on this book when ordered in bulk quantities. For more information, write to Special Sales/Professional Marketing at the headquarters address above.

Current printing (last digit):
10 9 8 7 6 5 4 3 2 1

PRINTED IN THE UNITED STATES OF AMERICA

Preface

Food Processing Operations Modeling is intended as a resource on modeling various fundamental mechanisms in food processing. Its broad scope makes it a useful volume for scientists, graduate and undergraduate students, and practicing engineers. Students who learn the concepts introduced herein will be able to incorporate simpler versions of the mathematical models in their senior design or term projects. The book could also be used as an excellent reference in any numerical methods course offered in departments of agricultural, biological, biosystems, chemical, or mechanical engineering. Additionally, engineers and scientists working outside of food science could use this book as a reference to understand the application of numerical methods in food processing.

Considering that our readership spans various disciplines and programs, our primary goal is to engage the audience with an array of topics based on fundamental engineering principles. Because of this diversity, the book begins with a brief review of the physical properties of food materials and an introduction to modeling. After that introduction, the book proceeds with discussions of applications involving basic to complex problems encountered in processing and handling of food materials.

Each chapter first addresses the theory behind the process, and then discusses a complex case study that demonstrates how to obtain the model, numerical formulation, and solution. For each case study, the discussion explains the thermophysical properties involved and takes into account the modeling complexity and any nonlinearity in the material properties of the system. The many phenomena addressed include heat and mass transfer, fluid flow, electromagnetics, and stochastic processes. Operations discussed in the course of the book are drying, microwave heating, infrared heating, frying, electric resistance heating, aseptic processing, the neural network approach to modeling and process control, and stochastic process modeling of heat and mass transfer in food.

Food Processing Operations Modeling applies a variety of theories to solve practical problems relevant to research in and teaching of food process engineering. Unfortunately, this book cannot offer a complete catalog of modeling for the numerous operations used in food processing. However, working through the case studies provided, the reader will learn a conceptual framework that will enable him or her to understand and solve diverse problems that emerge in food processing operations.

I wish to acknowledge my parents for their encouragement and support in a variety of ways in making this book possible. I would also like to express special thanks to Mr. Hong Yang for his help in preparing the index.

Joseph Irudayaraj

Contents

Contributors

Vikrant Chauhan Processing Technology Program, Asian Institute of Technology, Bangkok, Thailand

Ashim K. Datta, Ph.D. Department of Agricultural and Biological Engineering, Cornell University, Ithaca, New York

Laurence J. Davies School of Chemical Engineering, University of Birmingham, Birmingham, United Kingdom

Josse De Baerdemaeker, Ph.D. Dept. of Agro-Engineering and -Economics, Katholieke Universiteit Leuven, Leuven, Belgium

Oladiran O. Fasina, Ph.D. U.S. Department of Agriculture–Agricultural Research Service and North Carolina Agricultural Research Service, North Carolina State University, Raleigh, North Carolina

Peter J. Fryer, Ph.D. School of Chemical Engineering, University of Birmingham, Birmingham, United Kingdom

Sundaram Gunasekaran, Ph.D. Department of Biological Systems Engineering, University of Wisconsin–Madison, Madison, Wisconsin

Dennis R. Heldman, Ph.D. Department of Food Science, Rutgers – The State University of New Jersey, New Brunswick, New Jersey

Joseph Irudayaraj, Ph.D. Department of Agricultural and Biological Engineering, The Pennsylvania State University, University Park, Pennsylvania

Vinod K. Jindal, Ph.D. Processing Technology Program, Asian Institute of Technology, Bangkok, Thailand

Rosana G. Moreira, Ph.D. Department of Agricultural Engineering, Texas A&M University, College Station, Texas

Kasiviswanathan Muthukumarappan, Ph.D. Department of Agricultural and Biosystems Engineering, South Dakota State University, Brookings, South Dakota

Bart M. Nicolaï Department of Agricultural and Applied Biological Sciences, Katholieke Universiteit Leuven, Leuven, Belgium

Virendra M. Puri, Ph.D. Department of Agricultural and Biological Engineering, The Pennsylvania State University, University Park, Pennsylvania

K. P. Sandeep, Ph.D. Department of Food Science, North Carolina State University, Raleigh, North Carolina

Nico Scheerlinck, M.D. Department of Agro-Engineering and -Economics, Katholieke Universiteit Leuven, Leuven, Belgium

Robert Thomas Tyler, Ph.D. Department of Applied Microbiology and Food Science, University of Saskatchewan, Saskatoon, Canada

Pieter Verboven, Ph.D. Department of Agro-Engineering and -Economics, Katholieke Universiteit Leuven, Leuven, Belgium

1

Prediction Models for Thermophysical Properties of Foods

Dennis R. Heldman
Rutgers – The State University of New Jersey, New Brunswick, New Jersey

1 INTRODUCTION

Properties of food and food ingredients are critical parameters in the design of a process used in the manufacturing of a food product. Although property magnitudes may be estimated based on published values for similar materials, improvements in process efficiency and the design of equipment used to perform the process, depend on more accurate property magnitudes. Thermophysical properties are unique and influence the design of any thermal process; a food manufacturing process involving changes in temperature of the ingredients and product. Thermophysical properties normally include *specific heat, density,* and *thermal conductivity*. Individually, these properties may have influence on process evaluation and design. For example, specific heat and density are important components of an analysis involving mass and/or energy balances. Thermal conductivity is the key property in the quantification of thermal energy transfer within a material by conduction. The combination of the three properties is *thermal diffusivity*, a key property in the analysis of unsteady-state heat transfer.

Thermophysical properties of food and food ingredients have been investigated for several years, with much of the emphasis on the measurement of property magnitudes for many foods as a function of temperature and composition. Although these published properties have and continue to contribute to the improved design of selected processes, the properties are not in a format for ideal input to most process design situations. All thermal processes involve changes in product temperature, and many involve changes in product composition. Most published thermophysical properties data do not accommodate situation in which property magnitudes change during the process.

More recently, prediction models for thermophysical properties based on product composition have evolved. Many of these models are based on property magnitudes for the basic compositional components of foods; proteins, fat, carbohydrates, ash, and water. Knowledge of the properties of these basic components as a function of temperature provides the opportunity to develop prediction models that will accommodate the needs of process design models. These thermophysical property models represent a significant opportunity to improve the efficiency of thermal processes for food and the ultimate design of equipment used for processing of foods.

The overall objective of the information to be presented is to discuss models for the prediction of thermophysical properties of food and food ingredients based on composition. The following are more specific objectives:

- To present and discuss models for the prediction of the specific heat of foods based on the composition of foods and food ingredients, with emphasis on the application of models to process design
- To present and discuss models for the prediction of density of foods based on the composition of foods and food ingredients, with emphasis on the influence of physical structure of the product.
- To present and discuss models for the prediction of the thermal conductivity of foods as a function of composition and temperature, with emphasis on the use of models that incorporate the influence of physical structure of the product.

As suggested, the emphasis will be on models and steps needed to use the models in process design. Reference to the appropriate thermophysical property data obtained from experimental measurements will be illustrated.

2 PREDICTION OF SPECIFIC HEAT

The specific heat of a food is defined as the quantity of thermal energy associated with a unit mass of the food and a unit of change in temperature. This thermophysical property is often referred to as heat capacity and is an essential component of a thermal energy analysis on a food product, a thermal process, or processing equipment used for heating or cooling of a food. A prediction model provides the opportunity to conduct analyses over defined ranges of temperature and composition for a given process used for a food product.

The influence of composition on the specific heat of foods is obvious in the earliest of prediction models from Siebel [1]:

$$c_p = 0.837 + 0.034(\text{moisture content, \%}) \tag{1}$$

This empirical expression is based on experimental data for high-moisture foods, and it is anticipated that the coefficients within the expression will vary with temperature. Several similar models have summarized recently by Sweat [2], An obvious dependence of the magnitude of the specific heat of a high-moisture food on moisture content is evident.

A relationship expanding on the dependence of the specific heat of a food on composition was suggested by Leninger and Beverloo [3] as follows:

$$c_p = (0.5M_f + 0.3M_s + M_w)4.18 \tag{2}$$

This equation contains the mass fractions of fat (M_f), nonfat solids (M_s), and water (M_w) and references the specific heat of water (4.18 kJ/kg) at 20°C. A very similar relationship was suggested by Charm [4]:

$$c_p = 2.094M_f + 1.256M_s + 4.187M_w \tag{3}$$

The Charm equation uses the specific heat of water at 75°C and the coefficients for the fat and nonfat solids are the same as Eq. (2) when the temperature is adjusted. The magnitude for specific heat of fat is 2.094 kJ/kg, when the fat is in a liquid state at 75°C. A value for solid fat would be 1.675 kJ/kg (at the appropriate lower temperature). An additional dimension of the dependence of specific heat on composition was suggested by Heldman and Singh [5]:

$$c_p = 1.424M_c + 1.549M_p + 1.675M_f + 0.837M_a + 4.187M_w \tag{4}$$

In this expression, the coefficients represent the specific heats of carbohydrates, proteins, fat, and ash at 20°C or less. As suggested earlier, the magnitude in the equation is for fat in the solid state.

As is evident, the models presented up to this point are entirely empirical or lack reference to the specific temperature for applications. A very general model for the prediction of specific heat of food was suggested by van Beek [6]:

$$c_p = \sum (c_{pi} M_i) \tag{5}$$

The general model indicates that the specific heat of a food can be predicted from knowledge of the composition and the specific heat of each component. More specifically, the predicted value is the summation of the product of the mass fraction of component (i) and the specific heat of component (i).

The successful use of the general model depends on two key inputs:

- Composition information on the food or food ingredient being considered. These types of data for an array of food products are found in USDA Handbook No. 8 [7]. The composition of an ingredient is likely to be established or measured, as would be the case for new product formulations. For processes where the composition changes during the process, information may be limited, but models of the type being presented should encourage monitoring of these changes during a process.
- Data on the specific heat of the key compositional components (proteins, carbohydrates, fat, ash) of food. It must be emphasized that the property magnitudes are for moisture-free components. It is recognized that foods contain many different types of proteins, carbohydrates, and fats. Data published to date suggests that differences in specific heat magnitudes for different proteins, carbohydrates, or fats are relatively small. These differences may be smaller than changes in specific heat magnitudes due to a phase change for the same component. It is very important for the specific heat data for these compositional components to be available over a range of temperatures associated with typical thermal processes for food. To date, the best and most complete data were published by Choi and Okos [8].

The data presented by Choi and Okos [8] are based on an extensive study and analysis of specific heat data for many liquid foods with different compositions and generally over a temperature range of 20–100°C. The results are summarized in Table 1.

The best approach to illustrating the use of the general specific heat model is in the form of an exarnple. The example will describe use of the general model [Eq. (5)] to predict changes in specifie heat of the product during a process in which both the product temperature and composition are changing in a defined manner during the process.

TABLE 1 Specific Heat Relationships for Food Product Components

Component	Temperature relationship	Standard error	Standard error (%)
Protein	$c_p = 1.9842 + 1.4733 \times 10^{-3}T - 4.8008 \times 10^{-6}T^2$	0.1147	5.57
Carbohydrate	$c_p = 1.54884 + 1.9625 \times 10^{-3}T - 5.9399 \times 10^{-6}T^2$	0.0986	5.96
Fat	$c_p = 1.9842 + 1.4733 \times 10^{-3}T - 4.8008 \times 10^{-6}T^2$	0.0236	1.16
Ash	$c_p = 1.0926 + 1.8896 \times 10^{-3}T - 3.6817 \times 10^{-6}T^2$	0.0296	2.47
Water (< 0°C)	$c_p = 4.0817 - 5.3062 \times 10^{-3}T + 9.9516 \times 10^{-4}T^2$	0.0988	2.15
Water (> 0°C)	$c_p = 4.1762 - 9.0864 \times 10^{-5}T + 5.4731 \times 10^{-6}T^2$	0.0159	0.38
Ice	$c_p = 2.0623 + 6.0769 \times 10^{-3}T$	0.0014	0.07

Source: Ref. 8.

2.1 EXAMPLE

A liquid food, with a composition of 3.5% protein, 4.9% carbohydrate, 3.9% fat, 0.7% ash, and 87% water, is heated from 20°C to 100°C, and the concentration of product solids increases to 30% during the process. The process requires 40 min when the heating medium temperature is 105°C. The changes in concentration and temperature as a function of time (t) are described by the following relationships:

$$\%\mathrm{TS} = \%\mathrm{TS}_0 \exp(0.021t)$$

where %TS is the percentage of total product solids, or total of protein, carbohydrate, fat, and ash within the product and expressed as a percentage, and

$$\frac{T - T_M}{T_0 - T_M} = \exp(-0.07t)$$

where T is temperature at any time (t), T_0 is the initial product temperature (at $t = 0$), and T_M is the heating medium temperature. Predict the specific heat of the product, as a function of time, during the process.

Solution

The results of the solution will be presented in the form of a table with predicted specific heat values at time increments during the process.

The solution will include steps required for prediction of each specific magnitude.

1. At $t = 0$: The specific heat of each product component is computed at 20°C.

   For protein,

   $$c_p = 2.0082 + (1.2089 \times 10^{-3})(20) - (1.3129 \times 10^{-6}(2)^2$$
   $$= 2.032\,\text{kJ/kg}$$

   For carbohydrate

   $$c_p = 1.5488 + (1.9625 \times 10^{-3}(20) - (5.9399 \times 10^{-6}(20)^2$$
   $$= 1.587\,\text{kJ/kg}$$

   For fat,

   $$c_p = 1.9842 + (1.4733 \times 10^{-3}(20) - (4.8008 \times 10^{-6}(20)^2$$
   $$= 2.012\,\text{kJ/kg}$$

   For ash,

   $$c_p = 1.0926 + (1.8896 \times 10^{-3}(20) - (3.6187 \times 10^{-6}(20)^2$$
   $$= 1.129\,\text{kJ/kg}$$

   For water,

   $$c_p = 4.1762 + (9.0864 \times 10^{-5}(20) + (5.4731 \times 10^{-6}(20)^2$$
   $$= 4.176\,\text{kJ/kg}$$

   The specific heat of the product at the beginning of the process is

   $$c_p = (2.032)(0.035) + (1.587)(0.049)$$
   $$+ (2.012)(0.039) + (1.129)(0.007) + (4.176)(0.87)$$
   $$= 3.868\,\text{kJ/kg}$$

2. At $t = 40$ min: For a temperature of 99.8°C:

Protein	$c_p = 2.116$ kJ/kg
Carbohydrate	$c_p = 1.685$ kJ/kg
Fat	$c_p = 2.083$ kJ/kg
Ash	$c_p = 1.245$ kJ/kg
Water	$c_p = 4/231$ kJ/kg

   At 40 min, the water content has decreased to 70% and the mass fractions of all other components have increased. Based on the adjusted composition and temperature change, the specific heat of the product becomes

$$
\begin{aligned}
c_p &= (2.116)(0.081) + (1.685)(0.1135) \\
&\quad + (2.083)(0.0903) + (1.245)(0.0162) + (4.231)(0.699) \\
&= 3.528\ \text{kJ/kg}
\end{aligned}
$$

3. At times between 0 and 40 min, the specific heat can be predicted using the same steps as illustrated previously. The results are presented in Table 2.

TABLE 2 Prediction of Specific Heat During a Process with Changing Temperature and Composition

Time (min)	Temperature (°C)	Total solids (%)	Specific heat (kJ/kg)
0	20.0	13.0	3.868
10	62.8	16.0	3.819
20	84.0	19.8	3.748
30	94.6	24.4	3.650
40	99.8	30.1	3.528

The results of the example illustrate that the specific heat of the product decreases during the process as product temperature increases and water content of the product decreases. The predictions indicate that the specific heat of product components increase with temperature. The influence of this increase is smaller than the influence associated with the change in product composition. As the concentration of product solids increases, the amount of water in the product decreases. Because the specific heat of the product solids is much lower than the specific heat of water, the higher mass fractions of the lower-specific-heat components result in a lower specific heat of the product at the end of the process.

3 PREDICTION OF DENSITY

There are only a limited number of models for predicting the density of a food product based on composition. The suggestions by Heldman [9,10] illustrate the influence of freezing on the density of a high-moisture food (Figure 1). These models are similar to the general model for density prediction as proposed by Choi and Okos [8]:

$$
\rho = \frac{1}{\sum (M_i/\rho_i)} \tag{6}
$$

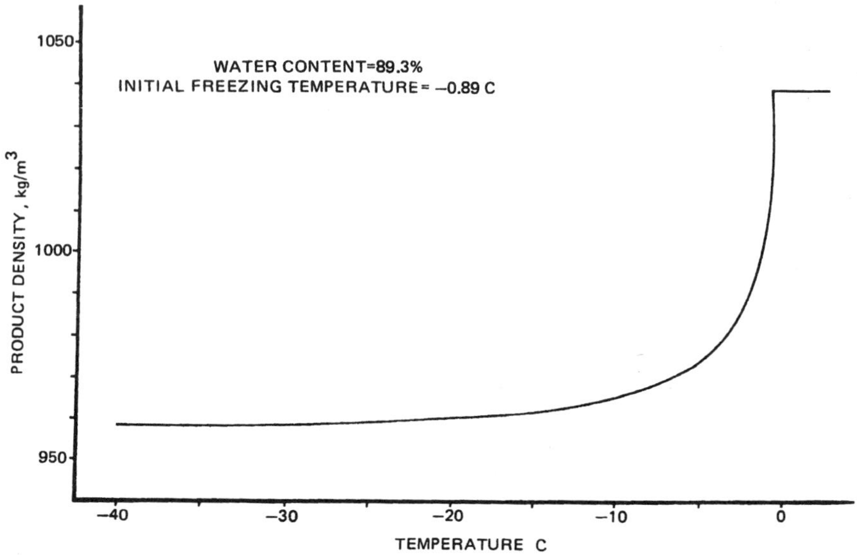

FIGURE 1 Influence of phase change on the density of strawberries. (From Ref. 9.)

The use of Eq. (6) is similar to specific heat and involves the use of product composition (M_i) for protein, fat, carbohydrate, ash, and water and the density (moisture-free) for each component (ρ_i). The density data to be used as inputs to the proposed model were published by Choi and Okos [8]. These data are summarized in Table 3.

The proposed model predicts the bulk density of a high-moisture food from typical composition information [7] and the density relationships

TABLE 3 Density Relationships for Food Product Components

Component	Temperature relationship	Standard error	Standard error (%)
Protein	$\rho = 1.3299 \times 10^3 - 5.184 \times 10^{-1} T$	39.9501	3.07
Carbohydrate	$\rho = 1.59919 \times 10^3 - 3.1046 \times 10^{-1} T$	93.1249	5.98
Fat	$\rho = 9.2559 \times 10^2 - 4.1757 \times 10^{-1} T$	1.2554	0.47
Ash	$\rho = 2.4238 \times 10^3 - 2.8063 \times 10^{-1} T$	2.2315	0.09
Water	$\rho = 9.9718 \times 10^2 + 3.1439 \times 10^{-3} T - 3.7574 \times 10^{-3} T^2$	2.1044	0.22
Ice	$\rho = 9.1689 \times 10^2 - 1.3071 \times 10^{-1} T$	0.5382	0.06

Source: From Ref. 8.

presented in Table 3. The output density magnitudes can reflect changes in composition and temperature as might occur during a food manufacturing process. There would be a lower limit on the proposed application based on moisture content. The exact magnitude of this limiting moisture content would be product dependent and will be discussed in more detail when discussing the prediction of density of low-moisture foods.

The density of a dry food is directly dependent on the structure of the product, with a gas phase (air) having a significant influence on the magnitude of the property magnitude. This relationship to a gas phase depends on many external factors (packing, pressure, etc.) and prevents the use of prediction models, before some type of reference measurement is accomplished. The following relationships provide the opportunity to predict the product density of a low-moisture food as a function of temperature and moisture content.

3.1 Dry Particle Foods

Many dry foods are in the form of particles created by the manufacturing process. For these types of product, the bulk density is dependent on particle density, as well as the magnitude of void space around the particles. Particle density is the mass of the particle per unit of particle volume. At a product moisture content of zero, the particle is a two-phase system and can be described in terms of volume fractions (e_i) as follows:

$$e_s + e_a = 1 \tag{7}$$

indicating that the particle volume is composed of product solids and air (gas phase). In addition, the particle density can be predicted by

$$\rho_p = \rho_s e_s + \rho_a e_a \tag{8}$$

with the density of product solids and air as inputs. It should be noted that the density of product solids could be predicted from the relationship, based on compositional components presented previously. Equations (7) and (8) can be used to obtain the following:

$$e_s = \frac{\rho_p - \rho_a}{\rho_s - \rho_a} \tag{9}$$

indicating that the volume fraction of solids (and for air) can be determined after measurement of the particle density of product at a moisture content of zero.

Based on the concept proposed by Sarma and Heldman (11), the initial addition of moisture to the low-moisture food results in the replacement of air space within the particle, and the volume fraction of solids is constant. The concept is illustrated in Figure 2. As the increase in moisture content

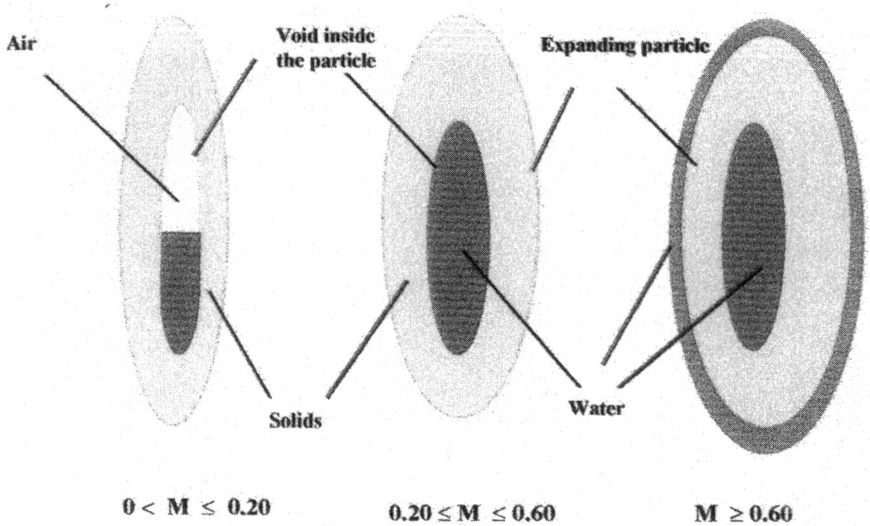

FIGURE 2 Proposed relationship of physical structure of food particles to moisture content for the range from zero to greater than 0.6 kg water/kg dry solids. (From Ref. 11.)

continues, the volume fraction of water increases until the magnitude is equal to the volume fraction of air at moisture content equal to zero. Within the low-moisture-content range, the particle density increases linearly with moisture content until the air space within the particle is replaced by water. The moisture content when the air space within the particle is replaced by water can be defined as the critical moisture content. When the moisture content of the product is increased above the critical moisture content, the particle density remains constant. The linear increase in the particle density and the maximum particle density at the critical moisture content has been illustrated for starch particles [11]. A particle density of 1476 kg/m^3 occurred at a dry-basis moisture content of 0.2 kg water/kg dry solids. Later, Sabliov and Heldman [12] have shown that the magnitude for casein particles was 1279 kg/m^3 at 0.25 kg water/kg dry solids.

Above the critical moisture content, the particle is a two-phase system; water and product solids. An increase in moisture content causes an expansion of particle volume. Over a range of moisture contents (above 20–25% dry basis), the particle density can be predicted from

$$\rho_p = \frac{\rho_w \rho_s (1 + M)}{\rho_w + M\rho_s} \tag{10}$$

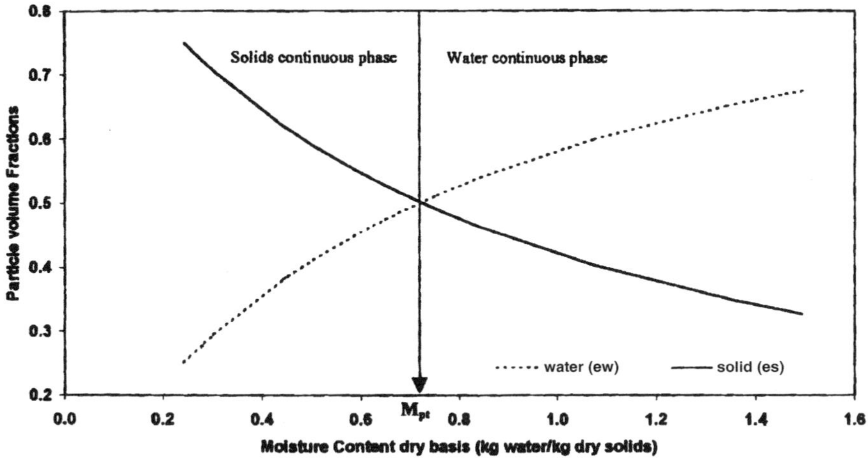

FIGURE 3 Influence of moisture content on volume fractions of water and product solids within a food particle. (From Ref. 12.)

where M is the dry-basis moisture content. Even though the particle volume expands with increasing moisture content, the volume fraction of water (as compared to the volume fraction of solids) increases. The moisture content when the two volume fractions are equal (at 0.5) can be predicted as illustrated by Sabliov and Heldman [12] in Figure 3.

When considering the prediction of bulk density, the total mass and volume of the product must be considered. By considering the product as a two-phase system [a particle phase and an air (gas) phase], the following relationship would apply:

$$E_p + E_a = 1 \tag{11}$$

where the volume fraction for the particle (E_p) and the volume fraction of air (E_a) are the total volume of product. Because the bulk density can be defined as

$$\rho_b = E_p\rho_p + E_a\rho_a \tag{12}$$

the volume fraction of the particle becomes

$$E_p = \frac{\rho_b - \rho_a}{\rho_p - \rho_a} \tag{13}$$

These relationships should be applied to nonparticulate dry foods by recognizing the product solids phase would replace the particle phase in the particle system. In these situations, the volume phase of product solids would be predicted by the previous equation. It should be noted that the solid phase for the nonparticle food would be composed of dry product solids and water.

For both of the previous situations, the bulk density of the dry food would be predicted as a function of moisture content and temperature using

$$\rho_b = E_p(\rho_p - \rho_a) + \rho_a \tag{14}$$

For such predictions, the volume fraction of particles (or solids) would be based on the measurement of bulk density at the critical moisture content. At moisture comments above the critical moisture content (20–25% db), the relationship would use inputs from the previous particle analysis to obtain the volume fraction of particles (E_p) and the particle density (ρ_p) as a function of moisture content and temperature. The expression for volume fraction can be used to predict the volume fraction of the particle (or solid) and for air, as a function of moisture content. As moisture content increases, the volume fraction of the particle increases and the volume fraction of air decreases. At a defined moisture content or bulk density, the two fractions become equal at 0.5. A typical relationship for the bulk fraction of starch has been presented by Sabliov and Heldman [12], as illustrated in Figure 4.

An additional analysis of the relationship between the volume fraction of particles (or solids) and moisture content indicates that the volume

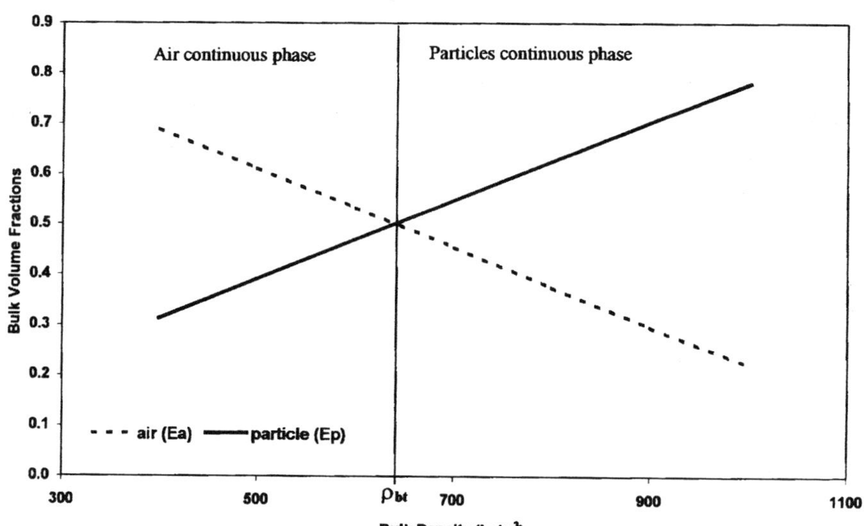

FIGURE 4 Influence of bulk density on volume fractions of air and product particles within a food system. (From Ref. 12)

fraction of air decreases to zero at a relatively high moisture content. At this upper limit of the moisture content of the product, the product becomes a two-phase system: product solids and water. Although the exact significance of this upper-limit moisture content, it would seem appropriate to use the general prediction models for foods above the upper-limit moisture content. An analysis of starch particles has been completed by Sabliov and Heldman [12] and indicates that the dry-basis moisture content at the upper limit is approximately 1.5 kg water/kg dry solids. It is anticipated that the changes in the various parameters would be modified slightly when the moisture content is decreased.

Based on the observations presented, six moisture content ranges can be defined. The approach to the prediction of the bulk density of the food product would vary with moisture content range. These ranges may be described in the following manner:

1. Very low moisture content: less than ~10% db, or where the volume fraction of water equals volume fraction of air, within the particle
2. Low moisture content: approximately 10% to ~20% db, or the moisture content when the volume fraction of air within the particle becomes zero
3. Low intermediate moisture content: 20% to ~65% db, or the moisture content when the volume fraction of water exceeds the volume fraction of product solids within the particle
4. Medium intermediate moisture content: 65% to ~90% db, or the moisture content when the volume fraction of particles exceeds the volume fraction of air within the product system
5. High intermediate moisture content: 90% to ~135% db, or the moisture context when the volume fraction of air become zero
6. High moisture content: above ~135% db, or the moisture content when the volume fraction of air is zero

The moisture content ranges identified are based on studies of starch and casein and will vary with food product and composition, as well as the process used to manufacture the product. In addition, the bulk density is a function of many factors associated with handling and packaging of the dry product.

4 PREDICTION OF THERMAL CONDUCTIVITY

Thermal conductivity is a basic thermophysical property of any material, the magnitude expressing the rate of thermal energy transfer within the material. Because the complexity of the physical structure of food, the prediction models for the thermal conductivity of food must account for characteristics of the physical structure. The development of prediction models for thermal conductivity of food should assist in the improvement of process design and the equipment used to accomplish the process. Because temperature and moisture content have a significant influence on thermal conductivity, the magnitude of the property changes significantly during the many processes used for the manufacturing of foods.

Historically, the models for the prediction of the thermal conductivity of foods have been empirical and based on experimental data. One of the earliest models was proposed by Riedel [13]:

$$k = (326.58 + 1.0412T + 0.00337T^2)(0.46 + 0.54M_w)(1.73 \times 10^{-3}) \tag{15}$$

This empirical expression is based on experimental data for fruit juices, sugar solutions and milk over a temperature range from 0°C to 180°C. Additional empirical models have evolved and have been summarized by Sweat [2]. These expressions include a model for the prediction of the thermal conductivity for fruits and vegetables, with water content above 60% [14]:

$$k = 0.148 + 0.439M_w \tag{16}$$

Later, Sweat [15] suggested a similar type of empirical model for meat, with moisture content between 60% and 80% (wet basis) and temperatures between 0°C and 60°C:

$$k = 0.08 + 0.52M_w \tag{17}$$

These empirical models do not account for the influence of thermal conductivity differences among the other compositional components in foods. Based on a detailed statistical analysis of thermal conductivity data for liquid foods, Choi and Okos [16] proposed the following prediction model:

$$k = 0.2051M_c + 0.2M_p + 0.175M_f + 0.135M_a + 0.61M_w \tag{18}$$

where the coefficients for each of the compositional components would represent the thermal conductivity of that component. A similar analysis was completed by Sweat [2] to create the following model:

$$k = 0.25M_c + 0.155M_p + 0.161m_f + 0.135M_a + 0.58M_w \tag{19}$$

As is evident, the coefficients for these models are similar, but the output from the model is influenced by the input data for the analysis. The prediction accounts for the temperature used during the collection of the experimental thermal conductivity. A more general model was proposed Choi and Okos [8]:

$$k = \sum k_i E_i \tag{20}$$

where the volume fraction (E_i) is estimated for each compositional component by

$$E_i = \frac{M_i/\rho_i}{\sum (M_i/\rho_i)} \tag{21}$$

The general model uses compositional data from measurements or the USDA Handbook No. 8 [7] and the thermal conductivity inputs from the relationships presented in Table 4.

The general prediction model is adequate for the prediction of the thermal conductivity of food as a function of temperature and product composition. The limits on the use of the model are moisture content and physical structure of the product. Because the experimental data used to generate the relationships in Table 4 were for high moisture foods, a lower limit on moisture content for the application of the relationships must be considered. Alternately, the best applications of the general model would be high-moisture foods, where water is the predominant component or continuous phase within the product. Physical structure is not considered in the general model and the influence of the orientation of product com-

TABLE 4 Thermal Conductivity Relationships for Components of Food

Component	Temperature relationships	Standard error	Standard error (%)
Protein	$k = 1.7881 \times 10^{-1} + 1.1958 \times 10^{-3}T - 2.7178 \times 10^{-6}T^2$	0.012	5.91
Carbohydrate	$k = 2.0141 \times 10^{-1} + 1.3874 \times 10^{-3}T - 4.3312 \times 10^{-6}T^2$	0.0134	5.42
Fat	$k = 1.8071 \times 10^{-1} - 2.7604 \times 10^{-3}T - 1.7749 \times 10^{-7}T^2$	0.0032	1.95
Ash	$k = 3.2962 \times 10^{-1} + 1.4011 \times 10^{-3}T - 2.9069 \times 10^{-6}T^2$	0.0083	2.15
Water	$k = 5.7109 \times 10^{-1} + 1.7625 \times 10^{-3}T - 6.7036 \times 10^{-6}T\,2$	0.0028	0.45
Ice	$k = 2.2196 - 6.2489 \times 10^{-3}T + 1.0154 \times 10^{-4}T^2$	0.0079	0.79

Source: Ref. 8.

ponents on the magnitude of the thermal conductivity is not a part of the model.

The influence of physical structure on the thermal conductivity prediction has been considered in several different models. These models have been reviewed by Sweat [2]; the first type of model is for thermal conductivity within an isotropic system with a discontinuous component homogeneously dispersed within a continuous second component of the product. The model derived and proposed by Kopelman [17] for the situation when the thermal conductivity magnitude of the continuous component is similar to the magnitude for the discontinuous component is

$$k = k_c\left(\frac{1 - E_d(1 - k_d/k_c)}{1 - E_d^2(1 - k_d/k_c)(1 - E_d)}\right) \tag{22}$$

where E_d is the volume fraction of discontinuous product component. A similar model for the condition when the thermal conductivity of the continuous component is much, much larger than thermal conductivity of the discontinuous component, would be

$$k = k_c\left(\frac{1 - E_d^{2/3}}{1 - E_d^{2/3}(1 - E_d^{1/3})}\right) \tag{23}$$

The application of this second model to foods is limited.

In many foods, the discontinuous component has a defined orientation within the continuous component. For this situation, Sweat [2] suggested the following simple model, when heat transfer is parallel to the orientation of the compositional components:

$$k = E_c k_c + E_d k_d \tag{24}$$

The more complex model for the anisotropic model was derived and presented by Kopelman [17] as follows:

$$k = k_c\left[1 - E_d\left(1 - \frac{k_d}{k_c}\right)\right] \tag{25}$$

The corresponding expression for heat conduction perpendicular to the orientation of the compositional components, as presented by Sweat [2] is

$$k = \left(\frac{E_c}{k_c} + \frac{E_d}{k_d}\right)^{-1} \tag{26}$$

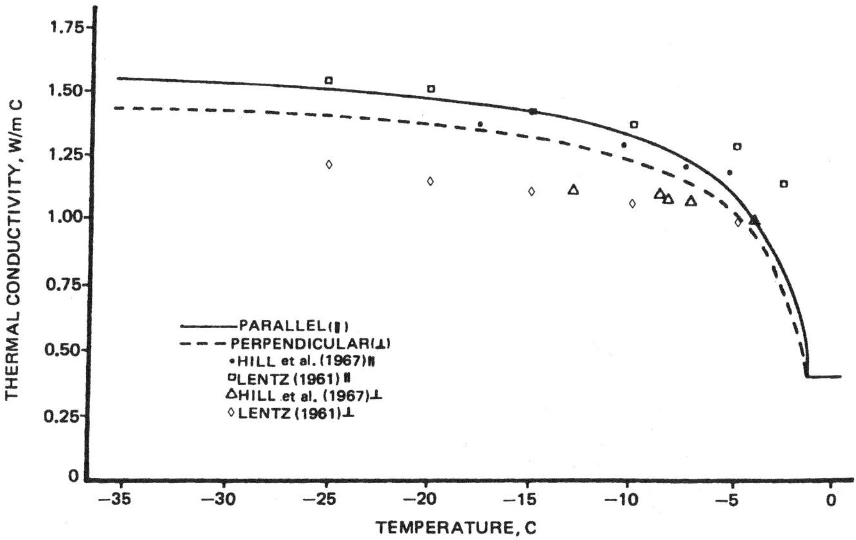

FIGURE 5 Influence of phase change and component orientation on thermal conductivity of lean beef. (From Ref. 18).

Kopelman [17] derived and presented the following model for heat conduction perpendicular to the orientation of compositional components:

$$k = k_c\left(\frac{1 - E_d^{1/2}[1 - (k_d/k_c)]}{1 - E_d^{1/2}[1 - (k_d/k_c)](1 - E_d^{1/2})}\right) \tag{27}$$

The use of the models to account for the influence of component orientation on thermal conductivity is illustrated in Figure 5, where the influence of phase change and muscle fibers in beef are evident. The influence of frozen water on the thermal conductivity of beef is dramatic. The thermal conductivity prediction for heat transfer parallel to the direction of muscle fibers is higher than for heat transfer perpendicular to the fibers and is consistent with experimental data.

Kopelman [17] derived and described a third model for prediction of thermal conductivity within an anisotropic, two-component system, where the two components are in layers within the system. The physical structures for all three of the models developed by Kopelman [7] are illustrated in Figure 6. For heat transfer parallel to the two component layers,

$$k = k_c\left\{1 - E_d\left[1 - \left(\frac{k_d}{k_c}\right)\right]\right\} \tag{28}$$

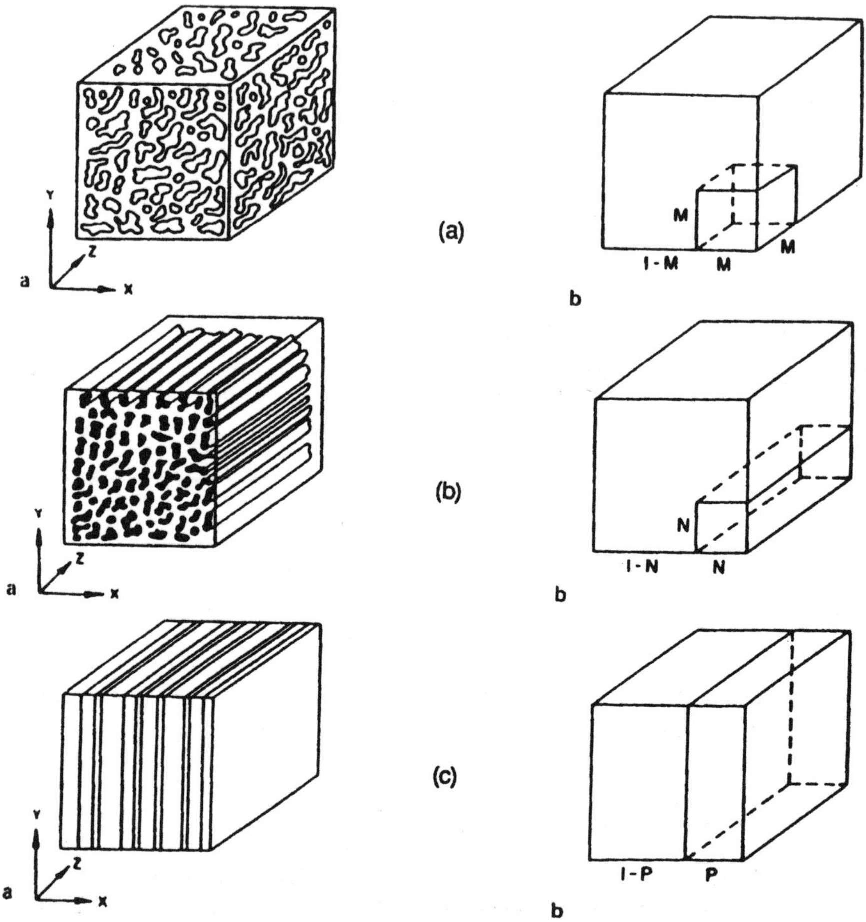

FIGURE 6 Physical model for heat transfer within two-component homogeneous food systems: (a) three-dimensional dispersion system: (b) two-dimensional fibrous system; (c) one-dimensional layered system. (Ref. 17)

The corresponding expression for heat transfer perpendicular to the component layers is

$$k = k_c\left(\frac{k_d}{E_d k_c + k_d(1 - E_d)}\right) \tag{29}$$

All of the models presented suggest that thermal conductivity of a food system can be predicted based on knowledge of the volume fractions and the thermal conductivities of the two components in the food system. The importance of defining the discontinuous phase is evident in all of the models. Application of the models to food systems requires careful definition of the components within the product. It is appropriate to refer to property magnitudes for food products as "effective thermal conductivities" to differentiate the property from the magnitudes for the components of the food.

The various models for prediction of effective thermal conductivity must be used in combination with the general model [Eq. (20)] to adequately describe heat transfer in typical food products. The key factors that differentiate among food products are physical structure and composition. These factors, in addition to temperature, are essential to predicting the changes in effective thermal conductivity of the product during a typical manufacturing process. The following steps have been demonstrated far the prediction of the thermal conductivity of food products at various moisture contents and temperatures.

A general component in application of the prediction models for effective thermal conductivity is the general model [Eq. (20)], in combination with the relationships in Table 4. These relationships are used to predict the thermal conductivity of the product solids component within the food structure. Overall, the influence of the product solids on the effective thermal conductivity of the food system is represented by the general model I and the relationships in Table 4. The influence of other components (water and/ or air) is described by the various models incorporating physical structure. The moisture content has the most significant influence on the use of the various models.

4.1 Very-Low-Moisture Foods

At moisture contents below the critical moisture content, as defined in discussions related to density prediction, the primary component influencing effective thermal conductivity of the product is the gas phase or air. Within this moisture content range, food particles will contain three components (product solids, water, air). Within the lower portion of the moisture content range [0 to $\sim$10% (db)], air will be the continuous component and the combination of water and product solids will be the discontinuous component. The homogeneous physical structure model [Eq. (22)] can be used to predict the thermal conductivity of the particle. The next step would be the prediction of the effective thermal conductivity using the same model, with air as the continuous component and product particles as the discontinuous component.

4.2 Low-Moisture Foods

At the higher-moisture-content end of the range below the critical moisture content, the volume fraction of water within the particle exceeds the volume fraction of air within the particle. It follows that the homogeneous model [Eq. (22)] is used to predict the thermal conductivity of particles, with the particles {product solids and water) as the continuous component. The final step in the prediction of the effective thermal conductivity of the product is that it would be the same as for the very low-moisture food.

4.3 Low-Intermediate-Moisture Foods

As indicated during the analysis of density of dry particle foods, the particle becomes a two-component (product solids and water) system in the moisture content range immediately above the critical moisture content [~20% (db)]. Within the lower portion of this moisture content range, the thermal conductivity of the particle is predicted by the homogeneous model [Eq. (22)]. The product solids are the continuous component and water is the discontinuous component. The effective thermal conductivity of the product is predicted using the homogeneous model (or appropriate model, based on the orientation of components) with air as the continuous component and product particles as the discontinuous component.

4.4 Medium-Intermediate-Moisture Foods

As the moisture content increases, the volume fraction of water within the particle exceeds the volume fraction of the product solids. The point at which the volume fractions are equal has been defined as the particle transition moisture content. At moisture contents higher than the particle transition moisture content, the thermal conductivity of the particle is predicted by the homogeneous model [Eq. (22)], with water as the continuous component and product solids as the discontinuous component. The effective thermal conductivity of the product is predicted using the homogeneous model (or appropriate alternative model), with air as the continuous component and product particles as the discontinuous component. The upper limit of this moisture content range occurs when the volume fraction of particles exceeds the volume fraction of air within the product system. This point has been defined as the transition bulk density.

4.5 High-Intermediate-Moisture Foods

At moisture contents above the transition bulk density, when the volume fraction of particle exceeds the volume fraction of air in the product, the thermal conductivity of the particle is predicted by using the homogeneous model. Within the model, the volume fraction of water becomes the continuous component and product solids are the discontinuous component. As the moisture content increases, the volume fraction of particle increases and the volume fraction of air decreases. Within this moisture content range, the effective thermal conductivity of the product is predicted by using the appropriate model, with product particles as the continuous component and air as the discontinuous component. The upper limit of this moisture content range occurs when the volume fraction of air becomes zero.

4.6 High-Moisture Foods

At moisture contents and bulk densities when the volume fraction of air is zero, the product becomes a two-component system of water and product solids. The model for the prediction of thermal conductivity depends on the orientation of product solids within the water component. The effective thermal conductivity is predicted by using water as the continuous component and the product solids as the discontinuous component in the appropriate model. The thermal conductivity of the product solids would be predicted from the composition of protein, carbohydrate, fat, and ash and the appropriate relationship for each compositional component.

4.7 Summary

The six moisture content ranges clearly illustrate the relationships between density of product and the appropriate models for the prediction of the effective thermal conductivity of the food product. The results in Figure 7 illustrate the agreement between the predicted effective thermal conductivity and experimental measurements for a significant range of moisture contents. The experimental verification of the proposed models was accomplished by Sabliov and Heldman [19]. During these investigations, the bulk density was measured as a part of the experimental determination of thermal conductivity. This experimental bulk density was used as an input to the prediction models. Although the application of the models have been for situations when moisture content is increasing, it is proposed that the same models would be used for decreasing moisture content. The differences in application would be created by differences in the physical structure of the product occurring during moisture removal from the product.

5 SUMMARY

The prediction of specific heat, density, and thermal conductivity of foods and food ingredients is accomplished based on composition and use of appropriate models to account for property magnitudes for each

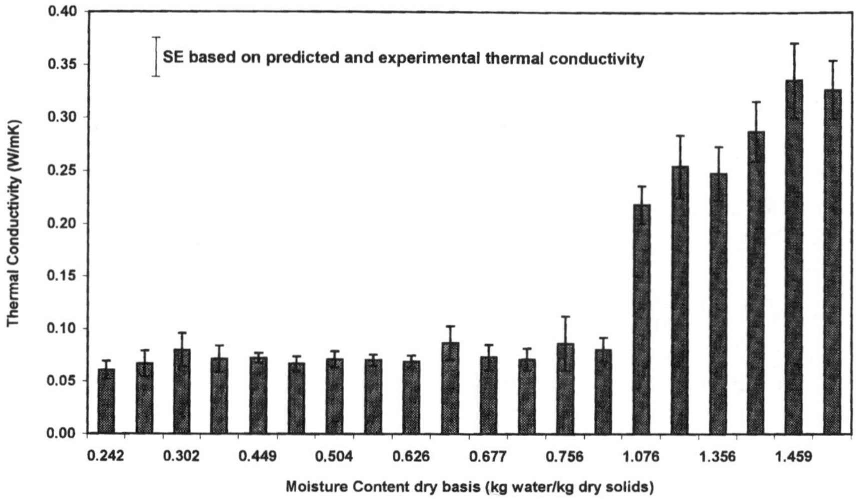

FIGURE 7 Experimental verification of the predicted effective thermal conductivity of casein at 25°C, over intermediate moisture content ranges. (From Ref. 19.)

component, temperature, and physical structure. The models for the prediction of specific heat are based entirely on composition and require knowledge of the specific heat of the moisture-free components of the food, as well as the composition.

The prediction models for density of a food product are equally straightforward at high-moisture contents. When considering lower-moisture contents when a gas phase or air is introduced into the product structure, the prediction models used must accommodate the influence of the low-density air on the product density. When the dry product structure includes particles, the prediction models incorporate the prediction of particle density.

The models for prediction of thermal conductivity of a food product are the most complex, due to the need to consider the influence of physical structure at all moisture contents. In general, the prediction models consider the influence of the orientation of the components on heat conduction within the product structure. At lower moisture contents when air becomes a component of the product structure, the prediction models are developed to accommodate the influence of a thermal conductivity component. At low moisture contents, the distribution of water and air within the physical structure of the product requires an understanding of the components representing the continuous and discontinuous components of the food product system.

REFERENCES

1. JE Siebel. Specific heat of various products. Ice Refrig 2:256, 1892.
2. VE Sweat. Thermal conductivity of foods. In: MA Rao, SSH Rizvi, eds. Engineering Properties of Foods, 2nd ed. New York: Marcel Dekker, 1995, pp 99–138.
3. HA Leninger, WA Beverloo. Food Processing Engineering. Dordrecht: Reidel, 1975.
4. SE Charm. The Fundamentals of Food Engineering. 3rd ed. Westport, CT: AVI Publishing, 1978.
5. DR Heldman, RP Singh. Food Process Engineering. 2nd ed. New York: Van Nostrand Reinhold, 1981.
6. G van Beek, Vade mecum, Koeltechnik-Klimaatregeling 3.1.1–3.3.4, 1976.
7. BK Watt, AL Merrill. Compositian of Foods. Agriculture Handbook No. 8. Washington, DC: US Department of Agriculture, 1975.
8. Y Choi, MR Okos. Effects of temperature and composition on the thermal properties of foods. In: M LaMaguer, P Jelen, eds. Food Engineering and Process Applications, Vol. 1, Transport Phenomenon, New York: Elsevier, 1986.
9. DR Heldman. Food properties during freezing. Food Technol 36(2):92, 1982.
10. DR Heldman. Food Freezing. In: DR Heldman, DB Lund, eds. Handbook of Food Engineering, New York: Marcel Dekker, 1992, pp 277–315.
11. SC Sarma, DR Heldman. An improved approach to prediction of thermal conductivity of a granular food. Unpublished Paper. University of Missouri, Columbia, 1996.
12. CS Sabliov, DR Heldman. Factors influencing thermal conductivity of a food over the range of moisture content from 20 to 120% (db). J Food Sci (in press).
13. L. Riedel. Measurement of the thermal conductivity of sugar solutions fruit juices and milk. Chem Ing-Tech 21:340–341, 1949.
14. VE Sweat. Experimental values of thermal conductivity of selected fruits and vegetables. J Food Sci. 39:1980–1083, 1974.
15. VE Sweat. Modeling the thermal conductivity of meats. ASAE Trans 18(3): 564–568, 1975.
16. Y Choi, MR Okos. The thermal properties of tomato juice concentrate. ASAE Trans 26(1):305–311, 1983.
17. IJ Kopelman. Transient heat transfer and thermal properties in food system. PhD dissertation, Michigan State University, East Lansing, 1966.
18. DR Heldman, DP Gorby. Prediction of thermal conductivity in frozen food. ASAE Trans. 18:156, 1975.
19. SC Sabilov, DR Heldman. A model for prediction of thermal conductivity of food as a function of moisture content and temperature. J Food Process Engr (in press).

2

Introduction to Modeling and Numerical Simulation

K. P. Sandeep
North Carolina State University, Raleigh, North Carolina

Joseph Irudayaraj
The Pennsylvania State University, University Park, Pennsylvania

1 INTRODUCTION

Mathematical modeling is a very useful tool for (relatively) quickly and inexpensively ascertaining the effect of different system and process parameters on the outcome of a process. It minimizes the number of experiments that need to be conducted to determine the influence of various parameters on the safety and quality of a process. Parametric analyses can be conducted to understand the relative effects of different parameters.

The use of approximate methods to solve problems described by partial differential equations has been employed for various reasons, including but not limited to the lack of availability of analytical solutions or empirical correlations, simplicity of solution technique, ability to quickly perform parametric analyses, and because it serves as a means for quickly honing in on the range of parameters to be used in experimental studies or for design purposes.

There are three main categories into which mathematical modeling falls: differential method, integral method, and stochastic method. The finite-difference method falls under the category of differential method. Under the integral method, we have the variational method, finite-volume method, and method of weighted residuals. The method of weighted residuals can be further divided into four categories: collocation method, subdomain method, Galerkin's method, and least squares method; the finite-volume (or control-volume) method can be categorized into two groups: cell-centered schemes and nodal point schemes. The variational method and the method of weighted residuals form the basis for the finite-element method. The boundary element method is a subset of the finite-element method in that it uses a similar approach, but for the surface or boundary under consideration. It can be used in conjunction with the finite-element or finite-volume method. The Monte Carlo method falls under the stochastic method. This is a computationally intensive and probabilistic method used primarily when the number of independent variables are large.

The finite-element method and the finite-difference methods are the most popular techniques used to solve problems associated with food processing. Relatively simple problems can be tackled with ease by commercially available software. Complicated problems require either modification of commercial codes or writing the code from scratch.

2 CLASSIFICATION OF PARTIAL DIFFERENTIAL EQUATIONS

Partial differential equations (PDEs) are classified as linear or nonlinear depending on whether there is a product of two terms containing either the dependent variable or its derivatives. If a PDE is linear in its highest-order derivative, but nonlinear in one or more of the lower-order derivatives, it is called a quasilinear PDE. The order of a PDE is the highest power of the derivative in the equation.

Consider the following second-order PDE:

$$A\frac{\partial^2\Phi}{\partial x^2}+B\frac{\partial^2\Phi}{\partial x\,\partial y}+C\frac{\partial^2\Phi}{\partial y^2}+D\frac{\partial\Phi}{\partial x}+E\frac{\partial\Phi}{\partial y}+F\Phi+G=0$$

The coefficients A, B, C, D, E, F, and G can be functions of x, y, or Φ.

The above PDE is said to be elliptic if $B^2-4AC<0$, parabolic if $B^2-4AC=0$, and hyperbolic if $B^2-4AC>0$ at all points in the domain.

Auxiliary variables are usually introduced to convert second-order PDEs to first-order PDEs, at least for the purpose of classification. This formulation may then be used for solving the system of equations too.

A PDE is said to be in conservative form (or conservation form or conservation-law form or divergence form) if the coefficients of all the derivative terms in the equation are either constant or, if variable, their derivatives do not appear anywhere in the equation. The schemes that maintain the discretized version of the conservation statement exactly (except for round-off errors) for any grid size over any region in the domain for any number of grid points is said to have the conservative property.

The nonconservative form of the continuity equation is

$$\rho\,\frac{\partial u}{\partial x}+\rho\,\frac{\partial v}{\partial y}+u\,\frac{\partial \rho}{\partial x}+v\,\frac{\partial \rho}{\partial y}=0$$

The conservative form of the same equation is

$$\frac{\partial}{\partial x}\,(\rho u)+\frac{\partial}{\partial y}\,(\rho v)=0 \quad \text{or} \quad \nabla\cdot(\rho\mathbf{V})=0$$

Equilibrium problems (or jury problems) are problems for which the solution of the PDE is required in a closed domain for a given set of boundary conditions. Equilibrium problems are boundary-value problems and are governed by elliptic PDEs.

Marching (or propagation) problems are transient or appear to be transient problems and the solution of the PDE is required in an open domain for a given set of initial and boundary conditions. Problems in this category are either initial-value or initial-boundary-value problems. Marching problems are governed by hyperbolic or parabolic PDEs.

3 NUMERICAL FORMULATION

Numerical formulations are based on the classification of the governing equation. When dealing with the non-steady-state heat equation or the scalar (linear or nonlinear) Burger's equation, formulations applicable to parabolic equations are used. When dealing with the wave equation, formulations for hyperbolic equations are used, and when dealing with Laplace's equation, formulations for elliptic equations are used. Formulations for all types of equations can be explicit or implicit. Explicit formulations are simple, but the number of computations and the instability of the formulation (addressed in the next section) are some of its drawbacks.

The Navier–Stokes equations are hyperbolic in the inviscid region and parabolic in the viscous region. For steady-state conditions, they are hyperbolic in the inviscid region and elliptic in the viscous region. The scalar equations which are similar to the Navier–Stokes equations, are the Burger's equations (linear and nonlinear). Thus, the starting point for solving the

Navier–Stokes equations involves understanding the methods employed to solve the Burger's equations.

Some of the commonly used explicit formulations for parabolic equations are the forward time central space (FTCS) method, Richardson's method, and the DuFort–Frankel method, whereas some of the commonly used implicit methods are the Laasonen method, Crank–Nicolson method, and the beta formulation. The five-point and nine-point methods are the commonly used methods to address elliptic problems. Euler's forward-time forward space (FTFS), Euler's FTCS, first upwind differencing, Lax method, midpoint leapfrog method, Lax–Wendroff method, Rusanov or Burstein–Mirin method, and Warming–Kutler–Lomax (WKL) method are some of the commonly used explicit methods for hyperbolic equations. Euler's backward-time central space (BTCS) and the Cank–Nicolson methods are two of the commonly used implicit methods for hyperbolic equations.

Multistep (or splitting) methods are usually used for nonlinear problems and sometimes for linear problems too. In this method, the finite-difference equations are written out at two or more time steps. The first step involves determination (or prediction) of the variable at an intermediate time step and the second step involves correcting it; hence, multistep methods are also called predictor–corrector methods. The Richtmyer formulation, Lax–Wendroff multistep method, MacCormack method, and the Warming and Beam (upwind) method are some of the commonly used multistep methods with hyperbolic equations.

4 CLASSIFICATION AND GENERATION OF GRIDS

In order to solve the partial differential equations that represent the physical problem, the domain of interest has to be divided into grid lines; the points of intersection of these grid Lines are called nodes. The accuracy of the solution depends on many factors, including grid spacing. Grids are classified as structured or unstructured depending on whether there is a set pattern of identification of nodes and if the solution process can proceed in an ordered sequential manner from one node to the next. The advantages of using the complicated unstructured grid system are that they can be used to fit irregular, singly-connected and multiply-connected domains. For two-dimensional geometries, the most common method of unstructured grid generation involves discretizing the domain into triangles (most flexible shape to fit various kinds of boundaries). The advancing front method and the Delaunay method are two of the commonly used techniques for triangulation of the domain.

The grid system used could be orthogonal [Cartesian, cylindrical, spherical (depending on the boundary configuration of system)] or non-orthogonal (such as triangular). Due to the complex geometries of the domain of interest and the possibility or necessity of having more grids close to boundaries, the physical domain is transformed into a computational domain (by twisting or stretching), where the grids are rectangular.

Grid generation can be divided into three main categories: algebraic (simple and fast—using one of many algebraic equations or interpolation techniques), partial differential equation (elliptic, hyperbolic, or parabolic), conformal mapping using complex variables. Grid systems are also classified as fixed (independent of solution and generated before solving the problem) or adaptive (grids move toward regions of steep gradients as the solution process proceeds).

Some of the desirable features of a grid system are (1) a mapping that ensures one-to-one correspondence with grid lines of the same family not intersecting, (2) grid point distribution that is smooth, (3) grid lines that are orthogonal or close to orthogonal and, (4) option for grid point clustering.

Grid point clustering (or grid embedment) is a technique used to increase the number of grid points around a specific grid point or around a grid line. It is performed by the appropriate choice of functions used in the transformation of coordinates. Two of the common ways of handling grid embedment is by the meshing of the grid method (where weighting factors are introduced to determine the relative influence of each point near the interface of the coarse and fine grids on the solution variable) and the separate regions method (in which there are two types of grids: interface and noninterface; at the fine-grid boundary, interpolation of the values at the coarse grid is performed to obtain values of the variable).

One of the easiest ways of obtaining staggered grids is by shifting the grid vertically or horizontally by half a grid space. This technique is used to improve the stability criterion by coupling of variables when the governing system of equations can be solved sequentially. Thus, there is a primary set and a secondary set of grids with different variables being specified on the primary and secondary grids. Consider the example of the incompressible Navier–Stokes equations and consider a grid point in the system where the pressure is specified. Immediately to the right and left of this point, the x component of the velocities are specified, and immediately to the top and bottom of this point, the y component of the velocities are specified. The Marker and cell method and the DuFort–Frankel method are two of the commonly used methods with staggered grids. Another technique used for coupling of equations is the multilevel (multigrid) method and it has been used for the diffusion, Poisson, and Navier–Stokes equations.

5 BOUNDARY AND INITIAL CONDITIONS

A boundary condition (BC) is said to be of the Dirichlet kind if the value of the dependent variable is given along the boundary. If the derivative of the dependent variable is given along the boundary, it is said to be a Neumann BC. If the BC at the boundary is given as a linear combination of Dirichlet and Neumann BCs, it is said to be a Robin BC. If the BC along a part of the boundary is of the Dirichlet type and another part is of the Neumann type, the overall BC is said to be a Mixed BC.

6 ERRORS, CONSISTENCY, STABILITY, COMPATIBILITY, AND CONVERGENCE

The errors associated with first-order accurate methods are known as dissipation errors (tend to decrease the amplitude of the wave) and that of second-order accurate methods are known as dispersion errors (tend to cause oscillation of the solution).

Truncation error: It is the error introduced by truncating terms in the finite-difference formulation. It is the difference between the PDE and the finite-difference formulation.

Discretization error: It is the error in the solution of a PDE due to transformation of the continuous problem to a discrete problem, and it is the difference between the exact solution of the PDE (without round-off error) and the exact solution of the finite-difference formulation (without round-off error). Thus, it is the error in the solution due to truncation and any errors due to the BCs.

Round-off error: It is the error associated with rounding off numbers in mathematical operations.

Consistency: It relates to the extent to which the finite-difference formulation approximates the PDE. A formulation is said to be consistent if the truncation error tends to zero as the mesh size tends to zero. Methods which are of the order Δt or Δx are consistent as the error tends to zero as the mesh size tends to zero. However, schemes that are of the order $\Delta t/\Delta x$ may potentially be inconsistent unless it is ensured that $\Delta t/\Delta x$ tends to zero.

Stability: A scheme is said to be stable if errors (round-off, truncation, etc.) do not grow as the scheme proceeds (or marches) from one step to another and is, hence, strictly applicable to marching problems only. In the solution of finite-difference equations, two types of errors exist: discretization or round-off (computational). It is important to control the growth of these errors so that the solution is stable. Two standard methods exist for stability analysis: discrete perturbation stability analysis and von Neumann (Fourier) stability analysis. The latter method is simpler and more commonly used.

Convergence: Usually, a consistent and stable scheme is convergent. Convergence relates to the solution of the finite-difference formulation approaching the solution to the PDE as the mesh size is refined. According to Lax's equivalence theorem, "Given a properly posed initial value problem and a finite-difference approximation to it that satisfies the consistency condition, stability is the necessary and sufficient condition for convergence." Although this theorem has not been proven for nonlinear PDEs, it is used for them too.

7 SOLUTION OF THE FINITE-DIFFERENCE EQUATIONS

Once the finite difference equations have been formulated and the stability criteria met, the set of equations have to be solved. Several direct and iterative methods exist for solving them, and they are discussed in the following sections.

7.1 Direct Methods

Cramer's rule: Simple, but extremely time-consuming; Number of operations $= (N + 1)!$, for N unknowns.

Gaussian elimination: It is an efficient means of solving algebraic equations, especially the tridiagonal system of equations. Approximately N^3 multiplications are required for solving N equations. To improve accuracy, equations are rearranged such that the largest coefficients occupy the diagonal (this process is called pivoting).

Some of the other direct methods include the LU decomposition method, error vector propagation (EVP) for the Poisson equation [1], odd–even reduction method [2], and the fast Fourier transform method [3,4].

Direct methods require an exorbitant number of arithmetic operations and they are usually restricted by one or more of the following: type of coordinate system (e.g., Cartesian), type of domain (e.g., rectangular), size of coefficient matrix, and type of BCs.

7.2 Iterative Methods

Usually, an initial solution is guessed, new values computed, and the process continued until convergence is obtained. If a formulation results in only one unknown, it is called a point iterative method and if the formulation involves more than unknown (usually three unknowns that result in a tridiagonal coefficient matrix), it is called a line iterative method. Some of the commonly employed iterative techniques are listed.

Alternating direction implicit (ADI) method for parabolic equations: The ADI method is a subset of the approximate factorization method (replacement of original finite-difference formulation by tridiagonal formulation). This method applies to two or three-dimensional cases.

Fractional step method for parabolic equations: This technique involves splitting of a multidimensional problem into a series of one-dimensional problems and solving them sequentially.

Alternating direction explicit (ADE) methods for parabolic equations: They do not require tridiagonal matrices to be inverted and can be used for one-dimensional equations also.

Jacobi method: Initial values of the variable are either prescribed or guessed (at the first iteration step) and the value of the variable at all grid points (at the previous iteration step are used) to solve for the variable at the grid point (i, j) at the new iteration step.

Point Gauss–Siedel method: This is an improvement of the Jacobi method. In this method, the values of the variable computed at the new iteration step are immediately used in the computation of the variable at all grid points at the new time step (as soon as they become available). It has a much higher convergence rate than the Jacobi method.

Line Gauss–Siedel method: This method is applied when there are three unknowns. The finite-difference equation, when processed under the same guidelines as the point Gauss–Siedel method results in a system of linear equations with a tridiagonal coefficient matrix. This method has a faster convergence rate than the point Gauss–Siedel method.

Successive overrelaxation (SOR): This is a technique used to accelerate any iterative procedure based on guessing the trend of a solution and modifying the solution appropriately. A parameter, ω $(0 < \omega < 2)$, is used to multiply a set of terms in the equation used for a method such as the Gauss–Siedel method. If $0 < \omega < 1$, it is called underrelaxation, and if $1 < \omega < 2$, it is called overrelaxation. Overrelaxation is similar to linear extrapolation (and is used usually for Laplace's equation with Dirichlet BCs), whereas underrelaxation is used when the solution is oscillating (usually used for nonlinear elliptic equations). Determining the optimum relaxation factor (ω_{opt}) greatly accelerates the convergence; expressions for the optimum value for various types of equations and BCs have been determined by researchers.

8 LINEARIZATION

Consider a nonlinear term such as $u(\partial u/\partial x)$. All the values at time j are known for a given location i and the values at $i+1$ are to be determined. Three of the commonly used linearization techniques are listed:

Lagging: The coefficient is used at the known value, i. Thus, the formulation for $u(\partial u/\partial x)$ becomes

$$u_{i,j}\frac{u_{i+1,j}-u_{i,j}}{\Delta x}$$

There is only one unknown $(u_{i+1,j})$ in this expression and the finite-difference formulation is linear.

Iterative: This method involves updating the lagged value until convergence is reached. The formulation for this method is

$$u_{i+1,j}^{k}\frac{u_{i+1,j}^{k+1}-u_{i,j}^{k}}{\Delta x}$$

For the first iteration, $u_{i+1,j}$ is the value at the previous location, $u_{i,j}$. Once $u_{i+1,j}$ is determined at $k+1$ the coefficient $u_{i+1,j}^{k}$ is updated and a new solution is obtained and this process is continued until the convergence criterion has been met.

Newton's iterative linearization: This method uses the technique of evaluating the change in a variable between two iterations and dropping second-order terms to arrive at the following expression for the nonlinear term:

$$\frac{2u_{i+1,j}^{k}u_{i+1,j}^{k+1}-(u_{i+1,j}^{k})^{2}-u_{i,j}^{k}u_{i+1,j}^{k+1}}{\Delta x}$$

9 INTRODUCTION TO THE FINITE-ELEMENT METHOD

9.1 How It Works

The finite-element discretization procedure reduces the given region into a finite number of elements. A collection of the elements is called the *finite-element mesh*. The elements are connected to each other at points called *nodes*. The nodes typically lie on the element boundary where adjacent elements are connected. In addition to boundary nodes, an element may also consist of a few interior nodes. The nodal points depict the field variable or the unknown defined in terms of approximating or interpolating functions within each element. The nodal values of the field variable and the interpolating functions for the elements define the behavior of the field variable within the elements. The nature of the solution and the degree of approximation depend not only on the size of the elements but also on the interpolating functions which should satisfy compatibility and continuity conditions.

The solution using the finite-element technique is obtained predominantly by *variational* or *weighted residual* method. The variational approach has its foundations in variational calculus and requires the use of a *functional*, whereas the weighted residual approach used the governing equations. The finite-element procedure consists of the following steps.

9.2 Discretization

This involves dividing the problem domain into subdomains. Generally, for a one-dimensional problem, this is very simple. However, the degree of complexity increases with the number of dimensions and the nonuniformity of the object in question. Discretization or division of the domain into smaller components can be accomplished by choosing a variety of different element shapes and nodes. The choice of the type of element and the number of nodes in an element are left to the engineer/scientist's discretion based on experience.

9.3 Interpolating Functions

Once the elements are defined, the next step is to assign nodes to each element to choose the appropriate interpolating function to represent the variation of the field variable over the element. Generally, interpolating functions are polynomials that can be easily integrated and differentiated subject to certain continuity requirements imposed at the element boundaries.

9.4 Element Matrix Formation to Obtain Global Matrix

Depending on the choice of the procedure (variational or weighted residual method), element matrices are calculated by transforming the elements from the global to a local coordinate system where integration and differentiation are performed and then back-transformed into the global matrix. Depending on the element connectivity or the nodes in the element, the element matrix is incorporated into the global matrix. Similar calculations are performed for each element and the *global matrix* is assembled using the element matrix.

9.5 Boundary Conditions

Before solving for the unknown variables, boundary conditions are imposed to the global matrix. The two types of boundary conditions are natural and essential boundary conditions. Natural boundary conditions are convective boundary conditions, whereas essential boundary conditions are constant or specified boundary conditions [4].

9.6 Solution of the System of Equations

The assembled equations consists of a set of simultaneous equations that can be solved using the matrix solvers. For time-dependent problems, the unknown nodal values are a function of time; hence, an appropriate finite-difference time-stepping scheme is generally chosen.

9.7 Summary of the Steps Involved in a Typical Finite Element Problem

1. Discretize the problem domain and construct the finite-element mesh
2. Derive element matrices for the system
3. Evaluate element equations and assemble element matrix to form the global matrix
4. Impose boundary conditions
5. Solve the system of equations using an appropriate solver
6. Postprocessing: graphics, calculation of gradients, and so forth

9.8 Future Applications

Most of the future growth expected will be in the application and validation of the finite-element results by experimental data. Further refinement of the existing finite-element procedures will also increase [5]. Appropriate solution procedures for solving problems with nonlinear and random material properties and boundary conditions will increase. Interest in the application of the finite-element method in biological systems and more direct integration of the technique with the actual design will also be given priority. Another crucial area that will demand attention is in solving microstructural problems in engineering and biological sciences. Other areas that will demand attention are adaptive finite elements and the application of parallel processing.

10 COMMERCIAL CODES AND RESOURCES AVAILABLE

Commercial codes available are often packaged with preprocessing and post-processing modules. Preprocessing involves transformation of the physical problem into computational domain and generating a grid mesh in the computational domain. Postprocessing involves presenting the data obtained by the code in graphical form.

There are many commercial codes available to solve most standard problems involving standard governing equations, boundary conditions, and relatively simple geometries. They are available on different platforms (PC, UNIX, SGI, etc.). There are many universities and research groups that offer codes or services, or would be interested in collaborative efforts. Some of the resources for preprocessing, processing, and postprocessing are as follows:

Adina R&D (http://www.adina.com)
Phoenics (http://www.cham.co.uk)
Fluent (http://www.fluent.com)—Gambit, FLUENT, FIDAP, POLYFLOW, NEKTON, IcePak, and MixSim software.
Innovative Research, Inc. (http://www.inres.com)
CFD Research Corporation (http://www.cfdrc.com)
Amtec (http://www.amtec.com)
http://icemcfd.com/cfd/CFD_homepages.html (or http://icemcfd.com/cfd/CFD_codes_o.html)
ANSYS (http://www.ansys.com/)

REFERENCES

1. PJ Roache. Computational Fluid Dynamics. Albuquerque, NM: Hermosa, 1972.
2. O Buneman. A compact non-iterative Poisson solver. Institute for Plasma Research SUIPR Report 294, Stanford University, 1969.
3. RW Hockney. A fast direct solution of Poisson's equation using Fourier analysis. J Asso Computing Machin 12:95–113, 1965.
4. RW Hockney. The potential calculation and some applications. Methods Comput Phys 9:135–211, 1970.
5. JN Reddy, An Introduction to the Finite Element Method. New York: McGraw-Hill, 1993.

3

Aseptic Processing of Liquid and Particulate Foods

K. P. Sandeep
North Carolina State University, Raleigh, North Carolina

Virendra M. Puri
The Pennsylvania State University, University Park, Pennsylvania

1 INTRODUCTION

Aseptic processing involves sterilization of a food product (in a direct or indirect contact heat exchanger), followed by holding it for a specified period of time (in a holding tube), cooling it, and, finally, packaging it in a sterile container. The use of high temperature for a short period of time (in comparison with conventional canning) in aseptic processing yields a high-quality product. The demand for high-quality shelf-stable products has been the driving force for commercialization of aseptic processing. Deaeration (prior to sterilization) is usually an integral part of aseptic processing, as removal of air enhances product quality and increases the shelf life of a product. It also stabilizes the product prior to processing. Care should be taken to ensure that all process calculations are performed after the deaeration stage and not based on the initial raw product. Another important part of an aseptic processing system is the back-pressure valve, which provides sufficient pressure to prevent boiling of the product at processing temperatures which can be as high as 125–130°C. An aseptic surge tank provides the means for the product to be continuously processed even if the packaging system is not operational due to any malfunction. It can also be used to package the sterilized product while the processing section is being resterilized.

Sterilization of the processing system, packaging system, and the airflow system prior to processing is of utmost importance. This is what is referred to as presterilization. The recommended heating effect for presterilization (using hot water) of the processing equipment for low-acid foods is the equivalent of 250°F for 30 min. The corresponding combination for acid or acidified products is 220°F for 30 min. This often involves acidification of the water (to below a pH of 3.5 for acid products) used for sterilization. Presterilization of an aseptic surge tank is usually done by saturated steam and not hot water because of the large volume associated with the surge tank.

Better product quality (nutrients, flavor, color, texture), less energy consumption, eliminating the need for refrigeration, easy adaptability to automation, use of any size package, use of flexible packages, and less expensive packaging costs are some of the advantages of aseptic processing over the conventional canning process. Some of the reasons for the relatively low number of aseptically processed products include slower filler speeds and higher overall cost. Aseptic processing also requires better quality control of raw products, better trained personnel, and better control of process variables and equipments. Some of the disadvantages of aseptic processing include increased shear rates, degradation of some vitamins (some vitamins are stable at pasteurization temperatures but not at sterilization temperatures), separation of solids and fats, precipitation of salts, and change in the flavor or texture of the product relative to what consumers are accustomed to. Minimization of the off-flavors produced can be accomplished by steam injection (short heating time) followed by flash-cooling. Thus, it can be seen that not all products can be aseptically processed to yield a high-quality product.

Because of some of the stringent regulatory requirements of aseptic processing, many processors adopt an aseptic process but package it in nonaseptic containers. This results in products that are called "extended-shelf-life products." Such processes are easier to adopt, require less monitoring (because the resulting product–package combination does not need to be sterile), and are easier to file with regulatory agencies. One such process involves ultrapasteurization of milk, wherein extended shelf-life can be obtained.

Notwithstanding the problems associated in producing aseptically processed foods, several companies have adopted this technology. Some of the products that are aseptically processed include fruit juices, milk, condensed milk, coffee creamers, puddings, soups, butter, gravies, and jelly. Some of the companies that deal with aseptic processing and packaging equipment are International Paper, Tetra Pak, Combibloc, Elopak, Cherry Burrell, Alfa Laval, ASTEC, VRC, APV, FranRica, Benco, Scholle, Bosch, and Metal Box.

The pH of a food product is a critical factor in determining the type of processing to be adopted and the class of viable microorganisms of concern. Foods are usually divided into three pH groups while designing a thermal process. They are high-acid foods with pH values less than 3.7, foods with pH values between 3.7 and 4.6 and the low-acid foods with pH values greater than 4.6. For low-acid foods, the anaerobic conditions that prevail in aseptic processing are ideal for the growth of some toxin-producing microorganisms such as *Clostridium botulinum*. To obtain a commercially sterile product, all pathogenic microorganisms must be destroyed during aseptic processing.

Bacteria are the primary organisms of concern in food processing. They multiply by the process called fission, wherein one cell splits into two cells. The growth of bacteria is generally divided into seven stages: lag phase (no growth or even a decrease in numbers), accelerated growth phase (rate of growth is increasing), logarithmic phase (most rapid and constant increase in numbers), deceleration phase (rate of growth is decreasing), stationary phase (numbers remain constant), accelerated death phase (rate of death is increasing), and final death phase (numbers decrease at a constant rate). In order to extend the shelf life of products, one of the techniques is to prolong the first two phases (lag and accelerated growth phase) of the growth of bacteria. Once bacteria reach the third stage (logarithmic phase), spoilage will occur rapidly. Some of the techniques to prolong the first two phases are refrigeration, freezing, drying, reduction in available oxygen, and reduction in the initial number of bacteria. These techniques must be accompanied by other practices such as the use of appropriate packaging and storage conditions. These are some of the techniques commonly used to extend the shelf life of products.

2 TYPES OF PROCESSING

Techniques to process and preserve foods range from retorting (canning) to frozen storage. Some of the other techniques of processing and preservation include hot-fill, refrigeration, and drying of foods. Not all products can be processed or preserved using the same technique. Feasibility of processing and the quality of the end product determine the type of processing and preservation technique employed for various foods.

The quality of canned foods is not very high because products are subjected to heat treatment for an extended period of time. On the other hand, the short processing times involved during aseptic processing leads to the production of a high-quality product. Recovery of heat from the heat exchangers used in aseptic processing also makes it more energy efficient. In addition, the products are shelf stable and, hence, do not require further control like refrigerated or frozen foods. Refrigerated foods (after pasteurization) require careful monitoring of the storage and distribution temperatures. They also have a shorter shelf life than aseptically processed products and, hence, their range of distribution is limited. The quality of frozen foods is generally high, but they need to be thawed and then cooked. The thawing process can result in uneven heating zones, especially if a microwave oven is used. In addition, depending on the storage period, the energy requirements for freezing can be a major portion of the total cost involved.

2.1 Critical Factors and Problems Associated with Processing

Some of the factors that affect the choice of the type of process include the viscosity of the product and the presence of large particles and/or low-thermal-conductivity particles. The simplest type of food product is a homogeneous low-viscosity liquid product. Direct heating by steam injection or steam infusion is a commonly employed method for heating such products. For higher-viscosity products, plate and tubular heat exchangers are employed. For extremely viscous products, a scraped-surface heat exchanger is usually used. When relatively high-viscosity products containing large particles and/or low-thermal-conductivity particulates are involved, dielectric (microwave) and ohmic heating are two commonly employed methods. The density of the particles is also an important issue to be considered and will be addressed in the section that deals with residence time distribution.

Heat transfer from the carrier fluid to the particle is a function of the boundary layer surrounding the particle, which, in turn, is a function of the thermophysical and rheological properties of the fluid and the relative velocity between the particle and the fluid. This boundary layer governs the convective heat transfer coefficient at the particle–fluid interface. In addition, the existence of a residence time distribution presents a problem of some particles being subjected to less thermal treatment than others. If the heating time is based on mean velocity, the faster-moving particles will be understerilized, whereas the slower-moving particles will be oversterilized. Knowledge regarding the spread of residence times for the fluid and particles is essential in determining the thermal treatment that any product has received.

2.2 Relevant Historical Background

The work of Olin Ball and the American Can Research Department laid the foundation of aseptic processing in the United States as early as 1927 when the HCF (heat, cool, fill) process was developed [1]. This was followed by the Avoset process in 1942 (steam injection of the product coupled with retort or hot-air sterilization of packages such as cans and bottles) and the Dole–Martin aseptic process in 1948 (product sterilization in a tubular heat exchanger, metal container sterilization using superheated steam at temperatures as high as 450°F because dry heat requires higher temperature than wet heat, followed by aseptic filling and sealing of cooled product in a superheated steam environment). The early 1960s was marked with the advent of a form–fill–seal package (tetrahedron package). The late 1960s saw the advent of the Tetra Brick aseptic processing machine and the late 1970s saw the advent of the Combibloc (blank carton) aseptic system. Soon, aseptic filling in drums and bag-in-box fillers were established. One of the major landmarks in the history of aseptic processing is the approval of use of hydrogen peroxide for the sterilization of packaging surfaces by the Food and Drug Administration (FDA) in 1981. In recent years, a major breakthrough for the aseptic processing industry was in 1997, when Tetra Pak received a no-objection letter from the FDA for aseptic processing of low-acid foods containing large particulates.

3 FLUID MECHANICS ASPECTS OF PROCESSING

The type of fluid, flow characteristics, and fluid properties are some of the fluid mechanics aspects that are important in designing an aseptic process. These parameters and the system configuration, in turn, are important factors that determine the choice of the pump to be used. The residence time distribution of the fluid elements and, more importantly, that of the particles are the factors that eventually are used in designing holding tubes.

3.1 Types of Fluids

Time, shear rate, temperature, and particle concentration are some of the factors that affect the viscosity of a fluid or suspension. Fluids are characterized as time dependent or time independent depending on whether the shear stress experienced by a fluid under a constant shear rate varies as a function of time. If the shear stress increases with time, it is called a rheopectic fluid, and if the shear stress decreases with time, it is called a thixotropic fluid. Time-independent fluids are divided into two categories: Newtonian and non-Newtonian. Newtonian fluids obey Newton's law of viscosity [shear stress (σ) and shear rate ($\dot{\gamma}$) are linearly related], whereas non-Newtonian fluids do not have a linear relationship of shear stress versus shear rate. The Herschel–Bulkley model is the most commonly used model to describe the flow behavior of most liquid food products:

$$\sigma = \sigma_0 + K(\dot{\gamma})^n$$

In this equation, σ_0 is the yield stress, K is the consistency coefficient, and n is the flow behavior index. For a Newtonian fluid, $\sigma_0 = 0$, $K = \mu$, and $n = 1$. A pseudoplastic fluid is one for which $\sigma_0 = 0$ and $n < 1$, whereas a dilatant fluid is one for which $\sigma_0 = 0$ and $n > 1$.

For a non-Newtonian fluid, the concept of apparent viscosity is introduced because the ratio of shear stress to shear rate is not a constant. Apparent viscosity is the ratio of the shear stress to shear rate and is always expressed along with the shear rate because it varies with shear rate. For a pseudoplastic fluid, the apparent viscosity decreases with an increase in shear rate, whereas for a dilatant fluid, the apparent viscosity increases with an increase in shear rate. When small particles of low concentration (Φ) are suspended in a fluid, the concept of effective viscosity (μ_e) comes into picture. One of the equations used to determine the effective viscosity of a suspension is

$$\mu_e = \mu(1 + 2.5\Phi + 14.1\Phi^2)$$

Temperature is a major factor that affects the viscosity of Newtonian and non-Newtonian fluids. For a Newtonian fluid, the Arrhenius model, given by the following equation is the most commonly used equation to determine the effect of temperature on the viscosity of a fluid:

$$\mu = Be^{-E_a/R_g T}$$

Thus, to determine the Arrhenius parameters B and E_a, a graph of $\ln(\mu)$ versus $1/T$ is made, the slope of which is the $-E_a/R_g$ and the intercept is $\ln(B)$. Thus, the flow behavior of the fluid as a function of temperature can be modeled.

3.2 Dimensionless Numbers Governing Flow

When a fluid flows through a tube at low velocities, the flow is characterized by a steady streamline flow and the flow is referred to as laminar flow. At higher flow rates, the flow becomes erratic and is referred to as turbulent flow. The Reynolds number is the nondimensional number that is used to characterize the type of flow and the generalized Reynolds number (valid for power-law fluids, in addition to Newtonian fluids) is defined as

$$N_{\mathrm{GRe}} = \frac{\rho \langle \bar{u} \rangle^{2-n} d^n}{K[(3n+1)/n]^n 2^{n-3}}$$

This expression reduces to the following form for a Newtonian fluid:

$$N_{\mathrm{Re}} = \frac{\rho \bar{u} d}{\mu}$$

For flow in a straight tube of circular cross section, laminar-flow conditions are said to exist if the Reynolds number is less than 2100 and the flow is said to be turbulent if the Reynolds number is greater than 10,000. In the intermediate Reynolds number region, the flow is said to be in transition. The Reynolds number is thus a convenient nondimensional quantitative measure of the type of flow in different flow systems (different pipe diameters, flow rates, etc.).

Laminar-flow conditions offer the advantage of simplicity in computations involving flow and heat transfer equations. However, the major drawback of laminar flow is the relatively low heat transfer coefficient. One way to enhance heat transfer coefficient without going into the turbulent regime is by using coiled tubes. Flow in coiled tubes is characterized by flow in the primary (axial) direction and also in the secondary (radial) direction. This is the result of the radial pressure gradient that develops due to the centrifugal force. The secondary flow is characterized by two counterrotating vortices in the cross section of the tube. The strength of these vortices depends on many factors, such as the tube-to-coil diameter ratio, flow rate, viscosity, and pitch of the coil. The nondimensional number that characterizes flow in a coiled tube is the Dean number (N_{De}):

$$N_{\mathrm{De}} = N_{\mathrm{Re}} \sqrt{\frac{d}{D}}$$

The use of helical holding tubes as a means of narrowing the residence time distribution (RTD) of particles has been suggested by several researchers in the past. The narrowing of the RTD was attributed to the development of secondary flow. Dean [2] was the first to analyze mathematically the phenomenon of secondary flow in helically coiled tubes. Dean obtained analytical expressions for the velocity profile valid for large radii of curvature ($\lambda_c \gg 1$) and low Dean numbers ($N_{\mathrm{De}} = N_{\mathrm{Rc}}/\sqrt{\lambda} \ll 17$). Dean [3] solved the Navier–Stokes equation and obtained an approximate expression for the velocity of the fluid as a function of position. Truesdell and Adler [4] obtained a numerical solution of the Navier–Stokes equations which are valid over a wide range of curvature and Reynolds numbers.

Taylor and Yarrow [5] found that secondary flow could stabilize laminar flow, providing transition Reynolds number as high as 6000–8000 in a curved tube as compared to 2100 in a straight tube. Koutsky and Adler [6] pointed out that the pressure drop in a tube formed into a helix can be up to four times as great as that in an identical straight tube. They also found that stable laminar flow can be maintained in helices at Reynolds numbers up to 8000 or more. Both of these facts imply the existence of strong secondary flow in helices. Secondary flow is known to increase the momentum, mass, and heat transfer and an increase in Reynolds number is also known to decrease the axial dispersion.

The results of some of the studies that have been conducted to determine the critical value of Reynolds number that separates laminar and turbulent flows are mentioned here. White [7] conducted experiments with oil and water for different curvatures of helical tubes. For $N_{\mathrm{Re}} > 100$ and $d/D = 1/50$, the curved pipe had a greater resistance than a straight pipe of the same diameter and length. The resistance to flow became 2.9 times that in a straight pipe at $N_{\mathrm{Re}} = 6000$ ($\sim N_{\mathrm{Re}}$ when flow becomes turbulent). When $d/D = 1/15$, turbulent flow was seen at $N_{\mathrm{Re}} \sim 9000$, and when $d/D = 1/2050$, turbulence was seen at $N_{\mathrm{Re}} \sim 2250$–$3200$. Many equations have been developed for predicting the critical Reynolds number that separates laminar flow from turbulent flow. One such equation developed by Srinivasan et al. [8] is

$$N_{\mathrm{Re}_c} = 2100\left[1 + 12\left(\frac{d}{D}\right)^{1/2}\right]$$

It is important to note that most equations similar to the above have a range of applicability. The limitations may be to the range of Reynolds number, tube-to-coil diameter ratio, pitch, or other factors.

It is known that for a given pressure gradient, the flow rate in a coiled tube is lower than that in a straight tube. Several correlations have been developed to predict the flow rate in a helical tube. One such correlation is presented in a nondimensional form [9]:

$$\frac{\dot{V}_c}{\pi R^2 \bar{u}} = 1 - 0.0306\left(\frac{N_{\mathrm{De}}^2}{288}\right)^2 + 0.012\left(\frac{N_{\mathrm{De}}^2}{288}\right)^4$$

where $\bar{u}$ is the average velocity in a straight pipe of the same radius under the same axial pressure gradient.

Thus, it is important to make comparisons of Dean numbers while dealing with flow in helical tubes, just like comparisons of Reynolds numbers are made in straight tubes. It can also be seen that decreasing the coil diameter enhances the extent of secondary flow. Optimization is performed to choose the appropriate coil diameter because decreasing the coil diameter results not only in enhanced mixing and heat transfer but also in an increase in the pressure drop.

3.3 Friction Factor

As a fluid flows through a pipe, friction impedes axial flow and creates a pressure drop between the inlet and outlet of the tube. The pressure drop depends on the type of flow (laminar, transition, or turbulent), type of fluid, and the type of pipe. As the fluid flows through a tube, there is loss in energy (E_f) and pressure (ΔP_f) due to friction and they are determined as follows:

$$E_f = \frac{\Delta P_f}{\rho} = 2f\,\frac{\bar{u}^2 L}{d}$$

In this equation, f is the friction factor and it varies with the type of pipe, flow conditions, and system geometry. The friction factor for laminar flow in a straight pipe is given by $f_s = 16/N_{\mathrm{Re}}$, and for turbulent flow, it is usually determined from the Moody [10] diagram. An alternative way to determine friction factor is to use the following equation by Colebrook [11] and perform an iterative analysis:

$$\frac{1}{\sqrt{f}} = -4\ln\left(\frac{1}{3.7}\,\frac{\varepsilon}{d} + \frac{1.255}{N_{\mathrm{Re}}\sqrt{f}}\right)$$

These are standard methods for determining friction factors in straight pipes. However, for curved pipes, there are many additional factors, such as coil diameter, pitch, and Dean number, that come into play and hence, there is no standard formula or procedure available. Manlapaz and Churchill [12] have presented a list of studies conducted for the determination of friction factors in curved pipes.

3.4 Pumps and Pumping Requirements

The choice of the pump for any food processing operation, including aseptic processing depends on many factors, including whether the product has particulates in it, the extent of slippage (if applicable), piping arrangement, fittings present, and the flow behavior of the fluid The first step in determining the type and rating of a pump required for any food processing operation involves the use of the Bernoulli's equation:

$$E_p = \Delta\text{PE} + \Delta\text{KE} + \frac{\Delta P}{\rho} + E_f$$

where ΔPE and ΔKE are the changes in potential and kinetic energies, respectively. E_p is the energy supplied by the pump and E_f is the total loss in energy due to friction. The above equation can also be written as

$$gZ_1 + \frac{\bar{u}_1^2}{2\psi} + \frac{P_1}{\rho} + E_p = gZ_2 + \frac{\bar{u}_2^2}{2\psi} + \frac{P_2}{\rho} + E_f$$

In this equation, ψ is a constant and is equal to 0.5 for laminar flow and 1.0 for turbulent flow. All the terms in Bernoulli's equation are energy per unit mass and hence have the units joules per kilogram, which is also the same as meters squared per seconds squared. The subscripts 1 and 2 refer to the intake port and delivery port, respectively. Once E_p is determined, the power rating of the pump is determined by

$$\text{Power} = \dot{m}E_p$$

Once the power of the pump is determined, the next step is to determine the type of pump to be used. Pumps are broadly classified into two categories: centrifugal and positive displacement. In a centrifugal pump, product enters the center of an impeller and, due to centrifugal force, moves to the periphery. At this point, the liquid experiences maximum pressure and is forced out into the pipeline. For a centrifugal pump, the volumetric flow rate is directly proportional to the pump speed; the total head varies as the square of the pump speed; and the power required varies as the cube of the pump speed. In a positive-displacement pump (rotary, reciprocating, axial flow pumps), direct force is applied to a confined liquid to make it move. Some of the factors involved in pump selection are as follows:

1. Flow rate of fluid.
2. Net positive suction head required (NPSHR)—depends on impeller design. It is required to maintain stable operation of pump including avoiding cavitation.
3. Net positive suction head available (NPSHA)—depends on absolute pressure, vapor pressure of liquid, static head of liquid above the center line of the pump, and friction loss in the suction system.
4. Properties of fluid (such as density or viscosity).
5. Characteristic pump curves (graph of head, power consumption, and efficiency versus volumetric flow rate).

3.5 Residence Time Distribution of Fluid Elements and Particles

The FDA only credits heat treatment experienced in the holding tube, which makes its design critical. An important factor to be taken into account in designing the holding tube is the fact that the residence time in the holding tube should be based on the specific volume of the product at the hold tube temperature and not on the displacement of the pump (because the pump is operating at a different temperature and specific volume varies with temperature). The velocity profile of the fluid in the holding tube is affected by the degree of its deviation from the behavior of a Newtonian fluid. The degree of deviation is characterized by the flow behavior index, n, for Ostwall–de Waale fluids. For a Newtonian fluid flowing under laminar conditions in a straight tube of circular cross section, the maximum velocity occurs at the center of the holding tube and its magnitude is twice the average velocity of the fluid. For pseudoplastic fluids ($n < 1$), differences between the maximum and average velocities becomes smaller as n decreases. In other words, the velocity profile becomes flatter. For the, extreme case ($n = 0$), the plug flow profile is attained. However, for most cases ($n > 0$), the maximum velocity occurs at the axis of the tube, which means that the minimum residence time corresponds to the residence time of particles located along the centerline of the tube. Consequently, these particles receive the least amount of heat treatment. Thus, the holding tube length required to achieve the required F_0 value (time–temperature effect) can be calculated based on the knowledge of this minimum residence time, but this will result in an overprocessed product. This is where the RTD of the particles comes into the picture. To understand RTD, we begin with the following equation which describes the velocity profile for flow of a Newtonian fluid under laminar conditions in a pipe of circular cross section:

$$u = 2\bar{u}\left[1 - \left(\frac{r^2}{R^2}\right)\right]$$

Thus, it can be seen that different fluid elements (at different radial locations) spend different amounts of time in the tube. For instance, a fluid element traveling at the center of the tube will travel twice as fast as the average fluid element. The distribution of times spent by various fluid elements within the tube is referred to as the RTD of the fluid elements. Similarly, when different particles are flowing through the tube, they spend different times in the tube, and the distribution of these times is the RTD of the particles. The RTD of the particles depends a great deal on the RTD of the fluid. It also depends on flow rate and viscosity of the carrier medium, and also the size, density, and concentration of particles. Analysis of particle RTD is relatively simple when there is only one type of particle in a system. However, when different types of particles (especially particles of different densities) are present in a product, the flow behavior is quite different from the situation when they are each present as the only particle type in suspension. For instance, in a mixture of two types of particle, denser particles (which travel slowly at the bottom of the tube when present alone) could be sped up by rarer particles due to collisions and, in turn, the rarer particles could get slowed down. Thus, an analysis has to be performed for each combination of particle types present in a system and direct inferences cannot be made from the RTD of each particle type separately.

The existence of a RTD for the particles results in some particles receiving more heat treatment than others in the holding tube. From a safety standpoint, the fastest particle is what is of concern and the holding tube length is based on the fastest particle residence time. Thus, it can be seen that if the particle RTD is narrow, the quality of the product would be high because the difference between the fastest and slowest particle residence time is not very high. The wider the RTD of the particles in the holding tube section, the more nonuniform the process. One of the techniques that can be used to narrow the RTD of the fluid and particles is the use of helical tubes.

When a non-Newtonian (power-law) fluid flows through a straight tube under laminar-flow conditions, the velocity profile is given by

$$u = \frac{3n+1}{n+1}\left\{\bar{u}\left[1 - \left(\frac{r}{R}\right)^{(n+1)/n}\right]\right\}$$

Thus, for a pseudoplastic fluid ($n < 1$), the maximum velocity is given by

$$u_{\max} = \frac{3n+1}{n+1}\,\bar{u}$$

Hence, it can be seen that the maximum velocity in the case of a pseudoplastic fluid is less than twice the average fluid velocity. Thus, the RTD of the fluid is narrower for a pseudoplastic fluid in comparison with that for a Newtonian fluid. Hence, the RTD of particles is also narrower when the carrier medium is a pseudoplastic fluid.

Studies on RTD of fluid elements and particles in conventional holding tubes have been conducted by several researchers in the past. Some of these studies include those of Dutta and Sastry [13], Palmieri et al. [14], Sancho and Rao [15], Sandeep and Zuritz [16], and Baptista et al. [17]. Several studies have been conducted to determine the RTD of particles in helical holding tubes too because the RTD in helical holding tubes is narrower than that in conventional holding tubes. Some of the studies include those of Chen and Jan [18], Tucker and Withers [19], Ahmad et al.

[20], and Sandeep et al. [21]. The general trend from these studies is that an increase in flow rate, particle size, or particle concentration results in a decrease in the RTD of particles, whereas an increase in viscosity results in an increase in RTD. However, it should be noted that in determining the fastest particle (to compute process lethality based on this particle), RTD studies should be conducted for that particular combination of particle sizes and concentrations involved, as experiments have shown that the fastest particle in a single-particle situation is the neutrally buoyant particle that travels through the center of the tube, whereas in mixed-particle-type situations, it is usually a particle of slightly higher or slightly lower density. It should also be noted that the fastest particle is not always the critical particle (the particle that receives the least heat treatment and, hence, lethality). Slower-moving particles of lower thermal diffusivities could very likely receive less heat treatment than faster-moving particles of higher thermal conductivities. This factor further complicates determination of process lethality.

3.6 Forces Acting on Fluid Elements and Particles During Flow

In order to solve for the trajectory and velocity of various particles during flow in a tube, we need to know the fluid-flow characteristics. The equations that govern the flow of a fluid are the continuity and momentum equations. As particles flow through a tube along with a carrier fluid, they experience various forces. These forces are responsible for the translation and rotation of the particles as they flow through the system. The density, size, and shape of the particle are important particle characteristics that affect the motion of the particles, whereas the viscosity, flow rate, and density of the fluid are the important fluid characteristics that affect the motion of the particles.

3.6.1 Equations of Motion of the Fluid

The motion of the fluid is described by the continuity equation and three momentum equations. The continuity equation is

$$\frac{\partial \rho_f}{\partial t} + \nabla \cdot (\rho_f \mathbf{u}) = 0$$

This reduces to

$$\nabla \cdot \mathbf{u} = 0$$

for an incompressible fluid. The three momentum equations for the fluid phase are given in vector notation as follows:

$$\rho_f \frac{D\mathbf{u}}{Dt} = \rho_f \mathbf{g} + \nabla : \tau$$

For non-Newtonian fluids, the Ostwald–de Waele model is used [22]:

$$\tau = -m[\tfrac{1}{2}(\Delta : \Delta)]^{(n-1)/2}\Delta$$

where $\frac{1}{2}(\Delta : \Delta)$ is the second invariant of the strain-rate tensor and is given by

$$\tfrac{1}{2}(\Delta : \Delta) = 2\left[\left(\frac{\partial u_f}{\partial x}\right)^2 + \left(\frac{\partial v_f}{\partial y}\right)^2 + \left(\frac{\partial w_f}{\partial z}\right)^2\right] + \left(\frac{\partial v_f}{\partial x} + \frac{\partial u_f}{\partial y}\right)^2 + \left(\frac{\partial w_f}{\partial y} + \frac{\partial v_f}{\partial z}\right)^2 + \left(\frac{\partial w_f}{\partial x} + \frac{\partial u_f}{\partial z}\right)^2$$

This takes into account the spatial variation of viscosity. Thus, the apparent viscosity of the fluid at any particular shear rate (at any particular location in the tube) can be determined for the corresponding temperature. The effect of temperature and concentration of particles on the effective viscosity of suspensions have been modeled by several researchers [23–25]. Thus, the above equations can be used to describe the flow behavior of a power-law fluid containing particles under nonisothermal conditions.

3.6.2 Linear Dynamic Equations for Particles

The three linear dynamics equations for the particles (in vector notation) are as follows:

$$m_p\left(\frac{d\mathbf{V}_{pk}}{dt}\right) = \sum \mathbf{F}_k$$

where m_p is the mass of a single particle and $\mathbf{V}_{pk}$ and $\mathbf{F}_k$ $(k = x, y, z)$ are the velocities of the particles and forces acting on the particle in the x, y, and z directions, respectively.

Particles suspended in a viscous fluid are subjected to the following forces [26]:

1. **Magnus Lift Force**: The Magnus lift force acts in a direction perpendicular to the direction of motion of the particle and it is this force that causes the curving of a spinning sphere. The expression to compute the Magnus lift force ($\mathbf{F}_{rk}$) is given by

   $$\mathbf{F}_{rk} = \pi\rho_f a^3 \mathbf{\Omega} \times (\mathbf{V}_p - \mathbf{V}_f)$$

where $\mathbf{\Omega}$ is the angular velocity of the particle. In the above expression, the difference in velocities $\mathbf{V}_p - \mathbf{V}_f$ is called the slip velocity or relative velocity ($\mathbf{V}_r$). The vector product $\mathbf{\Omega} \times \mathbf{V}_r$ is determined using

$$\mathbf{\Omega} \times \mathbf{V}_r = \varepsilon_{ijk}\mathbf{\Omega}_j\mathbf{V}_{rk} = \mathbf{i}(\Omega_y w_r - \Omega_z v_r) + \mathbf{j}(\Omega_z u_r - \Omega_x w_r) + \mathbf{k}(\Omega_x v_r - \Omega_y u_r)$$

Substituting this equation in the equation for computing the force results in the following three expressions for the Magnus lift force in the x, y, and z directions, respectively:

$$F_x = \pi a^3 \rho_f [\Omega_y(w_p - w_f) - \Omega_z(v_p - v_f)]$$

$$F_y = \pi a^3 \rho_f [\Omega_z(u_p - u_f) - \Omega_x(w_p - w_f)]$$

$$F_z = \pi a^3 \rho_f [\Omega_x(v_p - v_f) - \Omega_y(u_p - u_f)]$$

The experimental works of researchers [27,28] indicated that particles migrated radially even in the absence of rotation. Thus, there is some other force that contributes toward the lift forces experienced by particles.

2. **Saffman Lift Force**: Saffman [29] developed an expression for the lift force acting on a particle during an unbounded shear flow. The shear lift force is independent of the particle rotation unless the rotation speed is much greater than the rate of shear. For a freely rotating particle, $\Omega = \frac{1}{2}|\mathbf{K}|$, where Ω is the angular velocity of the particle and $|\mathbf{K}|$ is the magnitude of the vorticity vector. Oliver [27] and Theodore [28] found that if the particle velocity was smaller than the fluid velocity, the lift force acted toward the axis of the tube and if the particle velocity was greater than the fluid velocity, the lift force acted away from the axis of the tube, thereby moving the particle away from the axis. In vector notation, the Saffman lift force on a particle is given by

$$\mathbf{F}_s = 6.46 \rho_f a^2 \left(\frac{\nu}{|\mathbf{K}|} \right)^{1/2} \mathbf{K} \times (\mathbf{V}_p - \mathbf{V}_f)$$

where $\mathbf{K}$ is the curl of the fluid velocity, ν is the kinematic viscosity, and a is the radius of the particle. $\mathbf{V}_p$ and $\mathbf{V}_f$ are the velocities of particle and fluid, respectively. The expression for $\mathbf{K}$ in the above equation can be obtained as follows:

$$\mathbf{K} = \nabla \times \mathbf{V} = \varepsilon_{ijk} \frac{\partial V_k}{\partial x_j} = \begin{bmatrix} i & j & k \\ \dfrac{\partial}{\partial x} & \dfrac{\partial}{\partial y} & \dfrac{\partial}{\partial z} \\ u & v & w \end{bmatrix}$$

$$= \mathbf{i}\left(\frac{\partial w}{\partial y} - \frac{\partial v}{\partial z}\right) + \mathbf{j}\left(\frac{\partial u}{\partial z} - \frac{\partial w}{\partial x}\right) = \mathbf{k}\left(\frac{\partial v}{\partial x} - \frac{\partial u}{\partial y}\right)$$

We now define the relative or slip velocity, $\mathbf{V}_r$, as

$$\mathbf{V}_p - \mathbf{V}_f = \mathbf{V}_r = \mathbf{i}u_r + \mathbf{j}v_r + \mathbf{k}w_r$$

Evaluating the above expressions results in the scalar forms of the Saffman lift force in the x, y, and z direction, respectively, as follows:

$$F_{sx} = 6.46\rho_f a^2 \left(\frac{\nu}{|\mathbf{K}|}\right)^{1/2} \left[\left(\frac{\partial u}{\partial z} - \frac{\partial w}{\partial x}\right)(w_p - w_f) - \left(\frac{\partial v}{\partial x} - \frac{\partial u}{\partial y}\right)(v_p - v_f)\right]$$

$$F_{sy} = 6.46\rho_f a^2 \left(\frac{\nu}{|\mathbf{K}|}\right)^{1/2} \left[\left(\frac{\partial v}{\partial x} - \frac{\partial u}{\partial y}\right)(u_p - u_f) - \left(\frac{\partial w}{\partial y} - \frac{\partial v}{\partial z}\right)(w_p - w_f)\right]$$

$$F_{sz} = 6.46\rho_f a^2 \left(\frac{\nu}{|\mathbf{K}|}\right)^{1/2} \left[\left(\frac{\partial w}{\partial y} - \frac{\partial v}{\partial z}\right)(v_p - v_f) - \left(\frac{\partial u}{\partial z} - \frac{\partial w}{\partial x}\right)(u_p - u_f)\right]$$

The above expressions are valid if the tube Reynolds number is much greater than unity and the particle is not very close to the axis of the tube. However, it should be noted that the above requirements may not be met in many situations and, hence, the expression for Saffman force must be used with caution.

3. **Drag Force**: The expression for the drag on a particle in viscous fluid is given by:

$$\mathbf{F}_d = \tfrac{1}{2} C_d \rho_f \pi a^2 |\mathbf{V}_f - \mathbf{V}_p| (\mathbf{V}_f - \mathbf{V}_p)$$

where the drag coefficient, C_d, is obtained from the following equation [30]:

$$C_d = \frac{24}{\text{Re}_p}(1 + 0.15\text{Re}_p^{0.687}) \quad \text{for } 1 < \text{Re}_p < 1000$$

and the particle Reynolds number is defined by:

$$\text{Re}_p = \frac{(\rho)(2a)|\mathbf{V}_f - \mathbf{V}_p|}{\mu}$$

4. **Buoyancy Force (acting in the *y*-direction only)**: The expression to compute the buoyancy force exerted on the particle (acting only in the y direction) is given by

$$\mathbf{F}_b = (4/3)\pi a^3(\rho_f - \rho_p)\mathbf{g}$$

Substituting the above four equations into the linear dynamic equation for the particle results in

$$m_p \frac{d\mathbf{V}_p}{dt} = 6.46\rho_f a^2\left(\frac{\nu}{|\mathbf{K}|}\right)^{1/2}\mathbf{K} \times (\mathbf{V}_p - \mathbf{V}_f) + \pi a^3\rho_f\mathbf{\Omega} \times (\mathbf{V}_p - \mathbf{V}_f)$$
$$+ \tfrac{1}{2}C_d\rho_f\pi a^2|\mathbf{V}_f - \mathbf{V}_p|(\mathbf{V}_f - \mathbf{V}_p) + \tfrac{4}{3}\pi a^3(\rho_f - \rho_p)\mathbf{g}$$

The above equation can be rewritten in the following manner to obtain the expressions for the linear dynamic equations for the particles in the x, y, and z directions respectively (with the gravity force acting in the y direction):

x direction:

$$m_p \frac{du_p}{dt} = 6.46\rho_f a^2\left(\frac{\nu}{|\mathbf{K}|}\right)^{1/2}\left[\left(\frac{\partial u}{\partial z} - \frac{\partial w}{\partial x}\right)(w_p - w_f)\right.$$
$$\left. - \left(\frac{\partial v}{\partial x} - \frac{\partial u}{\partial y}\right)(v_p - v_f)\right]$$
$$+ \pi\rho_f a^3[\Omega_y(w_p - w_f) - \Omega_z(v_p - v_f)]$$
$$+ \tfrac{1}{2}C_d\rho_f\pi a^2|\mathbf{V}_f - \mathbf{V}_p|(u_f - u_p)$$

y direction:

$$m_p \frac{dv_p}{dt} = 6.46\rho_f a^2 \left(\frac{\nu}{|\mathbf{K}|}\right)^{1/2} \left[\left(\frac{\partial v}{\partial x} - \frac{\partial u}{\partial y}\right)(u_p - u_f) - \left(\frac{\partial w}{\partial y} - \frac{\partial v}{\partial z}\right)(w_p - w_f)\right] + \pi\rho_f a^3[\Omega_z(u_p - u_f) - \Omega_x(w_p - w_f)] + \tfrac{1}{2}C_d\rho_f\pi a^2|\mathbf{V}_f - \mathbf{V}_p|(v_f - v_p) + \tfrac{4}{3}\pi a^3(\rho_f - \rho_p)g$$

z direction:

$$m_p \frac{dw_p}{dt} = 6.46\rho_f a^2 \left(\frac{\nu}{|\mathbf{K}|}\right)^{1/2} \left[\left(\frac{\partial w}{\partial y} - \frac{\partial v}{\partial z}\right)(v_p - v_f) - \left(\frac{\partial u}{\partial z} - \frac{\partial w}{\partial x}\right)(u_p - u_f)\right] + \pi\rho_f a^3[\Omega_x(v_p - v_f) - \Omega_y(u_p - u_f)] + \tfrac{1}{2}C_d\rho_f\pi a^2|\mathbf{V}_f - \mathbf{V}_p|(w_f - w_p)$$

This accounts for the description of the translation of the spheres. However, the particles undergo rotation too, and to account for the rotational motion of the sphere, the angular momentum equations of the particle phase must be solved.

3.6.3 Angular Dynamic Equations for Particles

The three angular dynamics equations for the particles are

$$I\left(\frac{d\mathbf{\Omega}_k}{dt}\right) = \sum \mathbf{T}_k$$

where I is the moment of inertia [$I = (2/5)m_p a^2$ for a sphere] and $\mathbf{T}_k (k = x, y, z)$ is the local torque exerted by the viscous fluid on the surface of the particles. Substitution of the expression for the torque into the angular momentum equation results in the following sets of equations for the x, y and z directions respectively:

$$\frac{d\Omega_x}{dt} = \frac{15\mu}{\rho_p a^2}\left[\frac{\pi}{8}\left(\frac{\partial v}{\partial z} + \frac{\partial w}{\partial y}\right) - \Omega_x\right]$$

$$\frac{d\Omega_y}{dt} = \frac{15\mu}{\rho_p a^2}\left[\frac{\pi}{8}\left(\frac{\partial u}{\partial z} + \frac{\partial w}{\partial x}\right) - \Omega_y\right]$$

$$\frac{d\Omega_z}{dt} = \frac{15\mu}{\rho_p a^2}\left[\frac{\pi}{8}\left(\frac{\partial u}{\partial y} + \frac{\partial v}{\partial x}\right) - \Omega_z\right]$$

The linear and angular dynamics equations for the particles have to be solved simultaneously along with the four equations of motion for the fluid phase in order to completely describe the flow dynamics of the suspension.

3.7 Techniques to Determine Fluid and Particle Velocity

The average fluid velocity can be calculated once the volumetric flow rate of the product is known. To determine the distribution of fluid residence times, salt injections, dye tracers, and fine particles are used. Magnetic resonance imaging can also be used under certain circumstances to obtain a fluid-flow profile. Fluid-flow profiles, although important, are usually not the target, as the species of concern are the slow-heating particles. Particle residence times, residence time distributions, and velocities can be determined by using a stopwatch, digital image analysis, laser–Doppler velocimetry, and also with the aid of magnetically tagged particles.

4 HEAT TRANSFER ASPECTS OF PROCESSING

Some of the heat transfer aspects that are of importance in designing an aseptic process are the convective heat transfer coefficient, effect of temperature on the physical and thermal properties of the product, mode of heat transfer (conduction or convection), design of the heat exchanger, and the heat resistance of microorganisms, enzymes, and nutrients.

During heating or sterilization of a solid–liquid mixture by heat treatment using conventional means, the liquid part of the mixture gets heated first and then it transfers heat to the surface of the particles by convection. Further heating of the interior of the particles takes place by conduction. The difference between the bulk fluid temperature and the center temperature of particles can be due to the low convective heat transfer coefficient between the fluid and the surface of the particle or due to the low thermal diffusivity of the particles. Because there is not much that can be done to enhance the thermal diffusivity of particles (other than by reformulation), efforts have been geared toward measuring and enhancing the convective heat transfer coefficient between the fluid phase and particles.

4.1 Convective Heat Transfer Coefficient

The convective heat transfer coefficient has been described [31] as the thermal lag between the particle surface temperature and the fluid temperature for a particle being heated in a fluid. This is mathematically written as

$$Q = h_{\mathrm{fp}} A_p (T_{\mathrm{ps}} - T_f)$$

Some of the factors that affect the convective heat transfer coefficient between a fluid and particle (h_{fp}) include the shape of the particle, surface roughness, the position of the particle in the tube, particle concentration, type of flow, and the thermal properties of the fluid.

Heat transfer for laminar flow in a straight tube (heat exchanger or holding tube) is usually low because there is very little mixing in the radial direction. However, flow in a coiled tube is fully three dimensional and the rate of heat transfer can be much higher than that in a straight tube. This is due to the development of secondary flow (in the directions normal to the main direction of flow) due to the pressure gradient imposed as a result of the centrifugal forces present in the curved section. The secondary flow serves as a means of redistributing fluid elements in the radial direction, and thereby transferring heat more efficiently between the bulk of the fluid and the fluid elements near the tube wall.

4.2 Steam Quality

Steam quality refers to the amount of steam that is in vapor phase in saturated steam, with superheated steam having a quality of 1. The amount of energy given out by steam is given by

$$Q = \dot{m}_{\mathrm{st}}(H_s - H_c)$$

with

$$H_s = (X)H_v + (1 - X)H_c$$

In the above equations, H_s, H_v, and H_c are the enthalpies of steam, pure vapor, and pure condensate, respectively, and X is the steam quality. Thus, when steam quality is 1 (or 100%), all the steam is in vapor state and the enthalpy of steam is the same as the enthalpy of pure vapor, and when steam quality is 0, all the steam is in condensate form and the enthalpy of steam is the same as the enthalpy of pure condensate.

The enthalpy of pure vapor and pure condensate can be determined from saturated steam tables at the corresponding saturation temperature and pressure. The enthalpy of saturated steam can then be determined as a weighted mean of these enthalpies, once the steam quality is known. Thus, it can be seen that the higher the quality of steam, the higher the amount of energy transferred from steam to the product.

4.3 Dimensionless Numbers Governing Heat Transfer

Just like fluid-flow characterization is done by the use of a dimensionless quantity called Reynolds number, the convective heat transfer coefficient is represented by a dimensionless quantity called the Nusselt number. The

Nusselt number is a dimensionless temperature gradient at the surface and is given by

$$N_{\mathrm{Nu}} = \frac{hD_c}{k_f} = -\frac{(dT/dy)_{\mathrm{surface}}}{(T - T_{\mathrm{surface}})/D_c}$$
$$= \frac{\text{Temperature gradient at surface}}{\text{Avg. temp. gradient throughout the system}}$$

The Nusselt number also represents the ratio of the diameter of the tube to the equivalent thickness of the laminar boundary layer. Empirical correlations have been developed to determine the Nusselt number as a function of various other dimensionless quantities, including the Reynolds number under different flow and heat transfer conditions, including forced and free convection.

The Grashof number is the ratio of buoyancy to viscous forces and is of importance in free convection only (where buoyancy effects are significant). The expression for the generalized Grashof number (valid for power-law fluids, in addition to Newtonian fluids) is given by

$$N_{\mathrm{GGr}} = \frac{g\beta_f \rho_f^2 (T_{\mathrm{surface}} - T_\infty) D_{\mathrm{vertical}}^3 (4v^{1-n} D_{\mathrm{vertical}}^{n-1})^2}{\{K[(3n+1)/n]^n (2^{n-1})\}^2}$$

The quantity β is the coefficient of volumetric thermal expansion and is given by

$$\beta = \frac{1}{\langle V \rangle} \left(\frac{\partial \langle V \rangle}{\partial T} \right)_P$$

For the case of free convection, the Nusselt number is a function of the Grashof number. When dealing with Newtonian fluids, the expression for the generalized Grashof number reduces to

$$N_{\mathrm{Gr}} = \frac{g\beta_f \rho_f^2 (T_{\mathrm{surface}} - T_\infty) D_{\mathrm{vertical}}^3}{\mu^2}$$

The Prandtl number, which is the ratio of momentum diffusivity (μ/ρ) and thermal diffusivity ($k/\rho c_p$), comes into picture in the determination of the Nusselt number for both forced and free convection. The generalized Prandtl number (valid for power-law fluids, in addition to Newtonian fluids) is given by

$$N_{\mathrm{GPr}} = \frac{c_{p_f} k[(3n+1)/n]^n (2^{n-1})}{4v^{1-n} d^{n-1} k_f}$$

For Newtonian fluids, the above expression reduces to

$$N_{\mathrm{Pr}} = \frac{c_{p_f}\mu_f}{k_f}$$

The Biot number is the ratio of the internal (conductive) and external (convective) resistance offered to heat transfer in an object. It comes into picture for calculations involving non-steady-state heat transfer and is defined as

$$N_{\mathrm{Bi}} = \frac{hD_c}{k_s} = \frac{D_c/k_s}{1/h} = \frac{\text{Internal resistance}}{\text{External resistance}}$$

Non-steady-state heat transfer is characterized by another dimensionless quantity, which is the Fourier number and is defined as

$$N_{\mathrm{Fo}} = \frac{\alpha t}{D_c^2} = \frac{k(1/D_c)D_c^2}{\rho c_p D_c^3/t} = \frac{\text{Rate of heat conduction}}{\text{Rate of heat storage}}$$

4.4 Natural (Free) and Forced Convection

Convective heat transfer can take place by natural or forced means. In natural convection, flow occurs due to the differences in the density of the fluid as it comes into contact with a hot (or cold) surface, thereby resulting in buoyancy forces. The Nusselt number in this case is a function of the Grashof number (N_{Gr}) and the Prandtl number and it takes the form

$$N_{\mathrm{Nu}} = \frac{hL_c}{k} = c_1(N_{\mathrm{Gr}}N_{\mathrm{Pr}})^{c_2}$$

In the above equation, L_c is the characteristic length. For internal flows, the characteristic length is given by

$$L_c = 4\left(\frac{\text{Cross-sectional area}}{\text{Wetted perimeter}}\right)$$

In some situations, a combination of natural and forced convection takes place. Thus, the relative importance of the two has to determined. If $(N_{\mathrm{Gr}})(N_{\mathrm{Pr}}) < 8 \times 10^5$, the effect of natural convection can be neglected and forced convection governs the heat transfer. Another method of determining the relative importance of natural and forced convection is by determining the ratio of the Grashof number (measure of the buoyancy force) and the square of the Reynolds number (measure of the inertial force). If the ratio is close to unity, the effects of forced and free convection have to be taken into account. The magnitude of the Froude number also can be used to determine the relative importance of natural and forced convection.

4.5 Transient Heat Transfer Within Particles

Two basic approaches exist to solving the problem of heat transfer involving particles. The first method, called the lumped capacitance method, assumes that the entire particle is at the same temperature and the second method takes into account the temperature gradient within particles. The lumped capacitance method is valid only if the Biot number (N_{Bi}) is less than 0.1. If the lumped capacitance method is applicable, the following equation can be used to determine the temperature (T) within an object at any given time (t):

$$\frac{T - T_{\infty}}{T_i - T_{\infty}} = \exp\left(-\frac{hA}{\rho c_p \langle V \rangle} t\right)$$

It should be noted that when the lumped parameter method is valid, the resistance to heat transfer due to conduction is negligible in comparison to that due to convection. However, if the Biot number is greater than 0.1, the Heisler chart is used to determine the temperature at the center of an object. When the Biot number is greater than 40, the resistance to heat transfer due to convection is negligible in comparison to that due to conduction.

4.6 Hydrodynamic and Thermal Entrance Lengths

The flow behavior of fluids in a pipe depends a great deal on the temperature. As most products get heated, they become less viscous and flow much easier than when they are at room temperature. Thus, it is possible to have turbulent flow in the heat exchanger and holding tube (where the viscosity of the product is very low) and laminar flow in the cooling section; hence, care has to be taken in designing these components of the aseptic processing system depending on the product.

As a fluid flows through a straight pipe and convective heat transfer is taking place, the flow can be divided into three regions: entrance region, transient region, and fully developed region. The entrance region is the region where the velocity and temperature profiles are still developing. The length required for the flow (laminar) to become fully developed (hydraulic length or hydrodynamic entry length, l_h) is given by [32]

$$l_h = 0.05(N_{\mathrm{Re}})(d)$$

This equation is also referred to as the Langhaar equation. In the transient region, the velocity profile is fully developed, whereas the temperature profile is still developing. The length required for the temperature profile to become fully developed (thermal length, l_t) is given by

$$l_t = 0.036(N_{\mathrm{Pe}})(d)$$

where the Peclet number (N_{Pe}) is given by

$$N_{Pe} = \frac{\bar{u}_f d}{\alpha_f}$$

The Peclet number, given by this equation (for Newtonian and non-Newtonian fluids) is a product of the Reynolds number and the Prandtl number.

4.7 Heat Transfer Coefficient in Straight Tubes

When considering transfer of heat between a heating medium (such as water or steam) and a product for flow in a tubular heat exchanger, two convective heat transfer coefficients come into picture: convective heat transfer coefficient between the heating medium and the outer wall of the inner tube (outside heat transfer coefficient), and the convective heat transfer coefficient between the product and the inner wall of the inner tube (inside heat transfer coefficient). Different techniques exist to determine the heat transfer coefficient based on the state of the heating medium (liquid or gas) and whether the flow is in a tube or in an annulus. Some of the correlations used commonly are listed below.

The following equation [33] is used to compute the (outside) heat transfer coefficient between steam and the wall of the heat exchanger:

$$h_{o(\mathrm{hx})} = 0.725\left(\frac{k_{\mathrm{st}}^{3.0}\rho_{\mathrm{st}}^{2.0} g\lambda}{(T_{\mathrm{st}} - T_{w(\mathrm{hx})})d_{o(\mathrm{hx})}\mu}\right)^{1/4}$$

The (inside) heat transfer coefficient between the wall of the heat exchanger and the product is computed using the following equation [33]:

$$N_{\mathrm{Nu}} = 2.0\left(\frac{3n+1}{4n}\right)^{1/3}\left(\frac{\dot{m}c_{p(f)}}{k_f L_{\mathrm{hx}}}\right)^{1/3}\left(\frac{m}{m_w}\right)^{0.14}$$

The properties of the fluid are determined at the film temperature (T_{film}), given by

$$T_{\mathrm{film}} = \tfrac{1}{2}(T_{\mathrm{wall}} + T_{\mathrm{fluid}})$$

In the holding tube, the following equation is used to determine the (inside) heat transfer coefficient between the product and the wall of the holding tube [34]:

$$N_{\mathrm{Nu}} = 8.0\left(\frac{15n^3 + 23n^2 + 9n + 1}{31n^3 + 43n^2 + 13n + 1}\right)$$

In the cooling section, the following equations are used to determine the inside and outside heat transfer coefficients [33]:

Laminar flow:

$$N_{\mathrm{Nu}} = 2.0\left(\frac{3n+1}{4n}\right)^{1/3}\left(\frac{\dot{m}c_p}{kL}\right)^{1/3}$$

Turbulent flow:

$$N_{\mathrm{Nu}} = 0.023 N_{\mathrm{GRe}}^{0.8} N_{\mathrm{GPr}}^{1/3}$$

When determining the inside heat transfer coefficient, the inside diameter of the tube and the properties of the fluid undergoing processing have to be used. When determining the outside heat transfer coefficient, the outside diameter of the tube and the properties of the cooling water have to be used. The flow in the cooling section (both internal and external flow) is considered to be laminar if the generalized Reynolds number is less than 2100).

The surface heat transfer coefficient between the fluid and the particle is computed using the following correlation [35]:

$$N_{\mathrm{Nu}} = 2.0 + 28.37 N_{\mathrm{GRe}}^{0.233} N_{\mathrm{GPr}}^{0.143}\left(\frac{d_p}{d}\right)^{1.787}$$

Nusselt number expressions for heat transfer from a power-law fluid under laminar-flow conditions under an uniform wall heat-flux boundary condition have been presented [36] as a function of Peclet number and the local Graetz number, where the Graetz number (N_{Gz}) is given by

$$N_{\mathrm{Gz}} = \frac{\dot{m}c_p}{kx} = \frac{N_{\mathrm{Re}} N_{\mathrm{Pr}}}{x/d}$$

The most commonly used equations for determining the Nusselt number for laminar flow in horizontal pipes are given as follows:

for $N_{\mathrm{Re}} N_{\mathrm{Pr}}(d/L) < 100$

$$N_{\mathrm{Nu}} = 3.66 + \frac{0.085[N_{\mathrm{Re}} N_{\mathrm{Pr}}(d/L)]}{1.0 + 0.045[N_{\mathrm{Re}} N_{\mathrm{Pr}}(d/L)]^{0.66}}\left(\frac{\mu_b}{\mu_w}\right)^{0.14}$$

for $N_{\mathrm{Re}} N_{\mathrm{Pr}}(d/L) > 100$

$$N_{\mathrm{Nu}} = 1.86 \times \left(N_{\mathrm{Re}} N_{\mathrm{Pr}} \frac{d}{L}\right)^{0.33}\left(\frac{\mu_b}{\mu_w}\right)^{0.14}$$

Another commonly used correlation for laminar-flow heat transfer in pipes is given by the following equation [37]:

$$N_{\mathrm{Nu}} = 2.0\left(\frac{3n+1}{4n}\right)N_{\mathrm{Gz}}^{1/3}\left(\frac{\mu_b}{\mu_w}\right)^{0.14}$$

Thus, it can be seen from the above equation that the heat transfer coefficient will be higher for pseudoplastic fluids ($n < 1$) as compared to that for Newtonian fluids.

For transition flow, a graph of the Colburn j factor (j_{H}) versus the Reynolds number is provided by Perry and Chilton [38] to determine the convective heat transfer coefficient. The statement of the Colburn analogy between heat transfer and fluid friction is given by

$$j_{\mathrm{H}} = \tfrac{1}{2}f = 0.023N_{\mathrm{Re}}^{-0.2}$$

Researchers [39] determined the convective heat transfer coefficients between a fluid and a particle and developed the following correlation to determine the Nusselt number:

$$N_{\mathrm{Nu}} = 2.0 + 8.4703N_{\mathrm{GRe}}^{0.553}N_{\mathrm{GPr}}^{0.2716}\left(\frac{L_c}{d}\right)^{0.6272}\left(\frac{R-r}{R}\right)^{-0.1142}$$

They found that the convective heat transfer coefficient increased with decreasing particle size or viscosity and increasing flow rate. It was also found that the convective heat transfer coefficient was higher for a particle near the wall.

When dealing with a scraped surface heat exchanger, the convective heat transfer coefficient is determined using an equation such as [40]

$$N_{\mathrm{Nu}} = 1.2N_{\mathrm{re}}^{0.5}N_{\mathrm{Pr}}^{0.33}N_b^{0.26}$$

A more general equation has been developed [41] which takes into account the speed of rotation and the type of fluid:

$$N_{\mathrm{Nu}} = c_1\left(\frac{\rho u(D-D_s)}{\mu}\right)N_{\mathrm{Pr}}^{c_2}\left(\frac{D(\Omega/2\pi)}{u}\right)^{0.62}\left(\frac{D_s}{D}\right)^{0.55}(N_b)^{0.53}$$

For viscous liquids, $c_1 = 0.014$ and $c_2 = 0.96$, while for nonviscous liquids, $c_1 = 0.039$ and $c_2 = 0.70$.

Thus, depending on the situation presented, the appropriate equation to determine the convective heat transfer coefficient should be used. It should also be noted that each of the equations have a range of applicability and also assumptions involved.

4.8 Heat Transfer Coefficient in Helical Tubes

Researchers have conducted several studies on flow and heat transfer in helical tubes. These studies have included various flow and fluid types, tube diameters, and coil radii. Due to the complexities that are involved in flow and heat transfer in helical tubes, there is no simple correlation that can be applied to a wide range of process conditions. Nevertheless, different correlations are available for different situations and some of them have been presented below.

The following correlations have been developed [42] for constant wall heat-flux heat transfer in curved tubes:

$$N_{Nu} = 3.31 N_{De}^{0.115} N_{Pr}^{0.0108} \quad \text{for } 20 \leq N_{De} \leq 1200 \text{ and } 0.005 \leq N_{Pr} \leq 0.05$$

$$N_{Nu} = 0.913 N_{De}^{0.476} N_{Pr}^{0.2} \quad \text{for } 80 \leq N_{De} \leq 1200 \text{ and } 0.7 \leq N_{Pr} \leq 5$$

For constant wall temperature heat transfer in curved tubes, the following correlation was developed [43]:

$$N_{Nu} = 0.836 N_{De}^{0.5} N_{Pr}^{0.1} \quad \text{for } 80 \leq N_{De} \leq 1200 \text{ and } 0.7 \leq N_{Pr} \leq 5$$

Other researchers [44] developed the following Nusselt number correlations for different ranges of Dean numbers:

$$N_{Nu} = 1.7(N_{De}^2 N_{Pr})^{1/6} \quad \text{for } N_{De} < 20 \text{ and } (N_{De}^2 N_{Pr})^{0.5} > 100$$

$$N_{Nu} = 0.9(N_{Re}^2 N_{Pr})^{1/6} \quad \text{for } 20 < N_{De} < 100$$

$$N_{Nu} = 0.7 N_{Re}^{0.43} N_{Pr}^{1/6} \left(\frac{d}{D}\right)^{0.07} \quad \text{for } 100 < N_{De} < 8300$$

They concluded that the effect of d/D can be neglected for Dean numbers less than 100 in the fully developed thermal region. They also found that for all cases with $(N_{De}^2 N_{Pr}) > 100$, the Nusselt number in the fully developed thermal region was proportional to $N_{Pr}^{1/6}$ and that for the thermal entry region, the Nusselt number was proportional to $N_{Pr}^{1/3}$.

4.9 Heating Media and Equipment

Heating the product can be accomplished by direct contact of the hot medium and the product or by indirect contact. Direct contact heating is accomplished by steam injection (injecting steam into a product) or steam infusion (passing a product in a thin layer through a chamber of steam). Both of these methods are rapid means of sterilizing the product. The product is then cooled by evaporation in a vacuum chamber.

There are several types of indirect contact heat exchangers. Some of them are the plate, tubular, shell and tube, scraped surface, microwave, ohmic. Some of the nonthermal techniques include the use of high pressure, irradiation, and pulsed electric field. Plate heat exchangers have a large surface area; hence, rapid heat transfer can take place (due to the turbulent flow conditions) to liquid foods and liquid foods with small particles. Tubular and shell and tube heat exchangers can be used with relatively low-viscosity products with the limiting factor being the size of the particulates. Tubular heat exchangers can be double-tube, triple-tube, or corrugated-tube heat exchangers with flow in the cocurrent, countercurrent, or cross-flow mode. With viscous products, the blades of the scraped surface heat exchanger (SSHE) provide mixing and also prevent burning of products onto the wall of the heat exchanger. A SSHE is also suitable for products with large particulates. Microwave (dielectric heating) and ohmic (electrical resistance heating) heating results in rapid and simultaneous heating of the liquid and particulate phases of the product. However, they can also result in nonuniform and runaway heating. Some of the other techniques such as the use of radio frequency, pulsed electric fields, irradiation, membrane separation, and high pressure are currently under investigation for commercialization on a large scale. On a smaller scale, the use of pulsed light, ultrasound, and ultraviolet radiation have been attempted with limited success.

Some of the methods that handle liquids and particulates separately are the Jupiter system (particles are sterilized by steam in a double-cone pressure vessel), rotaholder (a tubular sterilizer in which particles are held back for extra time using forks), and the fluidized-bed system (particles are separately sterilized in a fluidized bed by steam and cooled by sterile nitrogen) and have been described in more detail elsewhere [45]. The disadvantages of these methods are the added costs for separating and recombining the liquid and particulate phases and the complexities introduced in the overall process.

4.10 Cocurrent and Countercurrent Heat Exchangers

Cocurrent and countercurrent heat exchangers are used abundantly in the food processing industry. Thus, it is important to be able to determine the overall heat transfer coefficient in these types of heat exchangers. The total energy transferred to the product can easily be determined by

$$Q = \dot{m} c_p (\Delta T)$$

where ΔT is the rise in temperature of the product. This information is then used to calculate the overall heat transfer coefficient in the heat exchanger, making use of equation

$$Q = UA_{\text{lm}}\Delta T_{\text{lm}}$$

where the logarithmic mean temperature difference (ΔT_{lm}) is given by

$$\Delta T_{\text{lm}} = \frac{\Delta T_1 - \Delta T_2}{\ln(\Delta T_1/\Delta T_2)}$$

with ΔT_1 and ΔT_2 being the difference between the temperatures of the hot and cold fluids at the inlet and exit of the heat exchanger, respectively. The quantity A_{lm} is the logarithmic mean area and is given by

$$A_{\text{lm}} = \frac{A_o - A_i}{\ln(A_o/A_i)}$$

with A_o and A_i being the outside and inside surface areas, respectively.

It is thus possible to compute the effectiveness of various heat exchangers based on the above outlined procedure to compute the overall heat transfer coefficient. For the same inlet conditions, it can be shown that the amount of energy transferred from the hot to the cold fluid is higher in the case of a countercurrent heat exchanger.

4.11 Governing Heat Transfer Equations and Energy Balance

The energy equation in spherical coordinates is

$$\rho c_p \frac{\partial T}{\partial t} = \frac{1}{r^2}\frac{\partial}{\partial r}\left(kr^2 \frac{\partial T}{\partial r}\right) + \frac{1}{r^2 \sin\theta}\frac{\partial}{\partial \theta}\left(k \sin\theta \frac{\partial T}{\partial \theta}\right) + \frac{1}{r^2 \sin^2\theta}\frac{\partial}{\partial \varphi}\left(k \frac{\partial T}{\partial \varphi}\right) + S$$

By symmetry,

$$\frac{\partial T}{\partial \theta} = \frac{\partial T}{\partial \varphi} = 0$$

Also, $S = 0$ for the case where there is no heat source term. Thus, the energy balance equation reduces to

$$\rho c_p \frac{\partial T}{\partial t} = \frac{1}{r^2}\frac{\partial}{\partial r}\left(kr^2 \frac{\partial T}{\partial r}\right)$$

For constant properties, this equation reduces to

$$\frac{\alpha}{r^2}\frac{\partial}{\partial r}\left(r^2 \frac{\partial T}{\partial r}\right) - \frac{\partial T}{\partial t} = 0$$

This is the equation that has to be then solved to determine the temperature distribution within spherical particles.

4.11.1 Energy Balance in the Heat Exchanger

In the heat exchanger, the steam loses heat to the fluid and the particles suspended in the fluid. The overall energy balance equation in the heat exchanger is

$$U_{\mathrm{hx}}A_{\mathrm{hx}}(T_{\mathrm{st}} - \bar{T}_f) = \dot{m}_f c_{p_f}(T_f' - T_f) + N_p h_{f_p} A_p(\bar{T}_f - \bar{T}_{\mathrm{ps}})$$

In this equation, $\bar{T}_f$ and $\bar{T}_{\mathrm{ps}}$ are the mean fluid and mean particle surface temperatures, respectively, and are given by

$$\bar{T}_f = \tfrac{1}{2}(T_f' + T_f) \quad \text{and} \quad \bar{T}_{\mathrm{ps}} = \tfrac{1}{2}(T_{\mathrm{ps}}' + T_{\mathrm{ps}})$$

The terms T_f' and T_{ps}' are the fluid and particle surface temperatures, respectively, at the new time step (or the new spatial location). Also, the surface area of the heat exchange surface, A_{hx}, is computed as

$$A_{\mathrm{hx}} = \pi d_{\mathrm{hx}} \Delta x$$

The following equation is used to determine the overall heat transfer coefficient in the heat exchanger (U_{hx}):

$$\frac{1}{U_{\mathrm{hx}}A_{\mathrm{lm(hx)}}} = \frac{1}{h_{o(\mathrm{hx})}A_{o(\mathrm{hx})}} + \frac{1}{h_{i(\mathrm{hx})}A_{i(\mathrm{hx})}} + \frac{x_{\mathrm{hx}}}{k_{\mathrm{hx}}A_{\mathrm{lm(hx)}}}$$

where the logarithmic mean area (A_{lm}) is given by

$$A_{\mathrm{lm}} = \frac{2\pi L(R_o - R_i)}{\ln(R_o/R_i)}$$

The terms $1/h_oA_o$, $\Delta r/kA_{\mathrm{lm}}$, and $1/h_iA_i$ represent the resistances to heat transfer from the steam to the outside wall of the heat exchanger, from the outside wall of the heat exchanger to the inside wall of the heat exchanger, and from the inside wall of the heat exchanger to the product, respectively. Thus, one can solve for the two unknowns, namely T_f' and T_{ps}'.

4.11.2 Energy Balance in the Holding Tube

In the holding tube, the fluid loses heat to the particles and also to the surroundings. The overall energy balance equation in the holding tube is

$$\dot{m}_f c_{p_f}(T_f - T_f') = U_{\mathrm{ht}}A_{\mathrm{ht}}(\bar{T}_f - T_{\mathrm{air}}) + N_p h_{\mathrm{fp}} A_p(\bar{T}_f - \bar{T}_{\mathrm{ps}})$$

The following equation is used to determine the overall heat transfer coefficient in the holding tube ($U_{\rm ht}$):

$$\frac{1}{U_{\rm ht}A_{\rm lm(ht)}} = \frac{1}{h_{o(\rm ht)}A_{o(\rm ht)}} + \frac{1}{h_{i(\rm ht)}A_{i(\rm ht)}} + \frac{x_{\rm ht}}{k_{\rm ht}A_{\rm lm(ht)}} + \frac{x_{\rm ins}}{k_{\rm ins}A_{\rm lm(ht)}}$$

A similar approach to that used in the determination of the fluid and particle temperatures in the heat exchanger was then used to determine the temperatures of the fluid and particles in the holding tube.

4.11.3 Energy Balance in the Cooling Section

In the cooling section, the fluid and particles lose heat to the cooling water. The overall energy balance equation in the cooling section is

$$\dot{m}_f c_{p_f}(T_f - T_f') = U_{\rm cs}A_{\rm cs}(\bar{T}_f - T_{\rm cw}) + N_p h_{\rm fp} A_p(\bar{T}_f - \bar{T}_{\rm ps})$$

The overall heat transfer coefficient in the cooling section ($U_{\rm cs}$) is computed using the same equation that was used to determine the overall heat transfer coefficient in the heat exchanger with all parameters for the heat exchanger being replaced by the corresponding parameters for the cooling section.

4.12 Fouling and Enhancement of Heat Transfer

As a product flows through a system, it tends to stick to the hot surface of the heat exchanger. This is referred to as fouling and can greatly impede the rate of transfer of heat. After a certain time, when the heat transfer rate becomes unacceptably low, the system has to be shut down and cleaned using a clean-in-place (CIP) solution. This translates to decreased productivity and increased costs. This is more of a problem with viscous, proteinaceous, and starchy foods and is predominant under laminar-flow conditions. The use of appropriate flow rate and temperature can minimize fouling, but not eliminate it totally. The use of scraped surface and helical heat exchangers also minimizes fouling. Another detrimental factor associated with fouling is the fact that with increased fouling, the cross-sectional area available for flow decreases, thereby increasing flow velocity and decreasing the residence time (in the heat exchanger or holding tube). This, in turn, could result in a decrease in the actual accumulated lethality which could potentially result in an unsafe product.

4.13 Techniques to Estimate the Temperature History of a Product

The most commonly used method to determine the temperature within a product is the use of thermocouples or RTDs in the flow regime, which may be installed in-line with the aid of clamps or compression fittings (such as Swagelok™). Another technique to determine temperature within a fluid is the introduction of tracer capsules (or data tracers) in the flow, retrieving it at the exit, and downloading the temperature data to a computer. Infrared imaging is a technique that can be used to obtain surface temperature information. In order to use infrared imaging for particles, the particles are retrieved at the exit, sliced, and imaged to determine variation of temperature at any cross section of the particle. Thermochromic dyes that change color with time and melting-point indicators (that melt at a specific temperature) are some of the other techniques that can be used to determine the temperature within fluids and also within particles. Thermoluminescent markers (that emit a certain wavelength of light depending on the temperature) can be used to determine the temperature within clear fluids on-line.

5 MICROBIOLOGICAL AND QUALITY CONSIDERATIONS

During processing, there are several factors that the processor takes into account. First and foremost comes the safety of the process and compliance with regulatory requirements. Other factors that come into picture are the extent of enzymatic inactivation and nutrient retention. Thus, the process is designed such that it is safe and results in maximum nutrient retention and the appropriate level of enzymatic inactivation.

5.1 Federal Regulations and HACCP

Unlike in European countries, where regulations are based on spoilage tests, the FDA requires microbiological tests to prove the safety of a process with sufficient latitude, for variability in process conditions. In the United States different regulatory agencies and rules apply to different products. For example, ultra–high temperature (UHT) milk processing is covered under title 21 (parts 108, 113, 114) of the code of federal regulations (CFR). The process should also adhere to the pasteurized milk ordinance (PMO). When meat is involved, the regulations are imposed by the United States Department of Agriculture. In addition to these regulations, certain states have state regulations imposed on certain processes. During the past few years, HACCP (Hazard analysis of critical control points) has gained tremendous importance and its implementation has been extended by the FDA to various products after its initial application to certain acidified and low-acid canned foods.

5.2 Kinetics of Microbial Destruction, Enzyme Inactivation, and Nutrient Retention

There are several techniques to estimate the extent of heat treatment received by various components in a food. The most direct technique to do this is by measuring the temperature at the desired locations. However, this is not easy in a continuous-flow situation. Thus, indirect mechanisms, such as the change in color of a dye or the extent of sucrose inversion, are used. As far as microorganisms go, one of the techniques to ascertain the extent of microbial destruction is by using an alginate particle with spores of *Bacillus stearothermophillus* embedded in it. The gel ensures that the microorganisms do not leak out and result in inaccurate degree of microbial destruction.

When vegetative cells of bacteria are subjected to harsh conditions (high heat or lack of nutrients), they form a hard proteinaceous coating outside the cell that can withstand the harsh conditions, and go into a passive stage, and the organisms in this state are called spores. Inactivating vegetative cells of bacteria can be achieved relatively easily (a few minutes at ~80°C), whereas inactivating the spores requires relatively high heat treatment (a few minutes at 120°C). The heat resistance of bacteria (vegetative cells and spores) is affected by previous events such as incubation temperature (resistance increases as incubation temperature is raised closer and closer to the optimum growth temperature), age (least resistant in logarithmic growth phase and most resistant in the last part of the lag phase and also in the stationary phase), growth medium (more nutritious the growth medium, more resistant the spore), and drying (some spores become more heat resistant after drying). Other factors affecting heat resistance are the presence of ionic species, oxygen content, water activity (moist heat is generally more effective than dry heat), pH (acid medium is usually more effective than alkaline medium, which is usually more effective than neutral medium), salts and sugars (high concentrations are effective in reducing their resistance), and proteins and fats (the presence of these materials increases the heat resistance). Thus, it is important to determine the heat resistance of the organisms of concern in the substrate of interest and under the appropriate processing condition. It should also be noted that bacteria which tend to clump together are generally more resistant to heat and care has to be exercised when dealing with them.

Most chemical and microbiological reactions encountered in thermal processing are first-order equations and are given by

$$\ln\left(\frac{c}{c_0}\right) = -k_T t$$

where c_0 is the initial concentration of the species under consideration, c is the concentration after time t, and k_T is the rate of the reaction. For microbial destruction, the concentration in the above equation is replaced by the number of viable microorganisms, and the rate of reaction is replaced by the decimal reduction time (D_T) to yield the following equation (expressed in base 10):

$$\log\left(\frac{N}{N_0}\right) = -\frac{t}{D_T}$$

with the decimal reduction time (in minutes) being related to the rate of the reaction (in seconds) by

$$D_T = \frac{2.303}{60k_T}$$

When a semilogarithmic plot is made between the number of viable microorganisms (on the y axis on a logarithmic scale) and time in minutes (on the x axis), a straight line is obtained. The slope of this line is equal to the negative reciprocal of the decimal reduction time. This graph is also referred to as the survivor curve, thermal death curve, inactivation curve or the thermal death time (TDT) curve.

The dependence of the rate of the reaction, k_T, on temperature, is given by the following equation by Arrhenius:

$$k_T = Be^{-E_a/R_gT}$$

with B being a constant which is referred to as the collision number or frequency factor and E_a being the activation energy. In the above equation, both B and E_a are assumed to be independent of temperature. However, there are other models that do not make this assumption. For *Clostridium botulinum*, researchers [46] determined the appropriate value of B to be $2 \times 10^{60}\ \text{sec}^{-1}$ and E_a to be 310.11 kJ/mol K.

Another commonly used technique to express the dependence of the rate of reaction on temperature is the quotient indicator method which defines a quotient indicator as the ratio of the reaction rates at two temperatures. When these temperatures are 10°C apart, the quotient indicator is then referred to as Q_{10} and is given by

$$Q_{10} = \frac{k_{T+10}}{k_T} = \frac{D_T}{D_{T+10}} = 10^{10/z}$$

The Q_{10} value for *Clostridium botulinum* is 10.

5.2.1 Process Lethality and Cook Values

The decimal reduction time of bacteria depends strongly on temperature and is given by

$$\log\left(\frac{D_T}{D_{\text{ref}}}\right) = -\frac{T - T_{\text{ref}}}{z}$$

where D_{ref} is the decimal reduction time at a reference temperature of T_{ref} and z is the temperature change required for an order-of-magnitude change in the decimal reduction time. The z value for *Clostridium botulinum* is 18°F (or 10°C).

The lethal rate (LR), which is a measure of the rate of inactivation of the microorganisms at any given temperature, is given by

$$\text{LR} = 10^{(T-T_{\text{ref}})/z} = \frac{D_{\text{ref}}}{D_T}$$

For a constant-temperature process, the above approach can be used. However, when the process temperature changes, the F value is used to calculate the total lethal rate:

$$F = \int_0^t (\text{LR})\, dt = \int_0^t 10^{(T-T_{\text{ref}})/z}\, dt$$

The F value at a reference temperature of 250°F (or 121.1°C) and a z value of 18°F (10°C) is referred to commonly as the F_0 value and is thus evaluated as

$$F_0 = \int_0^t 10^{(T-250)/18}\, dt$$

Another term that is commonly encountered in aseptic processing is lethality. Lethality is the ratio of the F_0 value of the process to the F_0 value required for commercial sterility. Thus, process lethality must be at least unity for commercial sterility.

An F_0 value of 5 min indicates that the process is equivalent to a heat treatment of 5 min at 250°F. It can be thus seen that many combinations of time and temperature can yield an F_0 value of 5 min. The appropriate combination of time and temperature that is used for processing is based on other factors such as nutrient retention and enzyme destruction. This is where the cook value (C) of a process comes into picture. The cook value is a measure of the extent of destruction of enzymes or nutrients and is given by

$$C = \int_0^t 10^{(T-T_{\text{ref}})/z_c}\, \mathrm{d}t$$

with z_c being analogous to z for microorganisms. Similar to F_0, C_0 is the reference cook value based commonly on a reference temperature of 100°C and the z_c value is much higher than that for microorganisms (e.g., 33°C for thiamine destruction). Graphical methods or optimization models are then used to determine the optimum time–temperature combination that renders the product safe and also retains the maximum possible amount of nutrients.

5.2.2 Commercial Sterility of the product

When insufficient data are available regarding cold spots in the product, a very conservative method is generally employed. The conservative approach involves the assumption that particles neither receive lethality nor any heat treatment in the heat exchanger. This approach is referred to as the "hold only" approach. The other two commonly used approaches are the "F_0 hold" and the "total system" approach. In the "F_0 hold" approach, it is assumed that particles gain heat treatment in the heat exchanger, but not lethality. In the "total system" approach, it is assumed that particles gain heat treatment and also accumulate lethality in the heat exchanger. The reason for not including lethality accumulated in the cooling section is that it is possible for particulates to break in the cooling section and thereby get cooled rapidly and, hence, not receive the assumed heat treatment and hence lethality.

6 FROM AN IDEA TO COMMERCIALIZATION

In order to commercially produce aseptically processed low-acid foods (pH > 4.6) containing large particulates (diameter >4.6 mm), there are several hurdles to overcome. It all begins with an idea for the product. Let us, for example, consider a product such as $\frac{1}{2}$-in carrot cubes (10% w/w) suspended in a 1% carboxymethylcellulose (CMC) solution. This is the product that we would like to aseptically process, package, and market as a high-quality shelf-stable product.

The first step is to identify a supplier who can provide high quality $\frac{1}{2}$-in carrot cubes (mildly blanched) and a supplier who can provide easy-to-dissolve, shear-stable CMC powder of satisfactory initial microbiological count. The next step is to determine the tentative process layout. This includes the choice of pump, heat exchanger, holding tube, cooling unit, and packaging equipment.

Let us begin by determining the appropriate pump to be used. Because we are dealing with large particulates, a piston pump such as the 20 hp Marlen twin-piston pump (Model 629A, Marlen Research Corp., KS) will be an appropriate choice. This type of pump will not only result in uniform product flow rate but also minimal damage to the particles. The product is batched in a 200-gal tank using an Admix Rotosilver submersible High-shear Mixing unit (Admix Inc., Londonderry, NH).

The next step is to identify the appropriate heating system. In situations where there are large particles, a scraped surface heat exchanger or a volumetric heating system is usually employed. Here, we select a scraped surface heat exchanger (SSHE) equipped with steam seals for aseptic processing applications (Model 6×9, Alfa Laval, Newburyport, MA) as the preheater (the speed of rotation of the rotor is set at 175 rpm and measured using an optical tachometer) and a 30-kW, 40.68 MHz continuous-flow radio-frequency (RF) heater (Model 464, Radio Frequency Co., Millis, MA) as the final heater. The SSHE serves to bring up the temperature of the product from room temperature to a certain elevated temperature and the RF heater is the finisher which has the effect of minimizing the difference in temperature between the fluid and particle because a low frequency (40.68 MHz) translates to a high depth of penetration of the electromagnetic waves.

The holding tube is one of the most important parts of the aseptic processing system, as this is where the product receives its heat treatment from a commercial sterility standpoint. A stainless-steel helical holding tube assembly (coil diameter 1 in.) is used. A helical holding tube results in the development of secondary flow and causes mixing of the solid–liquid mixture and, hence, translates to a relatively uniform heat treatment of the product.

The product is then cooled in a hydrocoil cooling unit (ASTEC, IA). A hydrocoil cooling unit will result not only in rapid cooling of the product but will also be gentle enough to the product so as to not cause disintegration of the particulates. The cooling medium is chilled water flowing through the system at a high flow rate (20 gpm of chilled water constantly flowing in and out of the system and 200 gpm of water continuously circulating through the system).

The packaging unit used in this case is a single-lane, two-head Metal Box cup filler (Model SL1-15, Metal Box, Reading, England) with a peroxidase spray system and an agitated Raque tank (100 gal PF-2-5-1A agitated filler, Food Systems Inc., Louisville, KY). The package used is a 12-oz cup with an aluminum foil used to heat-seal the top.

The data acquisition system consists of a datalogger (Model CR10, Campbell Scientific, Logan, UT) and multiplexer capable of handling 32 channels which is run using the software PC208W 3.0. The temperatures of the fluid are monitored at the entrance and exit of the pump, SSHE, RF heater, holding tube, and cooling section using type-T thermocouples (Omega Engineering, Stamford, CT) installed with sanitary fittings.

Back pressure is provided to the system by means of a lobe pump (Model 45U2, Waukesha-Cherry Burrell, Delavan, WI) along with a T junction with a manual control back-pressure valve in the vertical section and a wire mesh screen that ensured that only the fluid portion passed through the back-pressure valve and the particles passed through to the lobe pump.

The next step is to use mathematical modeling to design the length of the holding tube based on the properties of the product, process parameters, and other system parameters. For the heat exchanger and the holding tube sections, the theory of mathematical modeling has been presented in previous sections of this chapter. However, for modeling heat transfer in the RF heater, another model has to be used. Based on the modeling studies, the length of the holding tube is appropriately chosen (conservative estimate). During the modeling, care should be taken to account for the change in viscosity of the suspension as a function of time. To aid this, bench-top studies should be conducted to determine the rheological behavior of CMC as a function of shear rate, time, and the high temperatures encountered during processing. This can be done using a controlled-stress rheometer (Stresstech, ATS RheoSystems, Bordentown, NJ). During the modeling, conservative estimates of convective heat transfer coefficients and thermal diffusivities are used. Thermal conductivity is measured using the line heat source probe and specific heat is measured using a mixing calorimeter.

The first phase in experimental studies is to incorporate very small magnets (of different magnetic strengths) into several cube-shaped tracer particles in order to determine the residence times of the particles in various sections of the aseptic processing system (especially the holding tube section). Care is taken to compensate for the higher density of the magnets than the tracer particle because particle density is a major factor affecting the fastest particle residence time. Magnetic coils situated outside the tubes of the processing line pick up the signals produced by the motion of the magnets throughout the system, and this enables us to determine the residence times and, hence, the residence time distribution and also the fastest particle residence time. The magnets are of low enough strength to not affect the electromagnetic field created by the RF system. Based on statistics, it has been shown that the residence times of at least 299 particles must be determined in order to have a 95% confidence of collecting the fastest 1% of the particles.

The next step is to perform biological validation tests. These tests are performed at various stages of the process: just after start-up, during the middle of the run, and just before shutdown. These tests account for variations during the process and also for factors such as fouling. The validation tests are conducted at different temperatures to document a positive/negative result at the end of the process. This will aid in determining the minimum allowable process temperature that will result in a safe process. Microbiological validation tests are done using PA 3679 inoculated within alginate particles. Care should be taken to ensure that the spores do not leach out into the fluid. If the target for the process was a 5D process and an initial load of 10^5 spores per particle is used, a final count of <1 would indicate a safe process. The decimal reduction time of the organisms used is determined by means of thermal death time studies. Based on all of these tests, a process is designed and finally verification of the established process has to be conducted. During this process of verification, comparisons are made between actual temperatures and lethalities to the predicted temperatures and lethalities in order to ensure that the model results in a conservative prediction of process lethality.

Once verification is successful, all the process and system parameters are noted down and care should be taken to ensure that these parameters remain within an acceptable range. Some of the parameters include hydration time, mixing/batching time, temperatures at various locations, product flow rate, back pressure, and product properties. The final step in commercialization of the product involves process filing with the FDA using form 2541C. A comprehensive overview of the procedures and processes involved in process filing for a product such as the one discussed above has been given in a report elsewhere [47]. This is based on the workshops organized by the Center for Advanced Processing and Packaging Studies (CAPPS) and the National Center for Food Safety and Technology (NCFST).

7 CONCLUDING REMARKS

Aseptic processing has undergone a variety of changes since its inception as far as equipment, operating procedures, and critical points are concerned. With the advent of new technologies to inactivate microorganisms, some of the existing problems, such as slow heating of particles, can be overcome. Nevertheless, new technologies, such as the use of high pressure or pulsed electric field, have to be carefully studied, because the target microorganism, extent of enzymatic inactivation, and other factors might change. Despite the hurdles posed to aseptic processing, the high quality of the end product will make this technology more prevalent in the U.S. market as consumers are becoming more conscious about the nutritive value of foods and leading a healthy lifestyle.

NOMENCLATURE

a	Radius of particle, m
A	Surface area, m^2
B	Arrhenius parameter, Pa sec
c	Final concentration of species
c_0	Initial concentration of species
c_p	Specific heat, J/kg K
c_1, c_2	Constants

C	Cook value, min
C_d	Drag coefficient
d	Diameter of tube, m
D	Diameter of helical coil, m
D_c	Characteristic dimension, m
D_s	Diameter of shaft of SSHE, m
D_t	Decimal reduction time at temperature T, min
D_{vertical}	Vertical dimension, m
E_a	Activation energy, J/kg mol
E_f	Loss in energy due to friction, J/kg
E_p	Energy to be supplied by pump, J/kg
f	Friction factor
F	Force, N
F_0	F value when reference temperature is 250°F and z value is 18°F, min
F_b	Buoyancy force, N
F_d	Drag force, N
F_{rk}	Magnus lift force, N
F_s	Saffman lift force, N
g	Acceleration due to gravity, m/sec^2
h	Convective heat transfer coefficient, W/m^2 K
H_c	Enthalpy of condensate, J/kg
H_s	Enthalpy of steam, J/kg
H_v	Enthalpy of vapor, J/kg
I	Moment of inertia, kg m^2
j_H	Colburn j factor
k	Thermal conductivity, W/m K
k_T	Reaction rate constant at temperature T, sec^{-1}
K	Consistency coefficient, Pa sec^n
$\mathbf{K}$	Curl of velocity
l_h	Hydrodynamic entry length, m
l_t	Thermal entry length, m
L	Length of tube, m
L_c	Characteristic length, m
$\dot{m}$	Mass flow rate, kg/sec
m_p	Mass of particle, kg
n	Flow behavior index
N	Final bacterial count
N_b	Number of blades
N_{Bi}	Biot number
N_{De}	Dean number
N_{Fo}	Fourier number

N_{GGr}	Generalized Grashof number
N_{GPr}	Generalized Prandtl number
N_{GRe}	Generalized Reynolds number
N_{Gr}	Grashof number
N_{Gz}	Graetz number
N_{Nu}	Nusselt number
N_0	Initial bacterial count
N_p	Number of particles
N_{Pe}	Peclet number
N_{Pr}	Prandtl number
N_{Re}	Reynolds number
N_{Re_c}	Critical Reynolds number
P	Pressure, Pa
Q	Energy transferred, W
Q_{10}	Quotient indicator
r	Radial location, m
R	Radius of tube, m
Re_p	Particle Reynolds number
R_g	Universal gas constant, J/kg mol K
S	Source term, W/m^3
t	Time, sec
T	Temperature, K
T_i	Initial temperature, K
T_k	Local torque exerted by fluid, N m
T_∞	Free-stream temperature, K
u	Velocity, m/sec
$\bar{u}$	Average velocity, m/sec
u, v, w	x, y, and z components of velocity, respectively, m/sec
U	Overall heat transfer coefficient, W/m^2 K
$\dot{V}$	Volumetric flow rate, m^3/sec
$\mathbf{V}$	Velocity, m/sec
$\langle V \rangle$	Volume, m^3
X	Quality of steam
z	Temperature change required for an order of magnitude change in decimal reduction time, °C
Z	Height, m

Greek Letters

α	Thermal diffusivity, m^2/sec^2
β	Coefficient of volumetric thermal expansion, K^{-1}
ρ	Density, kg/m^3

ε	Roughness of pipe, m
σ	Shear stress, Pa
σ_0	Yield stress, Pa
$\dot{\gamma}$	Shear rate, sec^{-1}
λ	Latent heat of vaporization, J/kg
λ_c	Curvature
μ	Viscosity, Pa sec
μ_e	Effective viscosity, Pa sec
ΔP_f	Pressure loss due to friction, Pa
Δr, Δx	Thickness, m
ΔT	Temperature difference, K
ψ	Constant
Ω	Angular velocity, rad/sec
Φ	Particle concentration
τ	Shear stress, Pa

Subscripts

b	Bulk fluid
c	Coiled tube
cs	Cooling section
cw	cooling water
f	Fluid
fp	Fluid–particle interface
ht	Holding tube
hx	Heat exchanger
i	Inside
ins	Insulation
lm	Logarithmic mean
max	Maximum
o	Outside
p	Particle
ps	Particle surface
ref	Reference temperature
s	Straight tube
st	Steam
w	Wall

REFERENCES

1. JRD David, RH Graves, VR Carlson. Aseptic Processing and Packaging of Food: A Food Industry Perspective. Boca Raton, FL: CRC Press, 1996, pp 21–29.

2. WR Dean. Motion of fluid in a curved pipe. Philos Mag Ser 7 4(20):208–223, 1927,
3. WR Dean. The stream-line motion of fluid in a curved pipe. Philos Mag Ser 7 5:673–695, 1928.
4. LC Truesdell, RJ Adler. Numerical treatment of fully developed laminar flow in helically coiled tubes. AIChE J 16:1010–1014, 1970.
5. GI Taylor, FRS Yarrow. The criterion for turbulence in curved pipes. Proc Roy Soc London A124:243–249, 1929.
6. JA Koutsky, Rj Adler. Minimization of axial dispersion by use of secondary flow in helical tubes. Can J Chem Eng 42:239–246, 1964.
7. CM White. Streamline flow through curved pipes. Proc Roy Soc London A123:645–663, 1929.
8. PS Srinivasan, SS Napurkar, FA Holland. Pressure drop heat transfer in coils. Chem Eng 46(4):113–119, May 1968.
9. SA Berger, L Talbot, LS Yao. Flow in curved pipes. Ann Rev Fluid Mech 15:461, 1983.
10. LF Moody. Friction factors for pipe flow. ASME Trans 66:671–684, 1944.
11. CF Colebrook. Friction factors for pipe flow. Inst Civil Eng 11:133, 1939.
12. RL Manlapaz, SW Churchill. Fully developed laminar flow in a helically coiled tube of finite pitch. Chem Eng Commun 7:57–78, 1980.
13. B Dutta, SK Sastry. Velocity distributions of food particle suspensions in holding tube flow: Experimental and modeling studies on average particle velocities. Food Sci 55(5):1448–1453, 1990.
14. L Palmieri, D Cacace, G Dipollina, G Dall'Aglio. Residence time distribution of food suspensions containing large particles when flowing in tubular systems. J Food Eng 17:225–239, 1992.
15. MF Sancho, MA Rao. Residence time distribution in a holding tube. J Food Eng. 15:1–19, 1992.
16. KP Sandeep, CA Zuritz. Residence times of multiple particles in non-Newtonian holding tube flow: Effect of process parameters and development of dimensionless correlations. J Food Eng. 25:31–44, 1995.
17. PN Baptista, FAR Oliveira, SM Caldas, JC Oliveira. Effect of product and process variables in the flow of spherical particles in a carrier fluid through straight tubes. J Food Process Preserv 20:467–486, 1996.
18. W Chen, R Jan. The torsion effect on fully developed laminar flow in helical square ducts. J Fluids Eng 115:292–301, 1993.
19. GS Tucker, PM Withers. Determination of residence time distribution of non-settling food particles in viscous food carrier fluids using hall effect sensors. J Food Process Eng 17:401–422, 1994.
20. M Ahmad, SN Singh, V Seshadri. Distribution of solid particles in multisized particulate slurry flow through a 90° pipe bend in horizontal plane. Bulk Solids Handling 13(2):379–385, 1993.
21. KP Sandeep, CA Zuritz, VM Puri. Residence time distribution of particles during two-phase non-Newtonian flow in conventional as compared with helical holding tubes. J Food Sci 62(4):647–652, 1997.

22. RB Bird, WE Stewart, EN Lightfoot. Transport Phenomena. New York: John Wiley & Sons, 1960, pp. 91 and 103.
23. J Fichtali, FR van de Voort, GJ Doyon. A rheological model for sodium caseinate. J Food Eng 19(2):203–211, 1993.
24. A Ibarz, J Pagan, R Miguelsanz. Rheology of clarified fruit juices. II: Blackcurrant juices. J Food Eng 15:63–73, 1992.
25. Z Xuewu, L Xin, G Dexiang, Z Wei, X Tong, M Yonghong. Rgeological models for xanthum gum. J Food Eng 27:203–209, 1996.
26. SK Sastry, CA Zuritz. A model for particle suspension flow in a tube. ASAE Paper No.876537, 1987.
27. DR Oliver. Influence of particle rotation on radial migration in the Poiseuille flow of suspension. Nature 194:1269, 1962.
28. L Theodore. Sidewise force exerted on a spherical particle in a Poiseuille flow. Eng Doctoral thesis. New York University, 1964.
29. PG Saffman. The lift on a small sphere in a slow shear flow. J Fluid Mech 22(2):385–400, 1965.
30. R Clift, WH Gauvin. Motion of entrained particles in gas streams. Can J Chem Eng 49:439–448, 1971.
31. G Maesmans, N Hendrickx, S DeCordt, A Francis, P Tobback. Fluid-to-particle heat transfer coefficient determination of heterogeneous foods: A review. J Food Process Preserv 16:29–69, 1992.
32. EB Christiansen, DE Craig Jr. Heat transfer to pseudoplastic fluids in laminar flow. AIChE J 8(2):154–160, 1962.
33. WL McCabe, JC Smith, P Harriott. Unit Operations in Chemical Engineering. 4th ed. New York: McGraw-Hill, 1985, pp 294–354.
34. NJ Beek, R Eggink. In: WM Rohsenow, Developments in Heat Transfer. Cambridge, MA: MIT Press, 1962, pp 334.
35. CA Zuritz, SC McCoy, SK Sastry. Convective heat transfer coefficients for irregular particles immersed in non-Newtonian fluid during tube flow. J Food Eng 11:159–174, 1990.
36. I Filkova, B Koziskova, P Filka. Heat transfer to a power law fluid in tube flow: An experimental study. In: ML Maguer, P Jelen, eds. Food Engineering and Process Applications. 1: Transport Phenomena. Amsterdam: Elsevier Applied Sciences, 1986, pp 259–272.
37. WL Wilkinson. Non-Newtonian Fluids. London: Pergamon, 1960, pp. 104.
38. RH Perry, CH Chilton. Chemical Engineer's Handbook. New York: McGraw-Hill, 1973.
39. KB Zitoun, SK Sastry. Determination of convective heat transfer coefficient between fluid and cubic particles in continuous tube flow using noninvasive experimental techniques. J Food Process Eng 17:209–228, 1994.
40. H Weisser. Untersuchungen zum Warmeubergang im Kratzkuhler. PhD thesis, Karlsruhe Universitat, Germany, 1972.
41. AHP Skelland, DR Oliver, S Tooke. Br Chem Eng 7(5):346, 1962.
42. CE Kalb, JD Seader. Heat and mass transfer phenomena for viscous flow in curved circular tubes. Int J Heat Mass Transfer 15:801–817, 1972.

43. CE Kalb, JD Seader. Fully developed viscous-flow heat transfer in curved circular tubes with uniform wall temperature. AICHE J 20(2):340–346, 1974.
44. LAM Janssen, CJ Hoogendoorn. Laminar convective heat transfer in helical coiled tubes. Int J Heat Mass Transfer 21:1197–1206, 1978.
45. EMA Willhoft, Aseptic Processing and Packaging of Particulate Foods. London: Blackie Academic and Professional, 1993, 6–7.
46. SG Simpson, MC Williams. An analysis of high temperature short time sterilization during laminar flow. J Food Sci 39:1047–1054, 1974.
47. CAPPS and NCFST. Case study for condensed cream of potato soup. Aseptic Processing of Multiphase Foods Workshops, 1995 and 1996.

4

Modeling Moisture Diffusion in Food Grains During Adsorption

Kasiviswanathan Muthukumarappan
South Dakota State University, Brookings, South Dakota
Sundaram Gunasekaran
University of Wisconsin–Madison, Madison, Wisconsin

1 INTRODUCTION

Food grains are hygroscopic and hence adsorb or desorb moisture depending on the environment. Moisture diffusivity is a physical property of measurement, which aids in studying the moisture diffusion mechanism. Moisture gradients prevalent within a food grain due to the moisture adsorption/desorption phenomenon may lead to the development of internal stresses [1,2]. The internal as well as external stresses cause grain kernels to fissure. Fissured or stress-cracked kernels are objectionable because they are quite susceptible to breakage during handling and cause problems in storage, shipping, and processing [3]. If the stresses developed within the kernels can be calculated accurately, better processes can be designed to reduce fissure development. However, such an estimation requires accurate determination of moisture diffusivity of the grain components and a description of moisture adsorption/desorption mechanisms.

Adsorption and desorption are different mechanisms and there exists a hysteresis between them; that is, the equilibrium moisture content attained by grains via desorption is higher than that via adsorption for a particular temperature and relative humidity condition. There have been many theories to explain this hysteresis. Chung and Pfost [4] postulated that more sorption sites or polar sites are available to water vapor for the desorption process than for the adsorption process; that is, the moisture-transport mechanisms of desorption and adsorption are different. Desorption and adsorption processes are subjected to the same physical laws and, thus, can be treated analogously. Variations in the rate of desorption and adsorption (diffusion) occur due to the boundary conditions at the medium interface and may cause the apparent hysteresis in the sorption isotherms. It should be emphasized that during low-temperature deep-bed drying, whereas some parts of a large mass lose moisture (desorption) and others simultaneously gain moisture (adsorption). Thus, models that cover both desorption and adsorption processes are needed.

Extensive research work has been done on drying of different grains with a primary focus on modeling diffusion of moisture [5,6] and determining moisture diffusivities of major grain components. However, only limited information is available on diffusion of moisture in grains during adsorption [7,8]. In general, the moisture diffusivity of grains during adsorption is lesser (at least one order of magnitude) than during desorption. For example, rice kernels had a moisture diffusivity of 1.3×10^{-7} during desorption and $1.2 \times 10^{-8}\ m^2/hr$ during adsorption. Experimental methods of diffusivity determination, collecting moisture content data at various points inside the corn kernel over a time period, require sophisticated sensors and are cumbersome. Mathematical models, based on physical principles, can potentially predict with reasonable accuracy the moisture distribution inside the kernel during adsorption. However, for improved accuracy, the mathematical models require the moisture diffusivities of kernel components during adsorption. Moisture diffusivity of grain components during adsorption is also needed to better understand the moisture transport in grain conditioning, storage, deep-bed drying, and aeration processes.

For efficient processing operations, quantitative and predictive models relating the physical properties of food to transient time–moisture profiles that determine product quality are needed. In this chapter, different mathematical models to determine the moisture diffusivity of individual components of any heterogeneous food grain are described. As an example, the developed models were validated using the moisture adsorption data in a corn kernel. Moisture diffusivity of individual components of a corn kernel, namely pericarp, germ, and soft and hard endosperms, were determined using the finite-difference, analytical, and finite-element methods, respectively. The developed finite-element model was also used to predict the moisture distribution inside the corn kernel.

2 MOISTURE DIFFUSION IN FOOD GRAINS

2.1 Various Moisture-Transport Mechanisms

The mechanisms of moisture movement within a product can be primarily summarized as water-vapor-transport mechanisms and liquid-water-transport mechanisms (Figure 1). The water-vapor-transport mechanism consists of Knudsen diffusion, Stefan diffusion, mutual diffusion, Poiseuille flow, and condensation–evaporation. On the other hand, the liquid-water-transport mechanisms consists of capillary flow, liquid diffusion, and surface diffusion [9]. Among these, diffusion is the dominant mechanism.

2.1.1 Knudsen Diffusion

One of the water-vapor-transport mechanisms within a product may be explained in terms N of the Knudsen diffusion mechanism as explained in Figure 1. This type of diffusion occurs in gas-filled solids with small pores, or under low pressure when the mean free path of molecules is more than the pore size and the molecules collide with the walls more often than among themselves. Molecule reflection from the walls is normally diffuse. In this case, the water flux is a function of the vapor density and the Knudsen vapor diffusivity within the product. The size and amount of pores, tortuosity, and the geometry of the solid matrix affect the water-vapor flux as represented in the equation column.

2.1.2 Stefan Diffusion

In the case of Stefan diffusion, the water flux is a function of the vapor pressure, total pressure, and the Stefan vapor diffusivity within the product. A constant diffusivity is assumed.

2.1.3 Mutual Diffusion

Mutual diffusion is predominant in solids with large pores, whose size is much more than the free path of the diffusing vapor molecules. The roles Knudsen and mutual diffusions perform are commensurable within a certain range of pore sizes and gas pressures.

TYPE	PICTURE	EQUATIONS
KNUDSEN DIFFUSION	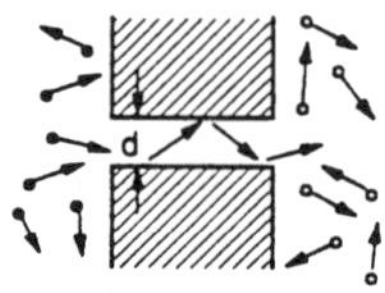	$n_w^d = -\varepsilon\tau\beta D_{kw} \nabla\rho_w^l$ $D_{kw} = \frac{2}{3} d \left(\frac{2}{\Pi}\frac{RT_k}{M_w}\right)^{\frac{1}{2}}$
STEFAN DIFFUSION	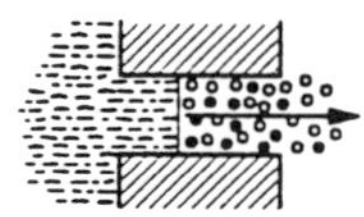	$n_w^d = -\varepsilon\tau D_{wg}\frac{P}{P-P_w}\frac{M_w}{RT_k}\nabla P_w$ $D_{wg} = 2.5x10^{-5}\ [m^2/s]$
MUTUAL DIFFUSION	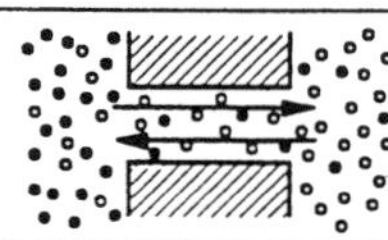	$n_w^d = -\varepsilon\tau D_{wg} \nabla\rho_w^l$
POISEUILLE FLOW	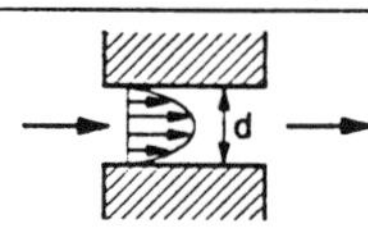	$n_w^d = -\varepsilon\tau \frac{d^2\rho}{32\mu} \nabla P$
CONDENSATION -EVAPORATION	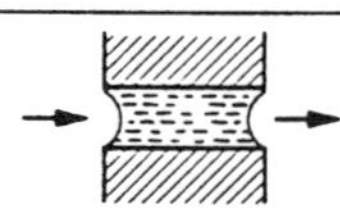	$n_w^d = n_w (\nabla T_k, \ldots.)$
CAPILLARY FLOW	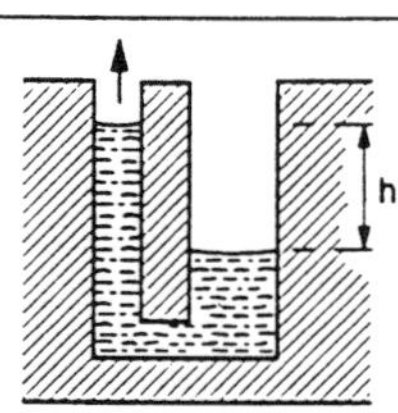	$n_w^l = -\rho_l \chi \nabla\theta$
LIQUID DIFFUSION	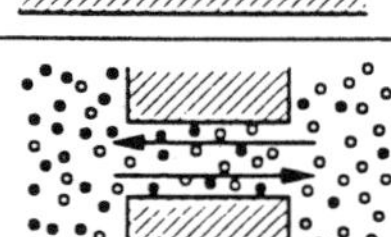	$n_w^l = -\rho_l D_{we} \nabla\omega_w$
SURFACE DIFFUSION	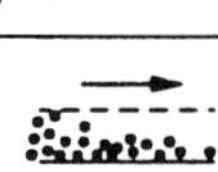	$n_w^l = -\rho_l D_{su} \nabla\omega_w$

FIGURE 1 Various mechanisms of moisture transport in porous materials. (From Ref. 9.)

2.1.4 Poiseuille Flow

Poiseuille flow is pressure-induced flow in a long duct. It is also called channel flow. In this case, it is assumed that there is laminar flow of an incompressible Newtonian fluid of viscosity μ induced by a constant positive pressure difference or pressure drop ΔP in a pipe of length L and diameter $d \ll L$.

2.1.5 Condensation–Evaporation Theory

Water vapor within the solid is condensed near the surface. This assumes that the rate of condensation is equal to the rate of evaporation at the surface of the solid and allows no accumulation of water in the pores near the surface. This theory takes into account the simultaneous diffusion of heat and mass, which assumes that the pores are a continuous network of spaces in the solid.

2.1.6 Capillary Flow

Moisture which is held in the interstices of solids, as liquid on the surface, or as free moisture in cell cavities moves by gravity and capillarity, provided that passageways for continuous flow are present. In drying, liquid flow resulting from capillarity applies to liquids not held in solution and to all moisture above the fiber-saturation point, as in textiles, paper, and leather, and to all moisture above the equilibrium moisture content at atmospheric saturation, as in fine powders and granular solids, such as soil, sand, and clays.

2.1.7 Liquid Diffusion

The movement of liquids by diffusion in solids is restricted to the equilibrium moisture content below the point of atmospheric saturation and to systems in which moisture and solid are mutually soluble. This applies to the drying of clays, wood, soaps, and pastes.

2.1.8 Surface Diffusion

Surface diffusion is observed during adsorption of a diffusing substance by a solid. Because the equilibrium surface gas concentration increases with an increase in partial pressure of the adsorbed spices, a surface concentration gradient of a diffusing substance appears in the surface layer of a pore. Under certain conditions like high temperature, this may enhance the total flow of a diffusing component.

The mechanisms described above refer only to a single-component diffusion. Multicomponent diffusion in porous solids is very complex. Because of its complex nature, it has not been adequately investigated.

2.2 Coupled Heat and Moisture Transport

The behavior of any food material during drying and rewetting depends on the heat and mass transfer characteristics of the product being dried or wetted. Knowledge of the temperature and moisture distribution in these products is vital for equipment and process design, quality control, and choice of appropriate storage practices. Wang and Hall [10] stated that if temperature distribution within the medium is uniform, the assumption of moisture concentration as the driving force is adequate. This assumption is reasonable because grains respond to temperature differences more rapidly than to moisture differences [11]. Further, Sharaf Eldeen et al. [12] reported that the body temperature of grains approached the drying air temperature in a small fraction of the total drying time; thus, temperature could be omitted in the model. Young [13] described a mathematical model for drying of a porous sphere using the diffusion equation for both moisture and heat transfer, assuming that moisture diffusivity is a linear function of moisture content. He defined a modified Lewis number and suggested that the moisture diffusion equation alone is sufficient if the Lewis number is greater than 60 (negligible temperature gradient).

About a decade ago, coupled heat and mass transfer equations have been solved for an isotropic sphere with constant material properties [14]. In 1992, Irudayaraj et al. [15] developed a comprehensive model that described the heat and mass transfer in a wide range of food grains (soybean, barley, and corn kernels) with varying material properties. Their simulated results from the heat and mass transfer models agreed well with the experimental results. Recently, Irudayaraj and Wu [16] developed models incorporating heat, mass, and pressure transfer equations to describe the moisture diffusion process in a barley kernel during soaking. The results obtained from the heat, mass, and pressure transfer show a marked difference from the results obtained from the heat and mass transfer model. This indicated that a pressure gradient exists during the soaking process, causing additional moisture movement due to filtration effect. Coupling the effect of moisture and temperature may be important for accurately modeling the drying process. However, for adsorption, the coupling effect may not be important because the adsorption process takes much longer (48–50 hr) than the desorption process (6–10 hr).

2.3 Characterization of Shape for Modeling Moisture Diffusion in Grains

Thin-layer models can be divided into three groups: (1) empirical models, (2) semiempirical models, and (3) theoretical models. Among these, theoretical models provide the most information about the moisture transport inside grains.

Young and Whitaker [17] and Whitaker and Young [18] evaluated different diffusion equations for a plane sheet, finite and infinite cylinders, and a sphere and an empirical model during drying of peanuts. They found that the diffusion equation better represented the drying than did the empirical model. Moreover, they concluded that the finite-cylinder model best fit the experimental data. Eckhoff and Okos [19] qualitatively explored the diffusion path of gaseous sulfur dioxide into yellow dent corn. They showed that sulfur dioxide (SO_2) enters at the tip cap, moves up through the area between the pericarp and seed coat, and then diffuses into the endosperm. Their observations clearly indicate that the pericarp of corn acts as a diffusion barrier even to gaseous SO_2. Modeling diffusion in cereal grains using the spherical diffusion model is thus inappropriate because the model does not adequately reflect the true nature of the diffusional processes via the tip cap. Recently, Eckhoff and Okos [20] modeled the gaseous SO_2 sorption by corn as an insulated cylinder with one end open for diffusion. Walton et al. [21] cautioned that the diffusion coefficient that is determined with one geometric shape could not be used with another geometric shape. Muthukumarappan and Gunasekaran [22–25] evaluated the effect of different shapes in determining the moisture diffusivity of corn samples and found that the infinite-slab model fitted the experimental adsorption data.

3 MODELING MOISTURE DIFFUSION IN FOOD GRAINS

3.1 Theoretical Considerations

A typical kernel is irregular in shape. Therefore, three geometries, namely an infinite slab, an infinite cylinder, and a sphere, were considered. The corresponding solutions of Fick's law of diffusion developed by Crank [26] were used. The differential equation with initial and boundary conditions to describe the system are

$$\frac{\partial M}{\partial t} = \frac{\partial}{\partial x}\left(D_m \frac{\partial M}{Px}\right) m = 1, n \tag{1}$$

$$\frac{\partial M}{\partial x} = 0, \quad x = 0, t \geq 0 \tag{2}$$

$$M = M_e, \quad x = \pm l_2, \quad t > 0 \tag{3}$$

$$M = M_0, \quad -l_2 < x < l_2, t = 0 \tag{4}$$

The following assumptions were made in solving Eq. (1):

1. Initial moisture content is uniform throughout the kernel.
2. Grain is isothermal during adsorption (i.e., the heat transfer equations may be neglected).
3. Moisture diffusivity is constant throughout the adsorption process.
4. Grain components are homogeneous and isotropic.
5. Expansion of the kernel during adsorption is negligible.

3.2 Boundary Condition

The boundary condition given in Eq. (3) implies that the moisture content at the kernel surface reaches equilibrium with the environment instantaneously. Newman [27] disputed this assumption during drying. Shivhare et al. [28] assumed that the surface moisture reaches equilibrium exponentially during microwave drying of corn. Muthukumarappan [8] found, during adsorption, that the numerical model with assumption of exponentially varying surface moisture fitted the experimental adsorption curve better than the model with assumption of instantaneous equilibrium surface moisture. Walton et al. [21] assumed a boundary condition, which is a function of the convective mass transfer coefficient in developing a drying model for corn. However, the experimental convective mass transfer coefficient values at different temperature and relative humidity conditions during adsorption are not currently available. This makes it difficult to predict surface moisture content as a function of adsorption time. Following these investigations, the surface moisture content was assumed to vary exponentially with adsorption time. The boundary condition describing the moisture content at the kernel surface (M_s, %) and the surface moisture ratio (MR_s) can be written as

$$M_s = [1 - \exp(-Kt)](M_e - M_0) + M_0 \tag{5}$$

$$\mathrm{MR}_s = \frac{M_s - M_0}{M_e - M_0} = 1 - \exp(-Kt) \tag{6}$$

The moisture ratio (MR) of a multicomponent system can be modeled as

$$\mathrm{MR}_i(t) = \sum_{i,j} X_{ij}\mathrm{MR}_j(t) \tag{7}$$

where i is the grain type 1 = soft and 2 = hard) and j is the grain component.

Instead of geometrically modeling the whole kernel, individual components (namely pericarp, soft and hard endosperms, and germ) can be modeled independently using Eq. (7). The number of components are those components that have differing properties, namely pericarp, soft and hard endosperms and germ, for a corn kernel. Similarly other grains, namely rice, wheat, and soybean, can be modeled using the above approach.

The Cartesian coordinates were used to represent the grain as a two-dimensional body. The general diffusion equation, which describes the moisture transport, has the form

$$\frac{\partial M}{\partial t} = \nabla(D\nabla M) \tag{8}$$

In two dimensions, it becomes

$$\frac{\partial M}{\partial t} = \frac{\partial}{\partial x}\left(D\,\frac{\partial M}{\partial x}\right) + \frac{\partial}{\partial y}\left(D\,\frac{\partial M}{\partial y}\right) \tag{9}$$

The initial and boundary conditions are

$$M = M_0, \quad t = 0 \tag{10}$$

and

$$M_s = [1 - \exp(-Kt)]^*(M_e - M_0) + M_0, \quad t > 0 \text{ on } \Omega \tag{11}$$

where Ω constitutes the complete boundary surface for the body.

3.3 Numerical Formulation

The element equations were developed by transforming the governing differential equations using the Galerkin's weighted residual approach. After the formulation, the element equations can be written in a simplified form as:

$$\sum_{j=1}^{n} \dot{M}_j[C_{ij}] + \sum_{j=1}^{n} \dot{M}_j[K_{ij}] = 0 \tag{12}$$

where the element moisture capacitance matrix

$$[C_{ij}] = \int_A N_i N_j \,\mathrm{d}x\, dy \tag{13}$$

and the element moisture conductance matrix

$$[K_{ij}] = \int_A \left(\frac{\partial N_i}{Px}\frac{\partial N_j}{\partial x} + \frac{\partial N_i}{\partial y}\frac{\partial N_j}{\partial y}\right) D\, dx\, dy \tag{14}$$

Assembling the element matrices in Eq. (12) using Eqs. (13) and (14), the global matrix equation can be written as

$$[C]\{\dot{M}\} + [K]\{M\} = 0 \tag{15}$$

where $[C]$ and $[K]$ are the global moisture capacitance and conductance matrices. The solution of Eq. (15) will result in moisture values at every time step in the domain of interest.

For the transient case under consideration, an implicit technique (backward difference scheme), which is unconditionally stable, was used. The final system of equations incorporating the known boundary conditions, had the form

$$([C] + \Delta t[K])M_{t+\Delta t} = [C]M_t + \Delta t F_{t+\Delta t} \tag{16}$$

4 MOISTURE DIFFUSION IN A CORN KERNEL DURING ADSORPTION

4.1 Corn Structure

Corn is a complex cereal grain. The kernel is flattened, wedge shaped, and relatively broad at the apex of its attachment to the cob. The kernel is composed of germ, soft (floury) and hard (horny) endosperms, and pericarp. The pericarp surrounds the kernel and is strongly adherent to the seed coat. The hard endosperm is found on the sides and back of the kernel and bulges in toward the center at the sides. The soft endosperm fills the crown (upper part) of the kernel and extends downward to surround the germ. The pericarp, the outermost part of the kernel and a major part of what the millers know as hull, is composed of several layers. Most importantly, the outer layer, the epidermis, is more or less cutinized on its outer surface. Cutin is relatively impervious to moisture, so the cutinized surface of the epidermis acts as a barrier to moisture movement during adsorption. Typically, a soft corn kernel is composed of 5% pericarp, 10% germ, 48% soft endosperm, and 37% hard endosperm. A hard kernel is composed of 4% pericarp, 9% germ, 21% soft endosperm, and 66% hard endosperm [8].

4.2 Moisture Diffusivity Determination

Chittenden and Hustrulid [29] reported that the mean diffusivity of shelled corn varied linearly with the initial moisture content; they concluded that actual diffusivity should depend also on moisture content at any point within the kernel. Steffe and Singh [5] verified that liquid diffusivity of rice components did not vary with the initial moisture content. However, Hsu et al. [30] demonstrated that, during soaking of soybeans, the moisture diffusivity is strongly dependent on the moisture content of the seeds and that the diffusion equation with constant diffusivity is inadequate in describing the water absorption curve. More recently, Lu and Siebenmorgen [31] modeled the moisture diffusion in rough, brown, and milled rice during adsorption with constant diffusivity. Their predictions agreed well with the experimental adsorption data.

The fluid parameters affecting moisture diffusion are temperature and relative humidity. Diffusion of moisture is generally enhanced by the temperature of the fluid medium (air) and has an exponential relationship (Arrhenius type) with the inverse of the fluid temperature. This has been demonstrated during drying of rough rice [32–34], brown and milled rice [35], peanuts [18], corn [21], and wheat [36]. Chu and Hustrulid [37] reported the diffusion coefficient as a function of temperature of the fluid medium and moisture content of corn during drying. Recently, Lu and Siebenmorgen [31] described the dependency of diffusivity of rough, brown, and milled rice on temperature by an Arrhenius-type function during adsorption. Further, they agreed that the relative humidity had some influence on moisture diffusivity. Therefore, the diffusivity may depend on the moisture content of a material and follow an exponential relation with fluid temperature. However, the dependence of diffusivity of grains on relative humidity of the environment is not clearly understood.

4.2.1 Germ

Three types of samples, namely corn germ, soft corn (FR27 × MO17), and hard corn (P3576), were tested. Corn germ obtained from Archer Daniels Midland (ADM) Company was used. The ADM company used steam tube dryers to remove moisture from the corn germ. The germ had an oil content of 44–48% and initial moisture content of 3–4% [38]. Two types of corn, namely soft and hard, of different densities (1229 and 1327 kg/m^3, respectively) were used in this study. The soft corn was grown on the Agricultural Research Station Farm at the Purdue University, West Lafayette, IN and combine-harvested at about 27% moisture content during Fall 1990. The hard corn was obtained from Frito-Lay Inc., Sidney, IL at about 15% moisture content during Spring 1991. The corn samples were dried using natural air at a temperature of 23°C and relative humidity (RH) of 55%.

The dried corn samples were hand-cleaned to remove the broken kernels. The moisture content of the samples was determined by the oven method [38] to be about 9–10%. The samples were stored in a refrigerator maintained at 5°C and 58% RH until the experiments. The adsorption tests were conducted in a controlled-environment chamber (2.21 × 0.74 × 1.95 m) available in the Biotron at the University of Wisconsin–Madison. The environment for this experiment consisted of four air temperatures of 25°C, 30°C, 35°C, and 40°C with each at two RH values of 75% and 90%. The temperature and RH of the air in the chamber were maintained within 0.1°C and 1.0%, respectively. Air was circulated constantly at 0.5 m/sec during the tests.

Two 50-g samples of each variety of corn and 25-g samples of germ were placed in individual perforated wire-meshed containers. The individual sample depth was 10 mm. This depth was chosen in order to acquire thin-layer adsorption data. The containers were individually supported with a load cell (Omega Model No. LCL-227G; rated capacity of 227 g). The output signals from the load cells were transmitted to a high-speed analog I/O board (DASl6G1, National Instruments) and a personal computer (Diversified Systems 486) through an expansion multiplexer (EXP-16). The digital data from the output signal was acquired through a DAS516G1 interface card. The data were saved on the computer using the software EASYEST LX. The voltage values from the load cell were converted to the corresponding mass values after calibrating each load cell with a set of known masses. The weight measurements were taken at 15-min interval. The tests were conducted for 48–72 hr during which the sample moisture content reached near equilibrium with the chamber.

Fick's law of diffusion models considering the geometry of corn germ and corn kernel as an infinite slab, an infinite cylinder, and a sphere were used for diffusivity determination. The first 10 terms of each model were considered using the nonlinear, least squares multivariate secant method [39]. The moisture diffusivity was estimated by minimizing the sum of square deviations (SSD) between the experimental and theoretical corn adsorption data. The characteristic dimensions of the germ and corn kernel were determined at the initial moisture content. Further details of the dimension measurement can be found in Muthukumarappan and Gunasekaran [22]. As a preliminary analysis, the diffusivities of corn germ and corn kernels exposed to air at 25°C and 90% RH were determined using infinite-slab, infinite-cylinder, and sphere models. Based on the SSD values, it was found that the infinite-slab model best predicts the moisture adsorption behavior of corn germ and corn kernels. Therefore, the moisture diffusivity of corn germ and corn kernels at other humid conditions were determined using only the infinite-slab model. The characteristic dimensions used in the infinite-slab model were 0.99, 2.08 and 2.40 mm for corn germ, kernels of FR27 × MO17, and P3576 variety corn, respectively. A multifactor analysis of variance [40] was performed to study the effects of air temperature, relative humidity, and type of corn on moisture diffusivity.

The surface adsorption coefficient (K) in Eq. (6) indicates how fast the moisture at the kernel surface is approaching equilibrium with the environment. In selecting the K values, it was assumed that the moisture content at the surface reaches equilibrium halfway through the adsorption process (i.e., MR = 0.5). This assumption is reasonable because the diffusivity is traditionally determined by calculating the time taken to reach MR = 0.5 [26]. Moreover, the effect of varying MR (MR = 0.6, 0.7, 0.8, and 0.9) on the

Table 1 Surface Adsorption Coefficient (K, hr^{-1}) Values Used in Eq. (6)

Temp. (°C)	Germ	Corn without pericarp FR27 × MO17	Corn without pericarp P3576	Corn with pericarp FR27 × MO17	Corn with pericarp P3576
25	0.776	0.672	0.388	0.487	0.371
30	1.381	1.036	0.829	0.606	0.592
35	1.776	1.130	0.888	0.710	0.777
40	2.072	1.243	1.036	1.036	0.921

goodness of fit was also investigated. From the preliminary investigation, it was found that the MR = 0.5 assumption best fitted the data. Therefore, the time taken to reach the moisture ratio of 0.5 was determined from the experimental data. This time value was substituted in Eq. (6) to obtain the K value. At this time, the surface moisture ratio should be, theoretically, unity. However, a value of $MR_s = 0.998$ was assumed because $MR_s = 1$ only at $t = \infty$. Values of MR_s ranging from 0.99 to 0.999 were evaluated and it was found that $MR_s = 0.998$ gave the best fit to the adsorption data. The values of K for all the environmental conditions are presented in Table 1. Using the K values in Table 1, the surface moisture values were calculated as a function of time. These surface moisture values were then used to calculate the modified moisture ratio with time. The moisture diffusivity of corn samples with varying surface moisture content was estimated by minimizing the sum of square deviations between the experimental (modified moisture ratio) and the theoretical data using the nonlinear, least squares multivariate secant method [39].

The moisture ratio for corn germ and corn samples (FR27 × MO17) exposed to air at 35°C and 75% RH are presented in Figure 2. Due to the smaller size of corn germ compared to the whole kernel, the germ approached the equilibrium moisture content more rapidly than the corn. This is reflected by the higher moisture ratio for the germ than for the com samples. Compared to the experimental values for both corn germ and corn kernels, the model initially overpredicted the moisture ratio and then underpredicted.

The moisture diffusivity values obtained for corn germ and composite corn kernel at different humid air conditions are presented in Table 2. In general, the moisture diffusivity increased with increasing temperature. In classical theory [41], increased temperature is interpreted to mean an increase in the average energy for each mode of motion of vapor (translational, rotational, and vibrational motions). Therefore, an increase of

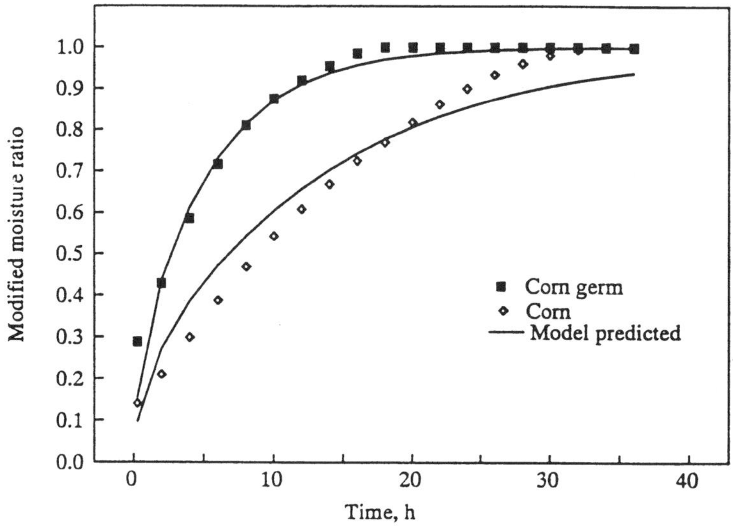

FIGURE 2 Modified moisture ratio for corn germ and corn kernels exposed to air at 35°C and 75% RH. The Fick's analytical model was used for an infinite slab with the varying surface moisture content assumption.

TABLE 2 Moisture Diffusivities (m^2/hr) of Pericarp, Germ, and Soft and Hard Endosperms of a Corn

Adsorption condition		Moisture diffusivity				
Temp. (°C)	RH (%)	Pericarp ($\times 10^{-8}$)	Germ ($\times 10^{-7}$)	Soft endosperm ($\times 10^{-7}$)	Hard endosperm ($\times 10^{-7}$)	Composite ($\times 10^{-7}$)
25	75	0.41	0.27	0.825	0.450	0.97
	80	0.34	0.17	0.733	0.320	0.68
	90	0.30	0.15	0.546	0.420	0.60
30	75	0.45	0.54	1.014	0.652	1.01
	80	0.42	0.20	0.997	0.566	0.90
	90	0.42	0.18	0.923	0.549	0.78
35	75	0.52	0.66	1.245	0.733	1.24
	80	0.49	0.33	1.142	0.680	1.20
	90	0.47	0.24	1.056	0.639	0.88
40	75	0.57	1.18	1.460	0.919	1.40
	80	0.53	0.40	1.221	0.687	1.32

temperature must mean an increase in the probability that the energy of the mode of motion required for interaction will attain a high value more frequently. The moisture diffusivity of germ decreased with increasing RH. During the adsorption tests, the concentration gradient is higher at higher RH. This higher concentration gradient should lead to higher moisture diffusivity values. However, the opposite trend was observed. At this time, there is no conclusive explanation for this trend. The moisture diffusivity values of corn germ were about two to five times lower than the effective moisture diffusivity of the composite corn kernels. This could be due to presence of oil in the germ. The difference was higher at low temperatures (25°C and 30°C) and lower at high temperatures (35°C and 40°C). From the multifactor analysis of variance tests, it was found that the differences in moisture diffusivity values obtained at different air temperatures, RH, and type of corn were statistically significant at the 0.05 level. The mean moisture diffusivity value of corn germ varied from 0.15×10^{-7} to $1.18 \times 10^{-7}\,m^2/hr$ for different air temperature and RH conditions.

The moisture diffusivity values of corn germ obtained during the adsorption study are lower than the published values during drying (desorption). For example, within the 10–24% moisture content range, reported diffusivity values of germ ranged from 5.974×10^{-7} to $34.731 \times 10^{-7}\,m^2/hr$ [6]. In general, the Fick's diffusion model better predicted the adsorption of corn germ than that of the composite corn kernel. One of the possible reasons is that the germ is more homogeneous than the composite corn kernel.

The temperature dependency of the moisture diffusivity of corn germ was modeled as an Arrhenius-type function used by Lu and Siebenmorgen [31]:

$$D = A \exp\left(-\frac{B}{T_a}\right) \tag{17}$$

The model coefficients A and B and the corresponding R^2 (coefficient of determination) values are summarized in Table 3. This model was satisfactory as evidenced by high R^2 values.

4.2.2 Pericarp

Two types of corn were used, namely soft (FR27 × MO17) and hard (P3576) with densities of 1229 and $1327\,kg/m^3$, respectively. To remove the corn pericarp, preliminary experiments were . conducted by soaking corn kernels in 25°C water for 15, 30, 60, and 120 sec. The pericarp was carefully removed using a razor blade. Corn kernels soaked for 30 sec absorbed less

TABLE 3 Coefficients of Arrhenius-Type Model [$D = A\exp(-B/T_a)$] for Temperature Dependency of Moisture Diffusivity of Corn Germ, Pericarp, and Soft and Hard Endosperms

Component	RH (%)	Coefficients		
		A	*B*	R^2
Germ	75	1.122×10^5	8645.3	0.964
	80	3.568	5727.4	0.956
	90	0.028	4313.9	0.981
Pericarp	75	4.917×10^{-7}	2116.1	0.991
	80	3.961×10^{-6}	2783.3	0.967
	90	3.293×10^{-4}	4137.1	0.928
Soft endosperm	75	9.338×10^{-3}	3463.7	0.996
	80	2.795×10^{-3}	3127.4	0.911
	90	0.418×10^2	6081.5	0.901
Hard endosperm	75	3.282×10^{-5}	1884.8	0.991
	80	0.2243	4657.7	0.903
	90	1.803×10^{-2}	3862.3	0.978

Note: D = moisture diffusivity (m^2/h); T_a = absolute temperature (K).

than 1% moisture and facilitated easy removal of the pericarp. The adsorption tests were conducted as explained in Section 4.2.1.

The corn kernel was modeled as a slab core (corn without pericarp composed of soft and hard endosperms and germ) surrounded by a slab shell (pericarp) as shown in Figure 3. First, adsorption of corn without pericarp (the slab core) was predicted numerically by solving the diffusion equation (in Cartesian coordinates) and the diffusivity of corn without pericarp was determined. Next, the adsorption of corn with pericarp (the slab core and shell) was predicted numerically. The diffusivity of corn pericarp was determined using the diffusivity of corn without pericarp and the experimental adsorption data for corn with pericarp.

The following assumptions were made in developing the adsorption models:

1. The mechanism of moisture transport is diffusion.
2. Corn is isothermal during adsorption (i.e., the heat transfer equations may be neglected).
3. Geometrically, the corn kernel is an infinite slab (i.e., the diffusion is one dimensional and end effects may be neglected).
4. Moisture diffusivity is constant throughout the adsorption process.

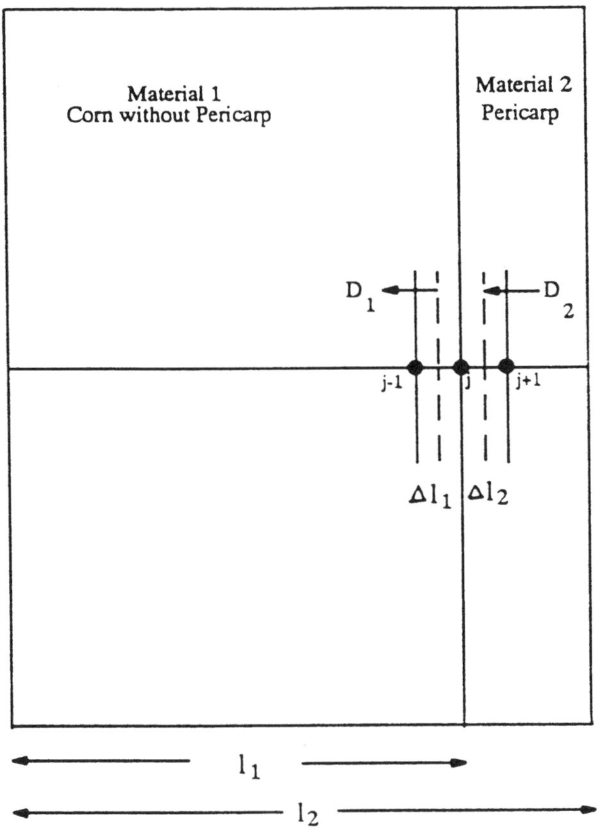

FIGURE 3 Schematic of a corn kernel modeled as a slab core (corn without pericarp) and a slab shell (pericarp).

5. Germ, soft and hard endosperms, and pericarp are homogeneous and isotropic.
6. Expansion of the corn kernel during adsorption is negligible.

The differential equation with initial and boundary conditions as shown in Eq. (1)–(4) were used with $n = 2$. The Crank–Nicolson finite-difference equations proposed by Crank and Nicolson [42] and summarized by Strikwerda [43] were used to solve the differential equations. Due to symmetry, only one-half of the whole system was considered for the analysis (Figure 3). Equation (1) should satisfy every point inside the system. Using the finite-difference formulations explained by Muthukumarappan [8], the nodal moisture contents were predicted.

The average moisture content of the corn can be obtained by volume averaging the nodal moisture content values. Therefore, the moisture ratio of corn during adsorption can be stated as a function of time (t):

$$\frac{\int_v M(x)\,dv - M_0}{M_e - M_0} = f(t) \tag{18}$$

Two FORTRAN programs were written to solve the finite-difference equations using the Thomas algorithm [43]. One program was used to determine the diffusivity of corn without pericarp and the other program was used to determine the diffusivity of corn pericarp. The listing of these programs can be found in Ref. 8. Space intervals of 0.01 mm and 0.001 mm were used for corn without pericarp and corn pericarp, respectively. An interval of 0.25 hr was used for the time marching. The thicknesses of corn with and without pericarp were determined for 50 kernels, each using a micrometer. They were 4.16 and 4.04 mm for the soft corn with and without pericarp and 4.80 and 4.66 mm for the hard corn with and without pericarp, respectively. The moisture ratio was calculated from the average moisture content. The moisture diffusivity was estimated by minimizing the sum of squares deviation between the experimental and predicted moisture ratio data. A multifactor analysis of variance [40] was performed to study the effects of air temperature, RH, and pericarp on moisture diffusivity.

The surface adsorption coefficient (K) was determined to represent the rate at which the kernels surface moisture approaches the equilibrium moisture as described in Section 4.2.1. The. values of K for all the environmental conditions (Table 1) were used to calculate the surface moisture contents as a function of time. These surface moisture values were then used in the numerical model as the time-varying boundary condition.

The variation of moisture ratio with time for corn samples (FR27 × MO17) exposed to air at 35°C and 75% RH is presented in Figure 4. The corn samples without pericarp attained higher moisture ratios than the samples with pericarp. Compared with the experimental data for corn samples without pericarp, the model overpredicted during the first 10 hr of adsorption and then underpredicted. However, the opposite trend was obtained for the samples with pericarp. This might be related to the resistance of the pericarp to moisture movement interacting with the boundary condition.

The moisture diffusivity values of pericarp at different humid air conditions are presented in Table 2. In general, the moisture diffusivity decreased with increasing RH and increased with increasing temperature. Similar trends were observed for corn germ and composite corn kernels

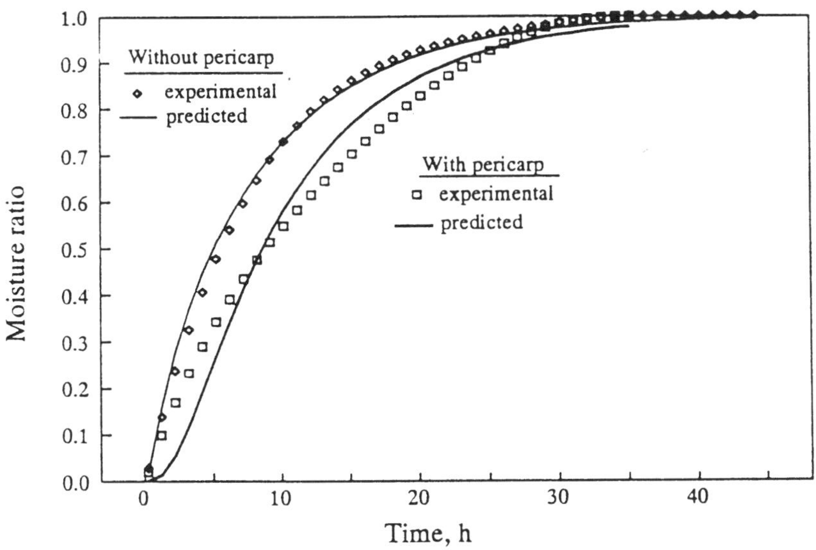

FIGURE 4 Experimental moisture ratio for corn samples (FR27 × MO17) exposed at 35°C and 75% RH and predicted moisture ratio of corn samples using finite-difference method.

during adsorption tests and possible explanations for these trends can be found in Section 4.2.1. The moisture diffusivities of composite corn samples were much higher (about two orders of magnitude) than the pericarp. This shows that the pericarp offers substantial resistance to moisture migration into corn kernels.

From the multifactor analysis of variance test, it was found that the differences in moisture diffusivity values among air temperatures, RH, and corn without pericarp and pericarp were statistically significant at 0.05 level. The mean moisture diffusivity of pericarp varied from 0.30×10^{-9} to $0.57 \times 10^{-9}\ m^2/hr$ for different air temperature and RH conditions. Diffusivity values of the pericarp obtained in this adsorption study are lower than the published values during drying (desorption). For example, within 10–24% moisture content, reported diffusivity values of pericarp varied from 0.568×10^{-7} to $3.299 \times 10^{-7}\ m^2/hr$ [6]. The moisture diffusivities of pericarp were much lower (about two orders of magnitude) than the germ (Table 2). The mean moisture diffusivity of corn germ varied from 0.15×10^{-7} to $1.18 \times 10^{-7}\ m^2/hr$ for different air temperature and RH conditions. This shows that the pericarp offers higher resistance to moisture movement into corn kernels than the germ.

The temperature dependency of the diffusivity of corn pericarp was fitted to the Arrhenius-type model shown in Eq. (17). The model coefficients and the corresponding R^2 (coefficient of determination) values for corn pericarp at all RH conditions are summarized in Table 3. In view of the high R^2 values, the Arrhenius-type model satisfactorily described the temperature dependency of diffusivity.

4.2.3 Soft and Hard Endosperms

The diffusivities of the soft and hard endosperms were determined using Eq. (10) along with the adsorption data for soft (FR27 × MO17) and hard (P3576) corn. The details of sample preparation and adsorption tests are presented in Section 4.2.1. The mass of individual components was determined by carefully breaking and weighing the component fragments from five kernels for each corn type. The amount of moisture in the individual components was estimated from the total amount of moisture in each corn type (S_i) and the component mass fraction (X_{ij}). The soft corn was composed of 5% pericarp, 10% germ, 48% soft endosperm, and 37% hard endosperm. The hard corn was composed of 4% pericarp, 9% germ, 21% soft endosperm, and 66% hard endosperm [8]. It was assumed that the differences in moisture diffusion between the two types of corn were due to different amounts of soft and hard endosperms in both types of corn. The amount of moisture in each component (M_j) was determined by normalizing the total mass of both types of corn to the component's total mass as

$$M_j = \frac{\sum S_i X_{ij}}{\sum X_{ji}} \tag{19}$$

The diffusivities of the soft and hard endosperm were determined using Eq. (19) along with the adsorption data for soft (FR27 × MO17) and hard (P3576) corn. The mass of individual components was determined by carefully breaking and weighing the component fragments from five kernels from each corn type. The amount of moisture in the individual components was estimated based on the component mass fraction (X_{11}, X_{12}, X_{13}, X_{21}, X_{22}, X_{23} for germ and soft, and hard endosperms of each type of corn). Then, the amount of moisture in each component was determined by normalizing the total mass of both types of corn to the component's total mass explained as follows:

If C_1 = total amount of moisture in corn type 1
C_2 = total amount of moisture in corn type 2
X_{11} = mass fraction of germ in corn type 1
X_{12} = mass fraction of soft endosperm in corn type 1

X_{13} = mass fraction of hard endosperm in corn type 1
X_{21} = mass fraction of germ in corn type 2
X_{22} = mass fraction of soft endosperm in corn type 2
X_{23} = mass fraction of hard endosperm in corn type 2

then the

$$\text{Amount of moisture in germ} = \frac{C_1X_{11} + C_2X_{21}}{X_{11} + X_{21}} \tag{20}$$

$$\text{Amount of moisture in soft endosperm} = \frac{C_1X_{12} + C_2X_{12}}{X_{12} + X_{22}} \tag{21}$$

$$\text{Amount of moisture in hardendosperm} = \frac{C_1X_{13}C_2X_{23}}{X_{13} + X_{23}} \tag{22}$$

This procedure can be used for other grains and food materials. Recently, Kang and Delwiche [44] used this approach for modeling the moisture diffusion in wheat kernels during soaking. In determining the soft and hard endosperm diffusivity values, the finite-element model described in Section 3.2 was used. Four-noded quadratic finite elements were used for discretization of the domain. Using the finite-element solutions developed by Muthukumarappan [8], the nodal moisture content of a corn kernel during adsorption was predicted. The average kernel moisture content (as distinguished from the nodal moisture values) is estimated to be the mass average value. Assuming constant density, the mass average moisture of a body ($\bar{M}$) is defined by Haghighi and Segerlind [14] as

$$\bar{M} = \frac{\int_V M(x, y)\,\mathrm{d}m}{\int_V dm} \quad \text{for every } \Delta t \tag{23}$$

A computer program for two-dimensional steady-state field problems written by Segerlind [45] was modified to solve the time-dependent diffusion problem. The modified computer program was written in Fortran77. This program can be used to (1) determine the diffusivity of individual components and (2) simulate the moisture adsorption of grains. The listing of the program is presented in Ref. 8.

The diffusivity values were estimated by optimizing the experimental and finite-element predicted moisture content data of the individual components. A corn kernel without pericarp was considered for the analysis. The cross section of the corn kernel with four distinct regions of germ, pericarp, and soft and hard endosperms is shown in Figure 5. The cross section presented is through the narrowest dimension of the kernel. The

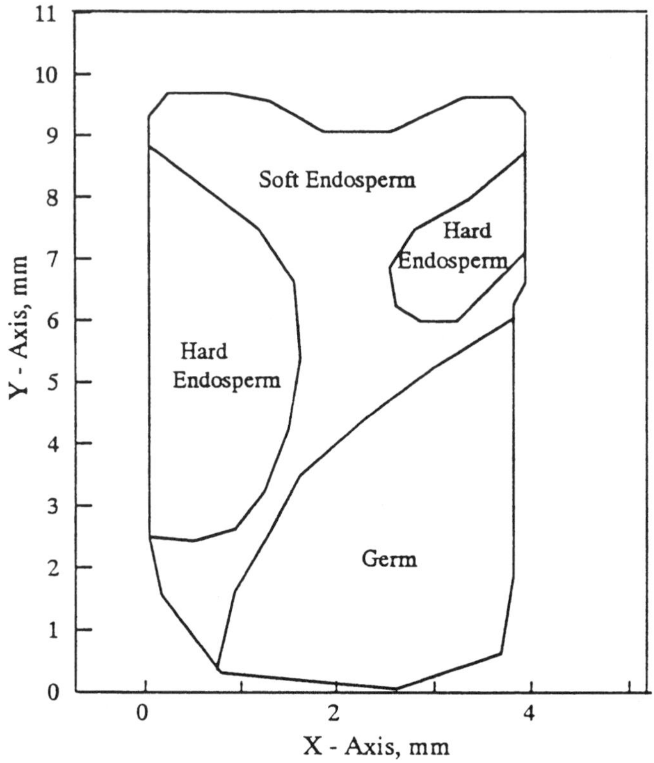

FIGURE 5 Cross section of a corn kernel through its narrowest dimension.

two-dimensional cross section was selected based on the average dimension of two hard corn kernels. Two hard corn kernels were cut through the narrowest dimension of the kernel. Then, the cut kernels were mounted on 10 × 10-mm aluminum cylindrical stubs using double-sided sticky tape. Further, silver paint was applied around the sides of the kernel. The mounted samples were sputtered with gold to a thickness of about 270 Å using a Bio-Rad Polaron Division Gold Coater (Model ESOOOM SEM Coater). The samples were examined in a scanning electron microscope (Model Hitachi S-570) at an accelerating potential of 10 kV and corresponding dimensions were determined. A finite-element discretization of the kernel is shown in Figure 6. The two-dimensional model in Cartesian coordinates consists of 53 elements.

The diffusivity values of the germ determined previously (Section 4.2.1) using the analytical model was used in the three-component finite-element

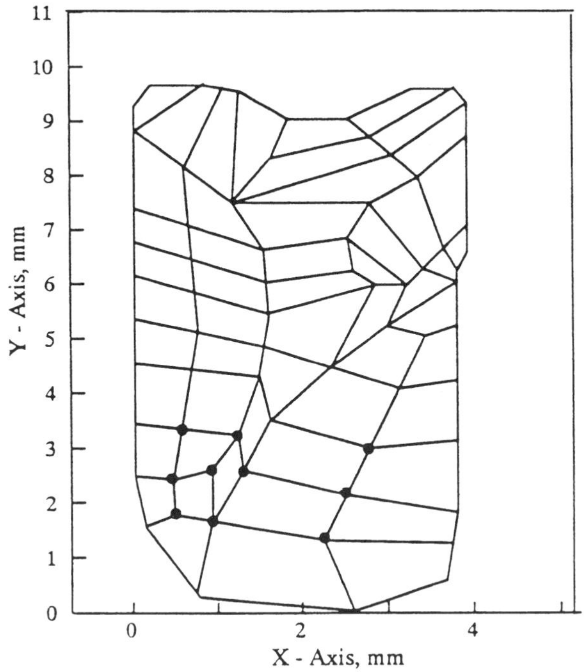

FIGURE 6 Finite element discretization of the corn kernel.

model. Because we have two variables to optimize (the diffusivity of soft and hard endosperms), two models were considered simultaneously. For the first model, called the "hard endosperm model," an initial diffusivity value of the soft endosperm was assumed and the hard endosperm diffusivity value was predicted. For the second model, called the "soft endosperm model," the previously estimated hard endosperm diffusivity was used and the new soft endosperm diffusivity was predicted. This procedure was repeated until the sum of square deviations (SSD) between the experimental and predicted moisture data was minimized. A subroutine based on the Gold Section search method [46] was used to optimize the diffusivity evaluation process.

The experimental and finite-element model predicted moisture ratios of soft and hard endosperms, exposed to air at 35°C and 90% RH, are presented in Figure 7. The moisture ratio of the soft endosperm was higher than that of hard endosperm at intermediate times. The difference in moisture diffusion rates between the soft and hard endosperms might be due to the packing of starch granules in both types of endosperm. The starch granules within the hard endosperm cells are small and tightly packed compared

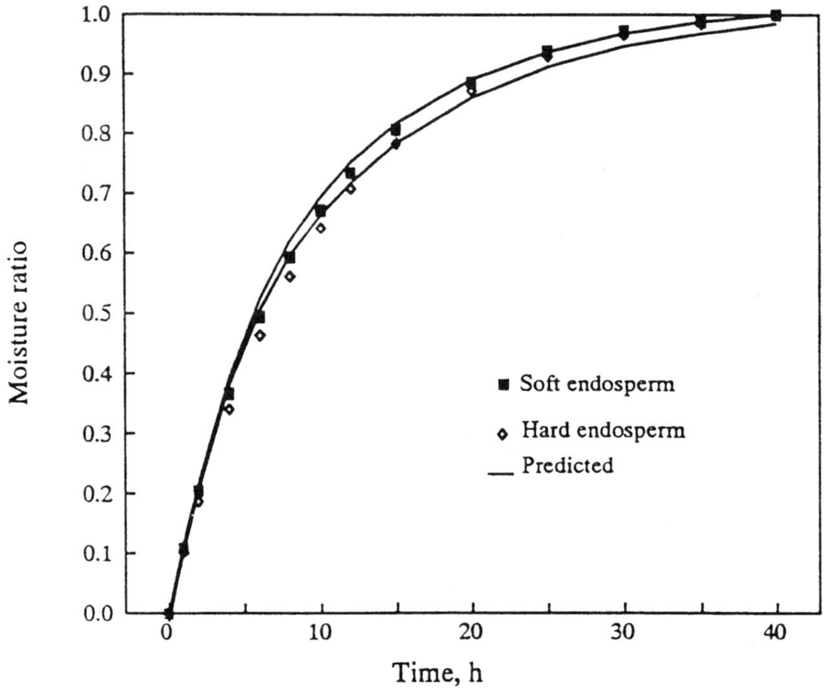

FIGURE 7 Moisture ratio of soft and hard endosperms exposed to air at 35°C and 75% RH using finite-element method.

to large and loosely organized granules in the soft endosperm cells. The soft endosperm might have more sorptive sites for water vapor compared to the hard endosperm. The moisture diffusivity values of soft and hard endosperm for all the humid air conditions are presented in Table 2.

The moisture diffusivity of hard endosperm was lower than the soft endosperm. The average diffusivity of soft and hard endosperms increased with increasing air temperature and decreased with increasing air relative humidity. These trends are similar to those for corn germ and corn pericarp (Section 4.2.1 and 4.2.2) and have been further explained in Section 4.2.1. From the multifactor analysis of variance [40] it was found that the differences in moisture diffusivity between soft and hard endosperms for different air temperatures and relative humidity were statistically significant at the 0.05 level.

The mean moisture diffusivity of the soft endosperm exposed to air at 35°C and 80% RH ($1.14 \times 10^{-7}\ m^2/hr$) is the largest, followed by hard endosperm ($0.68 \times 10^{-7}\ m^2/hr$), germ ($0.33 \times 10^{-7}\ m^2/hr$), and pericarp

(0.49×10^{-9} m^2/hr). From these results, it is evident that the pericarp offers the most resistance to moisture diffusion, followed by germ, hard endosperm, and soft endosperm. Based on these values, the diffusivity of composite corn is expected to be less than 1.14×10^{-7} m^2/hr. However, the mean moisture diffusivity of composite corn kernels exposed to air at 35°C and 80% RH is 1.20×10^{-7}. This value is higher than expected because it was obtained via one-dimensional analytical models rather than the finite-element analysis. Moreover, the analytical model assumed a regular infinite-slab geometry for a corn kernel and the finite-element analysis assumed an actual irregular shape. Muthukumarappan and Gunasekaran [22] compared the diffusivities of corn kernels for three different geometry representations. They found that the diffusivity values of corn kernels using the infinite-slab geometry were about 1–2.5 times the diffusivity values using the infinite-cylinder geometry. Thus, use of more nearly identical models would allow for a better comparison.

The temperature dependency of the diffusivity of corn endosperms were fitted to the Arrhenius-type model shown in Eq. (17) for corn germ and pericarp. The model coefficients and the corresponding R^2 (coefficient of determination) values for corn endosperm at all RH conditions are summarized in Table 3. In view of the high R^2 values, the Arrhenius-type model satisfactorily described the temperature dependency of diffusivity.

4.3 Finite-Element Simulation of Corn Moisture Adsorption

A corn variety of FR27 × MO17 was used. The finite-element model described in Section 4.2.3 was used to simulate the moisture diffusion into a corn kernel. A finite-element discretization of the kernel is shown in Figure 6. The two-dimensional model in Cartesian coordinates consists of 85 elements. The diffusivity values obtained from the experimental data (Sec. 4.2) were used along with the necessary initial and boundary conditions for the finite-element simulation.

The corn moisture adsorption was simulated using the finite-element model (FEM) and analytical model. The composite moisture diffusivity values reported in Table 2 were used in the analytical model to simulate the corn moisture adsorption. The experimental, analytical and FEM simulated moisture ratio of FR27 × MO17 corn samples exposed to air at 35°C and 75% RH are presented in Figure 8. In general, FEM simulated the experimental moisture ratio very well. The analytical model poorly overpredicted the finite-element model in the early stage of adsorption and underpredicted the model in the final stage of adsorption. The mean sum of squares deviation (MSSD) was used as an indicator to determine the prediction accuracy of the models studied. Based on the MSSD values,

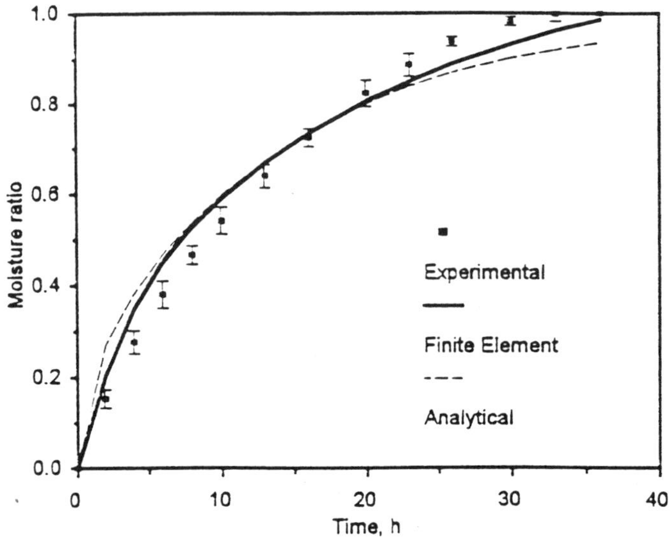

FIGURE 8 Moisture ratio of corn kernels exposed to air at 35°C and 75% RH.

the FEM predictions were clearly better than the corresponding analytical solutions. This may be because individual component moisture diffusivities of corn were considered for the FEM, whereas composite moisture diffusivity was considered for the analytical model. Ruan et al. [47] presented three-dimensional transient moisture profiles of corn kernels during a steeping process using a magnetic resonance imaging technique. From the images, they reported that the steepwater moved first into the corn kernel through the space between the germ and endosperm, and through the cross and tube cells of the pericarp layers. Then, it quickly diffused into the germ and slowly diffused into the endosperm. From these observations, it is clear that the moisture diffusion in a corn kernel is a complex phenomenon and more work is needed to better understand this behavior.

The FEM predicted nodal moisture contents were transformed to contour plots using Surfer software [48]. The moisture profiles for corn samples after 1 hr of exposure to air at 35°C and 90% RH is presented in Figure 9. The moisture gradient between the center and the surface of corn kernels during simulated moisture adsorption at 25°C, 30°C, and 35°C, each at 90% RH, is presented in Figure 10. In general, the moisture gradient inside a corn kernel during adsorption at 25°C was lower than at 35°C. The temperature effect on moisture gradient was significant during the early stage of adsorption (up to 5 hr). This is because of the different moisture

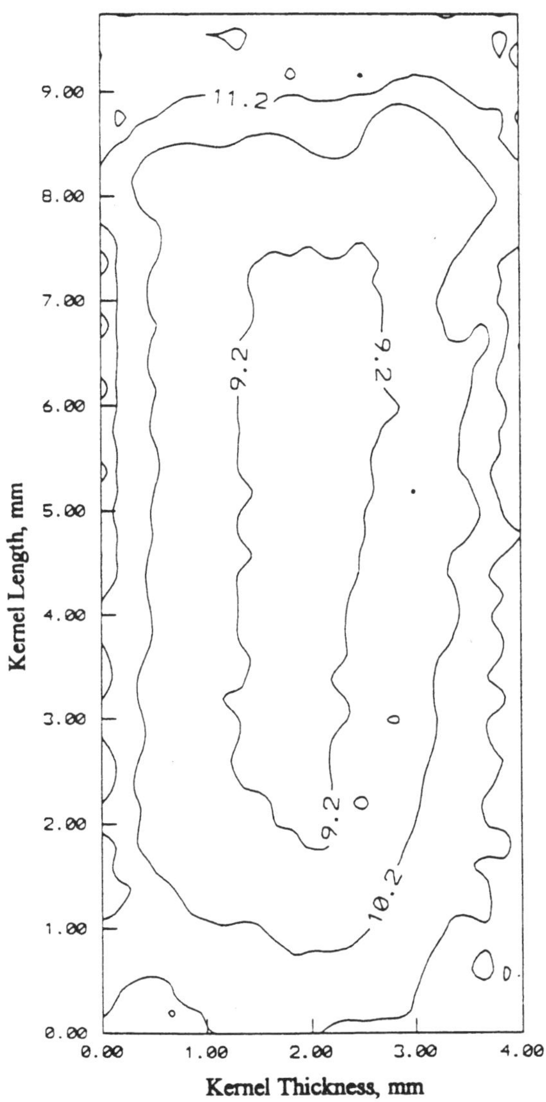

FIGURE 9 Moisture profile (% wet basis) within a corn kernel after 1 hr of exposure to air at 35°C and 90% RH during adsorption.

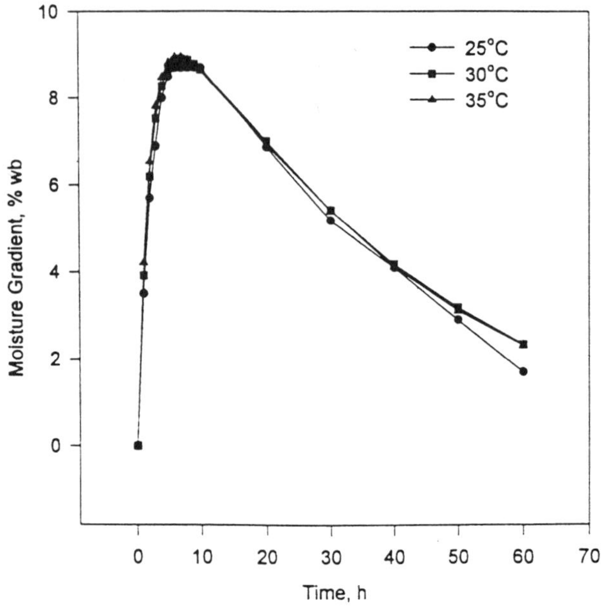

FIGURE 10 Moisture gradient (% wet basis) within a corn kernel with time when exposed to 25°C, 30°C, and 35°C each at 90% RH air condition.

diffusivity values and varying boundary condition used in the simulation model. The moisture gradient between the center and the boundary of a corn kernel exposed to air at 25°C, 30°C, and 35°C, each at 90% RH, was about 4% after 1 hr reaching a maximum of about 9% after 7.5 hr, and declined during subsequent adsorption. These times compare well with Sarwar and Kunze's [1] experimental observations; the corn samples took about 1 hr of exposure for fissures to start developing when exposed to 92% RH at 21°C. Further they reported that all the kernels exposed to 92% RH at 21°C fissured within 8 hr of adsorption. This shows that the difference in moisture gradient may cause the kernels to fissure. In addition, all the kernels may fissure when the moisture gradient was maximum.

5 RECOMMENDATIONS

Future research would be conducted in developing a model to predict possible failures in grains during adsorption and needs to be verified with experimental results. A simultaneous heat and mass transfer model could be developed to predict the temperature and moisture profiles inside a grain during adsorption at different temperature and humidity conditions. A storage model may be developed using the thin-layer moisture adsorption models presented in this chapter.

NOMENCLATURE

$[C]$	Global moisture capacitance matrix
$[C_{ij}$	Element moisture capacitance matrix
dm	Mass of an element
D	Diffusivity, m^2/hr
K	Surface adsorption coefficient, hr^{-1}
$[K]$	Global moisture conductance matrix
$[K_{ij}]$	Element moisture conductance matrix
$\bar{M}$	Mass average moisture content of a body
M	Moisture content at time t (hr), % wet basis
M_0	Average initial moisture content of the kernel, % wb
M_e	Equilibrium moisture content of the kernel, % wb
M_s	Surface moisture content of the kernel, % wb
t	Adsorption time, hr
x, y	Cartesian coordinates
Δt	Time step, hr
Ω	Boundary surface of the body
D_1	Diffusivity of corn without pericarp, m^2/hr
D_2	Diffusivity of pericarp, m^2/hr
D_m	Diffusivity of material m, m^2/hr
$j-1, j, j+1$	Spatial nodes defined in Figure 1
l_1	Half-thickness of corn without pericarp, m
l_2	Half-thickness of corn, m
m	Number of components included in the adsorption model
RH	Relative humidity, %
$\Delta l_1, \Delta l_2$	Space interval of corn without pericarp and pericarp, mm
F	Varying boundary condition
$\{M\}$	Nodal moisture values at time t, %
$\{\dot{M}\}$	Nodal moisture values at time $t + \Delta t$, %
M_j	Amount of moisture in each component j
$MR(t)$	Moisture ratio as a function of time, $(M - M_0)/(M_e - M_0)$
$MR_i(t)$	Moisture ratio of corn type i
$MR_j(t)$	Moisture ratio of the jth component
S_i	Total amount of moisture in each corn type i
T_a	Absolute temperature, K
X_{ij} fraction	Ratio of the mass of the jth component to total mass for each corn type i

REFERENCES

1. G Sarwar, OR Kunze. Relative humidity increases that cause stress cracks in corn. Trans ASAE 32(5):1737–1743, 1989.
2. GM White, IJ Ross, CG Poneleit. Stress crack development in popcorn as influenced by drying and rehydration processes. Trans ASAE 25(3):768–772, 1982.
3. S Gunasekaran, MR Paulsen. Breakage resistance of corn as a function of drying rates. Trans ASAE 28(6):2071–2076, 1985.
4. DS Chung, HB Pfost. Adsorption and desorption of water vapor by cereal grains and their products. Part I, Part III. Trans ASAE 10(4):549–551, 555–557, 1967.
5. JF Steffe, RP Singh. Liquid diffusivity of rough rice components. Trans ASAE 23(3):767–774, 782, 1980.
6. MA Syarief, RJ Gustafson, RV Morey. Moisture diffusion coefficients for yellow-dent corn components. Trans ASAE 30(2):522–528, 1987.
7. MK Misra. Thin-layer drying and rewetting equations for shelled yellow corn. PhD thesis, University of Missouri–Columbia, 1978.
8. K Muthukumarappan. Analysis of moisture diffusion in corn kernels during adsorption PhD thesis, University of Wisconsin–Madison, 1993.
9. S Bruin, KChAM Luyben. Drying of Food Materials: A Review of Recent Developments. Advances in Drying Volume 1. New York: Hemisphere, 1980, pp 155–215.
10. JK Wang, CW Hall. Moisture movement in hygroscopic materials: A mathematical analysis. Trans ASAE 4(1):33–36, 1961.
11. TB Whitaker, HJ Barre, MY Hamdy. Theoretical and experimental studies of diffusion in spherical bodies with a variable diffusion coefficient. Trans ASAE 12(5):668–672, 1969.
12. YI Sharaf Eldeen, JL Blaisdell, MY Hamdy. Factors influencing drying of ear corn-I: Mathematical description of the moisture history of fully-exposed ears of corn. ASAE Paper No. 78-6005, 1978.
13. JH Young. Simultaneous heat and mass transfer in a porous, hygroscopic solid. Trans ASAE 12(5):720–725, 1969.
14. K Haghighi, LJ Segerlind. Modeling simultaneous heat and mass transfer in an isotropic sphere—A finite element approach. Trans ASAE 31(2):629–637, 1988.
15. J Irudayaraj, K Haghighi, RL Stroshine. Finite element analysis of drying with application to cereal grains. J Agric Eng Res 53:209–229, 1992.
16. J Irudayaraj, Y Wu. Effect of pressure on moisture transfer during moisture adsorption. Drying Technology 13:1603–1617, 1995.
17. JH Young, TB Whitaker. Evaluation of the diffusion equation for describing thin-layer drying of peanuts in the hull. Trans ASAE 14(2):309–312, 1971.
18. TB Whitaker, JH Young. Simulation of moisture movement in peanut kernels: Evaluation of the diffusion equation. Trans ASAE 15(1):163–166, 1972.
19. SR Eckhoff, MR Okos. Diffusion of gaseous sulfur dioxide into corn kernels. Cereal Chem. 66(1):30–33, 1989.

20. SR Eckhoff, MR Okos. Sorption kinetics of sulfur dioxide on yellow dent corn. Trans ASAE 33(3):855–861, 1990.
21. LR Walton, GM White, IJ Ross. A cellular diffusion-based drying model for corn. Trans ASAE 31(1):279–283, 1988.
22. K Muthukumarappan, S Gunasekaran. Vapor diffusivity and hygroscopic expansion of corn kernels during adsorption. Trans ASAE 33(5):1637–1641, 1990.
23. K Muthukumarappan, S Gunasekaran. Moisture diffusivity of corn kernel components during adsorption Part I: Germ. Trans ASAE 37(4):1263–1268, 1994.
24. K Muthukumarappan, S Gunasekaran. Moisture diffusivity of corn kernel components during adsorption Part II: Pericarp. Trans ASAE 37(4):1269–1274, 1994.
25. K Muthukumarappan, S Gunasekaran. Moisture diffusivity of corn kernel components during adsorption Part III: Soft and hard endosperms. Trans ASAE 37(4):1275–1280, 1994.
26. J Crank. The Mathematics of Diffusion. 2nd ed. London: Oxford University Press, 1975.
27. AB Newman. The drying of porous solids: Diffusion and surface emission equations. Trans AICE 27:203–220, 1931.
28. US Shivhare, GSV Raghavan, RG Bosisio. Modeling of microwave-drying of corn through diffusion phenomena. ASAE Paper No. 91-3520, 1991.
29. DH Chittenden, A Hustrulid. Determining drying constants for shelled corn. Trans ASAE 9(1):52–55, 1966.
30. KH Hsu, CJ Kim, LA Wilson. Factors affecting water uptake of soybeans during soaking. Cereal Chem 60(3):208–211, 1983.
31. R Lu, TJ Siebenmorgen. Moisture diffusivity of long-grain rice components. Trans ASAE 35(6):1955–1961, 1992.
32. CY Wang, RP Singh. A single layer drying equation for rough rice. ASAE Paper No. 78–3001, 1978.
33. WJ Chancellor. Characteristics of conducted-heat drying and their comparison with those of other drying methods. Trans ASAE 11 (6):863–867, 1968.
34. JF Steffe, RP Singh. Diffusion coefficients for predicting rice drying behavior. J Agric Eng Res 27(6):489–493, 1982.
35. JF Steffe, RP Singh. Diffusivity of starchy endosperm and bran of fresh and rewetted rice. J Food Sci 45(2):356–361, 1980b.
36. HA Becker. On the absorption of liquid water by the wheat kernel. Cereal Chem. 37(3):309–323, 1960.
37. ST Chu, A Hustrulid. Numerical solution of diffusion equations. Trans ASAE 11(5):705–708, 1968.
38. SAE 5352.2 Moisture measurement—Unground grain and seeds, ASAE Standards. 35th ed. ASAE, St. Joseph, MI, 1990, p 53.
39. SAS Institute, SAS/STAT Guide for Personal Computers. Ver. 6 Edition. Cary, NC: SAS Institute, 1987.
40. STSC. STATGRAPHICS User's Guide. Ver. 5 Edition. Rockville, MD: STSC, Inc., 1991.
41. WM Clark. The laws of mass-action: Rates and reaction. In: WM Clark, ed. Topics in Physical Chemistry. Baltimore, MD: Williams & Wilkins, 1952.

42. J Crank, P Nicolson. A practical method for numerical evaluation of solutions of partial differential equations of the heat conduction type. Proc Camb Phil Soc 43:50–67, 1947.
43. JC Strikwerda. Finite Difference Schemes and Partial Differential Equations. Chapman & Hall, New York: 1998.
44. S Kang, SR Delwiche. Moisture diffusion modeling of wheat kernels during soaking. Trans ASAE 42(5):1359–1365, 1999.
45. LJ Segerlind. Applied Finite Element Analysis. 2nd ed., John Wiley & Sons, New York: 1984.
46. SLS Jacoby, JS Kowalik, T. Pizzo. Iterative Methods for Nonlinear Optimization Problems. Englewood Cliffs, NJ: Prentice-Hall, 1972.
47. R Ruan, JB Litchfield, SR. Eckhoff. Simultaneous and nondestructive measurement of transient moisture profiles and structural changes in corn kernels during steeping using microscopic NMR imaging. 1991 International Summer Meeting, 1991.
48. Surfer. Surfer Reference Manual. Ver. 4. Golden, CO: Golden Software, Inc., 1990.

5

Deep-Fat Frying of Foods

Rosana G. Moreira
Texas A&M University, College Station, Texas

1 INTRODUCTION

Deep-fat frying is the most common unit operation used in food preparation worldwide. Therefore, to optimize the process efficiency and the quality of the final product, a better understanding of the frying mechanisms is necessary.

Deep-fat frying is considered one of the oldest processes of food preparation. For decades, consumers have desired deep-fat fried products because of their unique flavor–texture combination. During deep-fat frying, a food product's surface is sealed when the product is immersed in hot oil so that all the flavors and the juices are retained in a crispy crust. Many foods have been successfully deep-fat fried; examples are potato chips, french fries, doughnuts, extruded snacks, fish sticks, and the traditional fried chicken products.

The frying technology is important to many sectors of the food industry, including suppliers of oils and ingredients, food service operators, food industries, and manufacturers of frying equipment. The amount of food fried and oils used at both the industrial and commercial levels are vast. The United States produces over 2.5 million metric tons (MMT) of snack food per year, the majority of which are fried [1]. There are more than 500,000 institutional and commercial restaurants in the United States that utilize approximately 1 MMT of frying fats and oils annually [2].

Deep-fat frying can be defined as a process of cooking and drying through contact with hot oil and it involves simultaneous heat and mass transfer. The quality of the products from deep-fat frying depends on the process conditions and on the type of oils and foods used during the process. The frying oil serves as a heat transfer medium between the food and the fryer and provides the food's texture and flavor characteristics.

1.1 Frying Equipment

The processes used to fry food products can be divided into two broad categories: (1) those that are static and smaller, classified as batch fryers used in the catering restaurants, and (2) those that fry large amounts of products in a moving bed, used in the food industry, classified as continuous fryers. These fryers can operate at atmospheric and low/high-pressure conditions.

1.1.1 Batch Fryers

A deep-fat fryer consists of a chamber where heated oil and a food product are placed. The frying oil is directly heated by means of electricity, gas, or fuel oil. The simplest heating system consists of gas flames directed placed against the bottom of a frying kettle. Some batch fryer designs use an infrared gas heater on open-pot fryers that have a V-shaped bottom. Others use tube-fired burners to heat the frying oil. Generally, the tube-type fryers are less efficient than the open-pot fryers. Tube-type fryers can accumulate food particles at the bottom of the vat, resulting in an insulating layer between the heat source and the oil, thus reducing heat transfer and accelerating fat breakdown due to local overheating and contamination.

Figure 1 shows a schematic of a gas-fired batch fryer. The fryer body is made of stainless steel. The product is placed in a basket and then lift down to the pan containing the frying oil. Some fryers have a basket lifting system that automatically raises the basket when the frying time is finished. A valve is used at the bottom of the fryer for easy a draining of the oil. Some models also include a built-in filtering system with automatic washdown of the oil. The main switches in the fryer consist of the oil-temperature setpoint, frying time, and the manual basket-raising button.

The frying oil temperature depends on the product to be fried and the food quality requirements. Typically, the temperature ranges from 160°C to 200°C. Thermostats or solid-state controllers are used to control the oil

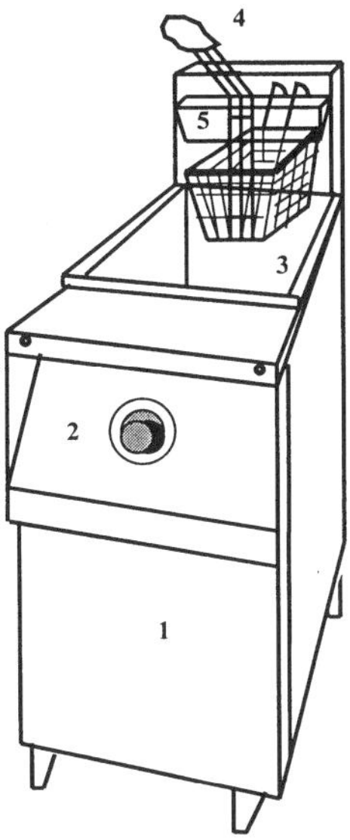

FIGURE 1 Schematic of a gas-fired batch fryer.

temperature in the fryer. In the case of an electric batch fryer, the heat source consists of an electric lift-out heater element.

Batch fryers can be of different types. Countertop are small capacity fryers with an oil capacity ranging from 8 to 11 L, and standard fryers are those that have a capacity ranging from 17 to 28 L of oil. The throughput can range from 12 to 35 kg/hr of finished product. Electric fryers are available with different energy inputs to produce the desired quantity of fried product [3].

Pressure batch fryers were developed primarily to be used in the food service industry for chicken frying. The time required to fry chicken parts is faster in a pressure fryer than in atmospheric (open) fryers and the food is moister and more uniform in color and appearance. Pressure batch fryers can be electric or gas fired and are available in many sizes. Oil capacity can range from 11 to 25 L and food capacity from 5.6 to 10.9 kg. The operating temperature of pressure fryers can vary from 160°C to 177°C and the frying time from 7 to 10 min. Two pressure ranges are used; some fryers employ 9–14 psi and others 28–32 psi [3].

1.1.2 Continuous Fryers

A continuous fryer system consists of at least five independent sets of equipment: (1) the kettle or tank containing the frying oil; (2) a heating unit with a

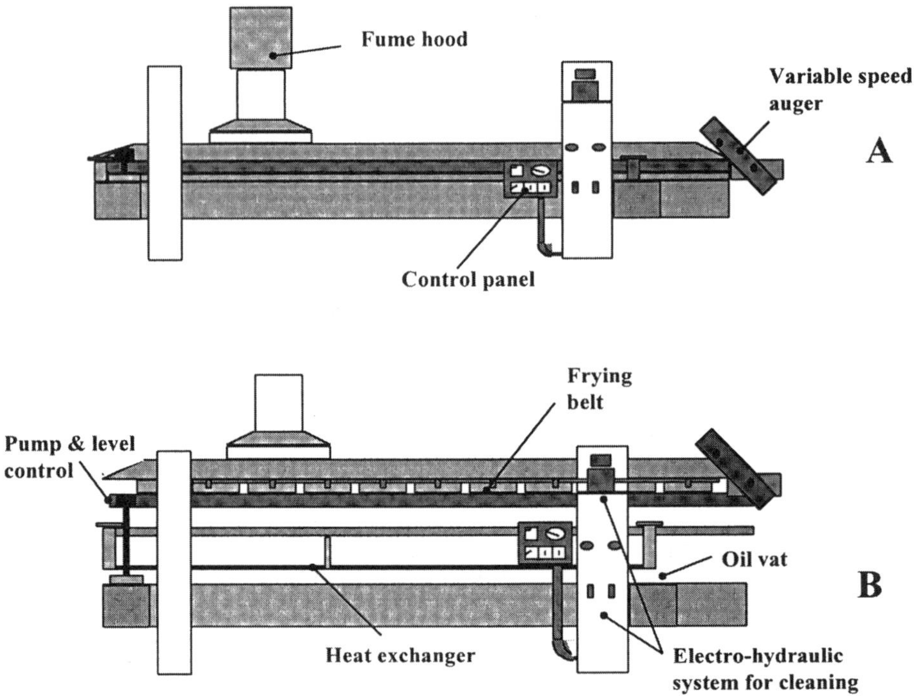

FIGURE 2 A schematic of a continuous fryer: (A) closed position; (B) open position. (Adapted from Ref. 3).

control system for generating thermal energy; (3) a conveying system for moving the product into, through, and out of the frying process; (4) a fat system, which pumps and filters the frying oil; (5) an exhaust system for removing the hot vapors emerging from the product. Figure 2 illustrates a schematic of a continuous deep-fat fryer in open and closed positions.

This is a "split-apart" type of fryer. The hood and the product conveyor system can be raised for cleaning and inspection. When in the open position, the heat exchanger remains in a fixed position between the hood and the kettle, making it easier for cleaning.

The output capacity of continuous fryers varies from 250 to 25,000 kg/hr for french fries, 100–2000 kg/hr for tortilla chips, and 100–2500 kg/hr for potato chips. A continuous frying system used for frying nuts can have a capacity from 500 to 3000 kg/hr [3].

Continuous vacuum frying was a concept developed by Florigo (H&H Industry Systems B.V., the Netherlands) in the early 1970s to produce high-quality french fries. Due to the improvement in quality of the raw materials and blanching techniques, the use of vacuum fryers almost died out with exception of one or two production companies who still insist in producing a nonblanching product. Today, the Florigo automatic continuous vacuum fryers are used mainly to produce fruit chips and very delicate snack products [3].

French fries processed in a vacuum fryer can achieve the necessary degree of dehydration without excessive darkening or scorching of the product (the frying oil temperature is much lower than in an atmospheric fryer). With the current preoccupation with lowering the fat contents of diets, vacuum frying would be an efficient way of reducing the oil content in fried snacks.

1.2 Factors That Affect the Frying Process

In deep-fat frying of foods, the factors that affect the process include the temperature of the heated oil, the frying time, and the fryer type (batch or continuous). The chemical composition of the frying oils, the physical and physicochemical constants, and the presence of additives and contaminants also influence the frying process. In addition, additives or contaminants can have a marked effect on the palatability, digestibility, and metabolic utilization of a fried food [4]. Finally, the food weight/frying oil volume and surface area/volume ratios determine how much fat penetrates the food.

1.2.1 The Process

During frying, rapid drying is critical for ensuring the desirable structural and texture characteristics of the final product. However, the loss of water results in a substantial absorption of oil by the product. A lower *oil temperature* results in lower oil content in the early stages of frying. Higher oil temperatures lead to a faster crust formation, thus favoring the conditions for oil absorption.

Oil content is temperature independent when products are fried at a specific temperature range. As the moisture is reduced with frying time, the ratio of oil content to the amount of water removed (OG/MR) becomes independent of the oil temperature, indicating that the oil content is not directly related to the oil temperature but to the remaining moisture present in the product [3].

1.2.2 Frying Time

Frying time is an important factor in processing fried products. Oil content increases, moisture content decreases, crust thickness increases, and the product becomes crispier with frying time.

Another important factor is if the product is fried in an isothermal frying condition. Isothermal frying conditions are obtained by maintaining a very low ratio of mass of product and the frying oil. Generally, this situation is found when frying a single piece of product. Nonisothermal conditions are encountered in commercial operations. During nonisothermal frying, the temperature gradient within the frying oil during the process can affect the residence time of the product in the fryer and, thus, the quality of the final product.

1.2.3 The Frying Oil

Surface appearance and texture are the most significant factors for acceptability for the majority of fried products. High-quality fried food requires a high-quality frying oil. Desirable frying oil must be low in free fatty acids and polar compounds and have a high breakdown resistance during continuous use. Hence, a thorough understanding of oil degradation and the effects of degraded oil on the quality of final products is important.

Frying oils degrade with continued use. During frying, the oil is exposed to the action of four agents which cause drastic changes in its structure: (1) moisture from the food, giving rise to oxidative alteration; (2) atmospheric oxygen entering the oil from the surface of the container, giving rise to oxidative alteration; (3) the high temperature at which the operation takes place (around 190°C) which results in thermal alteration; (4) contamination by food ingredients.

The type of oil used and the length of time the oil has been used for frying affect the desired flavor of fried foods. The method still most often used to determine when to discard frying oils is on sensory evaluation. In general, the food industry monitors product quality by the product's appearance, taste, and smell. The appearance of the fried product is usually monitored by color charts and taste panels. Food processors also incorporate shelf-life tests to determine flavor stability by using sensory panels [5].

1.2.4 The Product

The effect of raw material properties, composition, and prefrying treatment affect the quality of the final fried product.

For example, a high specific gravity of potatoes produces potato chips with lower oil content and good texture. Thicker chips also have less oil content. Prefrying treatment can affect the final oil content in fried foods, that is by reducing the initial moisture content of the product and partially gelatinizing the starch molecules in the sample can result in product with reduce oil content after frying [3].

The product composition or characteristics, such as particle size distribution of the masa flour in tortilla chips processing, can affect oil content after frying. The fine particles in the masa are responsible for most of the water uptake and viscosity development during mixing [6]. The function of the coarse particles is to produce fissures in the product that allow water to escape during frying and reduce oil absorption and to reduce the extent of pillowing during baking and frying. Therefore, a moderately coarse masa is desirable for fried corn snacks

Coatings for fried foods are produced from batters in which wheat flour is a major ingredient. The flour functionality in a batter system is based on two main components: protein and starch. Because of the water-binding characteristics of the gluten protein fraction in wheat flours, hard wheat flours require more water than soft wheat flours to produce the same batter consistency (viscosity). In a puff/tempura batter, gluten proteins help retain gases during leavening, resulting in the formation of a porous and crispy batter [7].

A higher level of protein increases the crispness of the fried product and produces a darker color [8]. Adhesion, a measure of coating ability to stick to the chicken skin, increases in batters made with higher protein flour (12.1%). Batter prepared with a fibrous ingredient, such as cellulose, reduces the oil uptake of fried products [9].

1.3 Deep-Fat Frying Process

Deep-fat frying involves simultaneous heat and mass transfer mechanisms. In addition, the physical properties of the material vary with changing temperature and moisture content. Therefore, theoretical treatment of the process is very complicated and a complete analytical solution to deep-fat frying processes is not available. Several simplified models have been developed, each of which is specific with regard to the material and boundary conditions.

Many physical, chemical, and nutritional changes occur in foods during deep-fat frying. Many of these changes are functions of oil temperature and quality, product moisture and oil contents, and product residence time in the fryer. Undesirable effects could be minimized, and the process could be better controlled if temperature, moisture, and oil distributions in food with respect to time could be accurately predicted.

In general, the frying process of a single product (french fries, for example) can be divided into four stages that are characterized by the following [10]:

- *Initial Heating*: short period; submersion of the food in the hot oil; product's surface heated to the boiling temperature of water; natural convection between oil and product's surface; no evaporation

- *Surface Boiling*: initial of evaporation; turbulence around the oil surrounding the product; forced convection between product's surface and hot oil; initial of crust formation at the product's surface
- *Falling Rate*: more internal removal of moisture from the food; rise in internal core temperature to the boiling point; increase of crust layer in thickness; decrease in vapor transfer to the surface
- *Bubble End Point*: reduction of the rate of moisture removal; decrease of bubbles leaving from product's surface; continued increase in the crust layer thickness.

2 HEAT AND MASS TRANSFER ASPECTS OF DEEP-FAT FRYING

Many attempts have been made to combine heat and mass transfer principles to describe the temperature and moisture content profiles in a product in deep-fat frying processes [10–13]. All of these models deal with the frying of a single piece of a product and assume constant physical properties. The practical importance of this information is limited because foods are seldom fried as individual pieces. Instead, tortilla and potato chips, French fries, and so forth are fried either in a stationary (batch fryer) or moving bed (continuous fryer). Depending on the fryer size, oil volume, batch size, and water content of the product, a temperature drop of 30–45°C of the frying oil can be observed in industrial operations [14]. Isothermal frying is only possible when frying a single piece of a product. In addition, the significant changes in the physical properties of a product during frying cannot be neglected.

2.1 Model Development

Figure 3 shows the cross section of a product being fried. Heat is transferred by convection from the oil to the surface of the product and by conduction to the center of the product. There is, however, a certain transfer of heat coupled to the transfer of water or vapor (i.e., the energy carried by the water vapor). Most of the water escapes from the product in the form of vapor during frying, and a small percentage of the frying oil also diffuses into the foodstuff [15]. Diffusion of moisture and diffusion of oil are in two opposite directions. The moisture content decreases, oil content increases, and the product becomes more porous during frying.

Two regions, crust and core, exist during frying. The crust–core interface moves toward the center of the product during frying. At the interface,

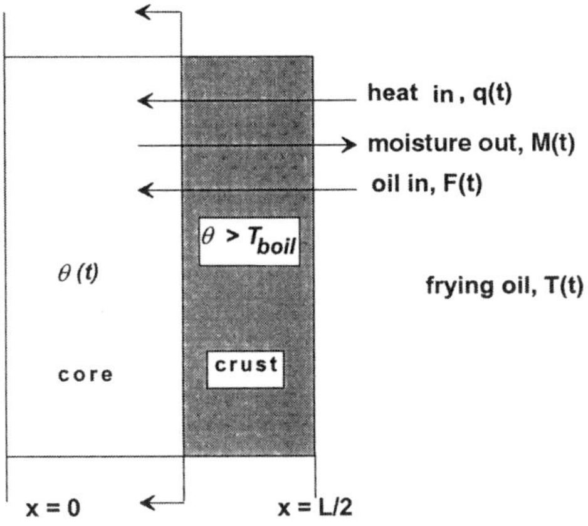

FIGURE 3 A product undergoing frying.

the temperature remains at the water boiling point for a short period of time to allow for the water presented in that region to evaporate.

Thermal and physical properties change greatly during frying. The bulk density of the food material decreases and the food becomes more porous. The thermal conductivity decreases as the porosity increases, the specific heat decreases as moisture content decreases, and oil content increases [12].

When a product is taken out of the fryer, it is covered with a thin layer of oil. As the temperature of the product decreases by natural convection with ambient air, the vapor pressure within the pores of the product decreases, forcing the surface oil to flow into the chips.

Moreira et al. [15] indicated that up to 80% of the oil is absorbed by the tortilla chips during the cooling period. Investigation of the cooling process is very important to fully understand the frying mechanism. The reader is referred to the work of Moreira and Barrufet [16].

2.1.1 Assumptions

The model development for a batch of tortilla chips deep-fat fried in a stationary batch fryer is described as follows [17]:

1. The product is considered to be initially isotropic and isothermal. The initial moisture and temperature distributions in the chip are uniform.
2. Because the thickness of the product is smaller than the other dimensions, an infinite-slab model is assumed in the study.

3. The heat required for chemical reactions (i.e., starch gelatinization, protein denaturation) is small compared to the heat required to evaporate the water.
4. Changes in the length of the product is negligible compared to changes in thickness during frying. Tortilla chips puff during frying (i.e., become thicker) [18].
5. Thermal and physical properties are functions of local temperature and moisture content during the frying process.
6. A "microscopically uniform" porous medium is formed after frying. The surface of the product is covered with a uniform layer of oil after frying and most of the oil diffuses into the product after frying during the cooling period.
7. The oil volume to product volume ratio is large so that only the oil that is in direct contact with the chips (in between the chips) is affected by the chips' moisture loss during frying, whereas the change in temperature of the surrounding oil in the kettle is negligible. It is assumed that the fryer consists of two oil zones, and the exchange of energy between these two zones is by convection only. This would not be true for the case of continuous frying or the volume of oil/volume of chips ratio is very small.

2.1.2 Governing Equations

Because of its symmetry, the computational domain may be simplified to a half-section of the product. The temperature (θ), moisture content (M), and oil content (F) in a batch of tortilla chips as well as the oil temperature (T) change dramatically during frying. Energy and mass balances are written on a differential volume located at an arbitrary position in a batch fryer containing tortilla chips (Figure 4).

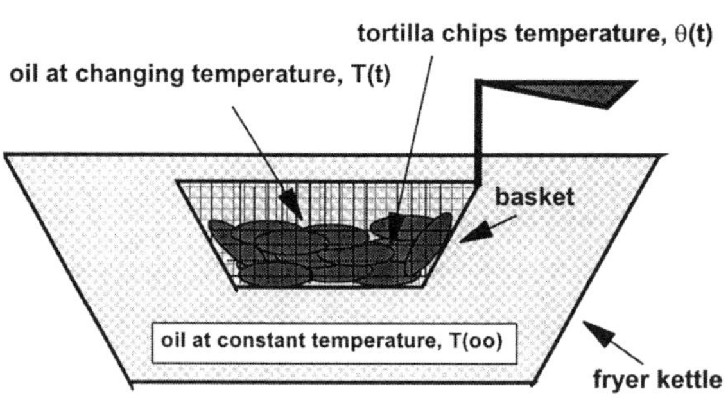

FIGURE 4 Batch fryer of tortilla chips.

The basic one-dimensional heat and mass diffusion equations were employed. Thus, four balances results in four equations. The explanation of the symbols are in Section 6.

1. The governing differential equation describing the temperature *change in the product* during frying is

$$\frac{\partial}{\partial x}\left(k\,\frac{P\theta}{Px}\right) - \frac{\partial(\Gamma C_{pw}\theta)}{\partial x} = \frac{\partial(\rho_b C_p \theta)}{\partial t} \tag{1}$$

The second term on the left-hand side of the equation represents the heat transfer caused by the diffusion of water vapor, where Γ is the water vapor flux,

$$\Gamma = -\frac{\partial(\rho_b D_w M)}{\partial x} \tag{2}$$

2. Fick's law of diffusion is used to calculate the *mass transfer rate in the product* in two different directions: Moisture (water vapor) diffuses from the chip to the oil and the oil diffuses from the surface to the center of the chip,

$$\frac{\partial}{\partial x}\left(D_w \rho_b\,\frac{\partial M}{\partial x}\right) = \frac{\partial(\rho_b M)}{\partial t} \tag{3}$$

$$\frac{\partial}{\partial x}\left(D_f \rho_b\,\frac{\partial F}{\partial x}\right) = \frac{\partial(\rho_b F)}{\partial t} \tag{4}$$

3. The temperature of the oil decreases significantly during the first seconds of frying when the product is dropped into the fryer. The change in enthalpy of the oil with respect to time in the void space (between tortilla chips) is equal to the sum of energy required for heating the product, for evaporating water from the chips, for heating the water vapor evaporated from the chips, and for exchanging energy to the surrounding oil. The equation for calculating the changes in the *temperature of the oil* is

$$\rho_{\text{oil}} C_p\,\frac{\partial T}{\partial t} = \left(-k\,\frac{\partial\theta}{Px} + h_{fg} D_w\,\frac{\partial(\rho_b M)}{\partial x} + \Gamma C_{pw}(\theta_{\text{sur}} - T)\right)\frac{A}{1-\phi} + h_s S(T_{\text{fo}} - T) \tag{5}$$

Equations (1)–(5) are used to describe temperature and moisture changes at all points in the chip (i.e., core and crust).

Two boundary conditions and one initial condition are needed for each governing equation. The first boundary condition is a symmetric condition at the centerline of the slab [Eq. (6)]. The second boundary condition is a convection surface condition [Eqs. (7)–(9)].

At $x = 0$ (centerline) for any time, no temperature, moisture, or oil gradient exists at the center of tortilla chips,

$$\frac{\partial \theta}{\partial x} = 0, \qquad \frac{\partial M}{\partial x} = 0 \qquad \frac{\partial F}{\partial x} = 0 \tag{6}$$

At $x = \pm L/2$ (surface) for any time, the energy transferred by convection from the oil to the chip's surface is equal to the sum of energy required for transferring heat to the center of the product by conduction, for evaporating water from the chips, and for heating the water vapor evaporated from the chips at temperature θ to the oil temperature T,

$$h(\theta_{\text{sur}} - T) = -k\,\frac{\partial \theta}{\partial x} + h_{fg}\Gamma + \Gamma C_{pw}(\theta_{\text{sur}} - T) \tag{7a}$$

The second term on the right-side of Eq. (7a) is eliminated when the temperature of the chip is above the boiling point of water,

$$h(\theta_{\text{sur}} - T) = -k\,\frac{\partial \theta}{\partial x} + \Gamma C_{pw}(\theta_{\text{sur}} - T) \tag{7b}$$

with two mass transfer boundary conditions,

$$k_d \rho_b (M_{\text{sur}} - M_{\infty}) = -\Gamma \tag{8}$$

where $M = M_{\text{e}} \simeq 0$ (i.e., the moisture content of the surrounding oil), and

$$k_f(F_{\text{sur}} - F_{\infty})) = -D_f\,\frac{\partial (F)}{\partial x} \tag{9}$$

where $F_{\infty} = F_e \simeq 1$ (i.e., oil content of the surrounding oil). The mass transfer coefficients (k_f and k_d) should be a function of the mass flux of water leaving the chip.

The initial conditions for any location x in the chip at time 0 are the following:

$$\theta(x,0) = \theta_0, \qquad M(x,0) = M_0, \qquad F(x,0) = F_0, \qquad T(x,0) = T_{\text{fo}} \tag{10}$$

The above 4 governing equations and 10 boundary conditions can be solved simultaneously by using the finite-difference technique. Heat and mass transfer equations are coupled by the transport properties and thermal properties, which are functions of moisture content and temperature. Equation (5) is used to calculate the temperature of the oil between the chips at each time step, and this value is then used in Eqs. (1)–(4) to obtain the temperature, moisture content, and oil content profiles in the chip during the batch frying process.

2.2 Solution of the Mathematical Model

The solution of the mathematical model is based on the work by Chen [19]. A control-volume formulation is used to discretize the governing equations, initial conditions, and boundary conditions. The continuous physical space (infinite slab) was divided into a number of nonoverlapping control volumes in the x direction such that there is one control volume surrounding each node. The differential equations are integrated over each control volume. Figure 5 shows a n-grid mesh in the x direction.

The control-volume formulation method obtains the finite-difference equation by applying conservation of energy to a control volume around each node. The most attractive feature of this method is that the resulting solution would imply that the integral conservation of energy is exactly satisfied over any group of control-volumes and over the whole calculation domain.

2.2.1 Discretization of the Governing Equations

Figure 6 shows a magnified interior node and its neighboring nodes for an infinite-slab tortilla chip.

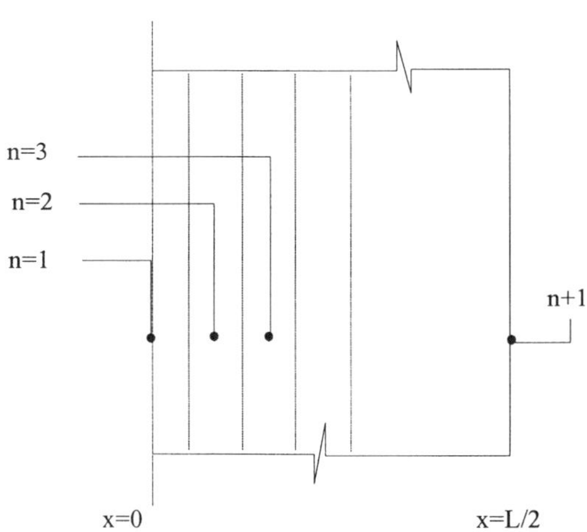

FIGURE 5 The grid generation for an infinite-slab tortilla chip.

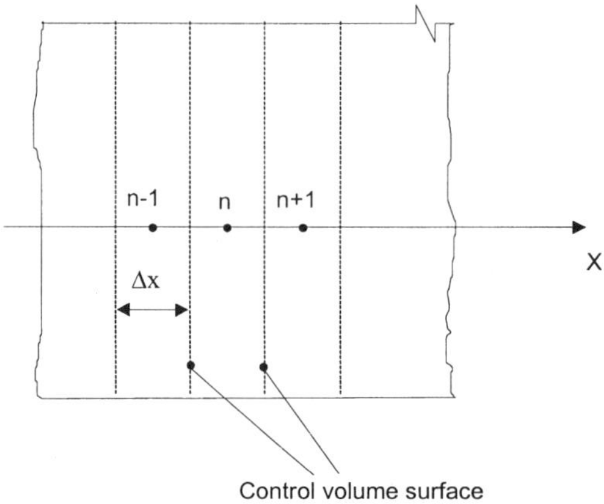

FIGURE 6 An interior node and its neighboring nodes.

Applying the energy balance to the control volume yields

$$k^p(n)\left(\frac{\theta^p(n-1)-\theta^p(n)}{\Delta x}\right)A - k^p(n)\left(\frac{\theta^p(n)-\theta^p(n+1)}{\Delta x}\right)A$$
$$+\tfrac{1}{2}[N^p(n-1)AC_p^p(n-1)\theta^p(n-1)+N^p(n)\theta^p(n)]$$
$$-\tfrac{1}{2}[N^p(n)AC_p^p(n)\theta^p(n)+N^p(n+1)AC_p^p(n+1)\theta^p(n+1)]$$
$$= A\Delta x\rho_p^p(n)C_p^p(n)\left(\frac{\theta^{p+1}(n)-\theta^p(n)}{\Delta t}\right) \tag{11}$$

Equation (11) is the discretized form of Eq. (1), where $n = 2, 3, 4, \ldots,$ $N-1$. The superscript p refers to the present time and $p+1$ refers to the next time step. Also, Δx is the distance step used in the calculation, Δt the time step used in the calculation, and A the cross-section area of the product.

Multiplying Eq. (11) by $1/(\Delta x \cdot \Delta)$, it is simplified to

$$k^p(n)\left(\frac{\theta^p(n-1)+\theta^p(n+1)-2\theta^p(n)}{\Delta x^2}\right)$$
$$+\frac{[n^p(n-1)C_p^p(n-1)\theta^p(n-1)-N^p(n+1)C_p^p(n+1)\theta^p(n+1)}{2\Delta x}$$
$$= \rho_b^p(n)C_p^p(n)\left(\frac{\theta^{p+1}(n)-\theta^p(n)}{\Delta t}\right) \tag{12}$$

Similarly, the principle of conservation of mass is applied to the interior node n to obtain the discretized moisture and oil transfer equations:

$$\rho_b^p(n)\left(\frac{M^p(n+1)+M^p(n-1)-2M^p(n)}{\Delta x^2}\right)$$
$$=\frac{1}{D_w^n(n)}\frac{M^{p+1}(n)\rho_b^{p+1}(n)-M^p(n)\rho_n^p(n)}{\Delta t} \tag{13}$$

$$\rho_b^p(n)\left(\frac{F^p(n+1)+F^p(n-1)-2F^p(n)}{\Delta x^2}\right)$$
$$=\frac{1}{D_f^p(n)}\frac{F^{p+1}(n)\rho_b^{p+1}(n)-N^p(n)\rho_n^p(n)}{\Delta t} \tag{14}$$

Next, the discretization method is applied to the nodes at the two boundaries of a chip to obtain the boundary conditions for the heat and mass transfer equations. Figure 7 is the magnified boundary node $n = 1$ with its neighboring nodes. Heat and mass transfer only occur in one surface and only half of the control volume is involved in each of the energy balance equations.

Similar to Eq. (12), the discretization form of energy balance at $n = 1$ can be expressed as

$$-k^p(1)\left(\frac{\theta^p(1)-\theta^p(2)}{\Delta x^2}\right)-\frac{[N^p(1)C_p^p(1)\theta^p(1)+N^p(2)C_p^p(2)\theta^p(2)]}{2\Delta x}$$
$$=\frac{1}{2}\rho^p(1)C_p^p(1)\left(\frac{\theta^{p+1}(1)-\theta^p(1)}{\Delta t}\right) \tag{15}$$

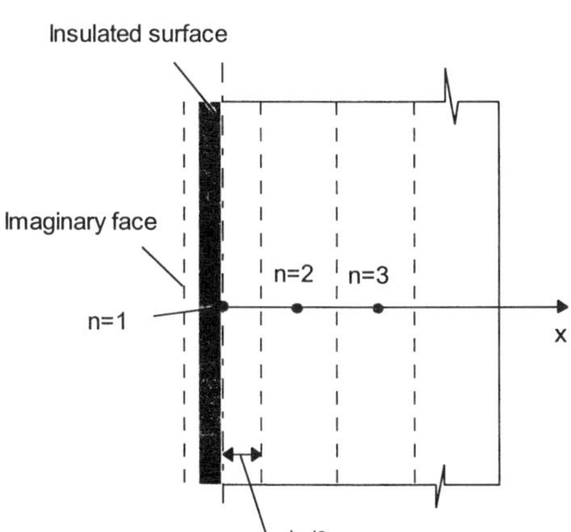

FIGURE 7 The boundary node at $n=1$ and its neighboring nodes.

The principle of conservation of mass was also applied to the boundary node $n = 1$ to obtain the discretized boundary conditions for moisture and oil transfer:

$$- \rho_b^p(1)\left(\frac{M^p(1) - M^p(2)}{\Delta x^2}\right)$$

$$= \frac{1}{2D_w^p(1)} \frac{M^{p+1}(1)\rho_b^{p+1}(1) - M^p(1)\rho_b^p(1)}{\Delta t} \tag{16}$$

$$- \rho_b^p(1)\left(\frac{F^p(1) - F^p(2)}{\Delta x^2}\right)$$

$$= \frac{1}{2D_f^p(1)} \frac{F^{p+1}(1)\rho_b^{p+1}(1) - F^p(1)\rho_b^p(1)}{\Delta t} \tag{17}$$

Figure 8 is the magnified boundary node $n = N$ with its neighboring nodes. Applying the energy balance to the boundary node and simplifying yields

$$k^p(N)\left(\frac{\theta^p(N-1) - \theta^p(N)}{\Delta x^2}\right) - h\left(\frac{\theta^p(N) - T}{\Delta x}\right)$$

$$+ \left(\frac{N^p(N_1)C_p^p(N-1)\theta^p(N-1) - N^p(N)C_p^p(N)\theta^p(N)}{2\delta x}\right)$$

$$= \frac{\rho^p(N)C_p^p}{2}\left(\frac{\theta^{p+1}(N) - \theta^p(N)}{\Delta t}\right) \tag{18}$$

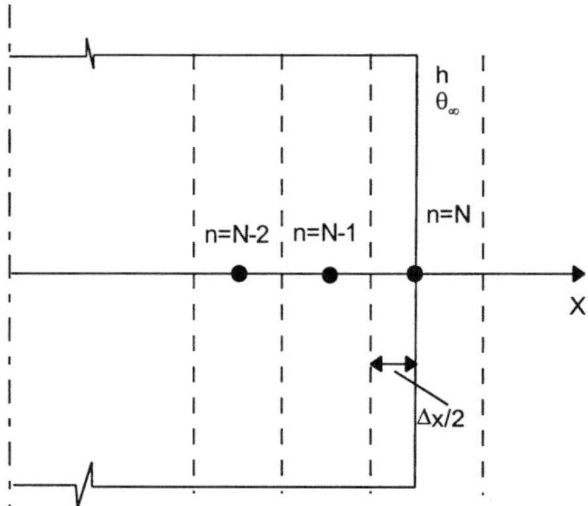

FIGURE 8 The boundary node $n = N$ and its neighboring nodes.

The principle of conservation of mass was applied to node $n = N$. The following equations were obtained:

$$\rho_b^p(N-1)\left(\frac{M^p(N-1)_M^p(N)}{\Delta x^2}\right) - \rho_b^p(N)h_d[M^p(N) - M_\infty]$$
$$= \frac{1}{2D^p(N)}\,\frac{M^{p+1}(N)\rho^{p+1}(N) - M^p(N)^p(N)}{\Delta t} \qquad (19)$$

$$\rho_b^p(N-1)\left(\frac{F^p(N-1)_F^p(N)}{\Delta x^2}\right) - \rho_b^p(N)h_d[F^p(N) - F_\infty]$$
$$= \frac{1}{2D_f^p(N)}\,\frac{F^{p+1}(N)\rho^{p+1}(N) - F^p(N)\rho^p(N)}{\Delta t} \qquad (20)$$

The moisture flux at a time step can be expressed as

$$N^p(n) = \rho^p(n)D^p(n)\,\frac{M^p(n+1) - M^p(n)}{\Delta x} \qquad (21)$$

For a batch frying process, Eq. (5) is discretized as

$$\rho_{\text{oil}}C_{pj}\,\frac{T^{p+1} - T^p}{\Delta t} = h[\theta^p(N) - T^p]\,\frac{A}{1-\phi} + h_s S(T_{\text{fo}} - T) \qquad (22)$$

The flowchart for the computational procedure of frying processes is shown in Figure 9. The computer program to solve the above problem was written using MATLAB (The MathWorks, Natick, MA) on a PG.

2.2.2 Grid Sensitivity and Stability

From the mathematical point of view, the development of the mathematical model could not be ended at this step. Even when the finite-difference equations have been properly formulated and solved, the results may still represent a coarse approximation to the exact solution. A numerical simulation enables the determination of the temperature at only discrete points that represents the average value of the surrounding region.

The finite-difference approximations can be made more accurate as the nodal network is refined. The cost is that the computer takes longer GPU time to complete the iterations. A grid sensitivity study is then performed to define when the computed results no longer depended on the choice of Δx and Δt.

An explicit finite-difference technique is used in this study. The temperature, moisture content, and oil content of any node at $t + \Delta t$ is calculated from the knowledge of temperatures, moisture contents, and oil

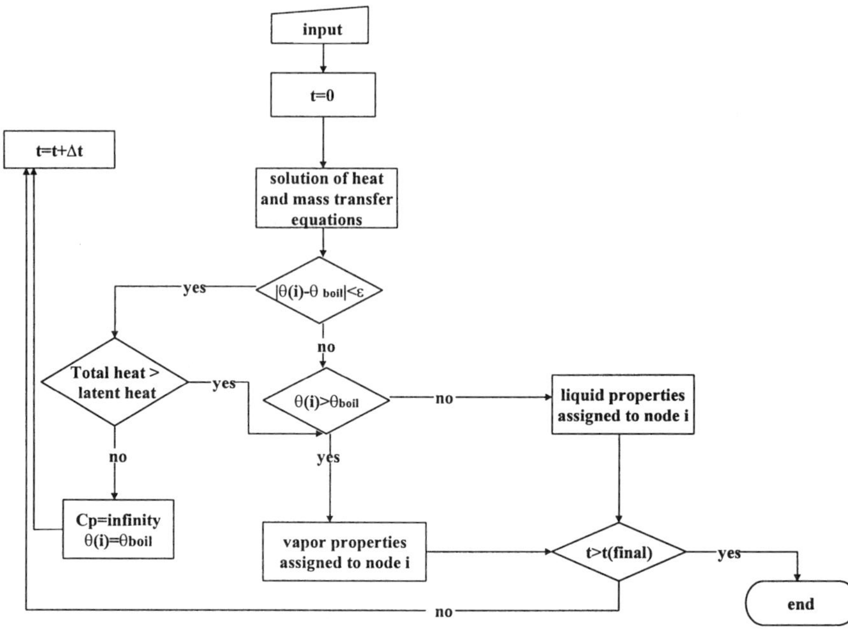

FIGURE 9 Flowchart of the frying process. (Adapted from Ref. 17.)

contents at the same and neighboring nodes for the preceding time. Hence, determination of a nodal temperature and moisture/oil content at some time is independent of temperatures and moisture/oil content at other nodes of the same time. In this method, the choice of Δx is based on a compromise between accuracy and computational time requirements, as mentioned earlier. Once this selection has been made, however, the value of Δt may not be chosen independently. Instead, it is determined by stability requirements. For a one-dimensional node, the following criterion is used to select the maximum allowable value of Fo and, hence, Δt to be used in the calculation

$$\mathrm{Fo}(1 + \mathrm{Bi}) \leq \tfrac{1}{2} \qquad (23)$$

where Fo is Fourier number, $F_0 = \alpha t/l^2$; α the thermal diffusivity, $\alpha = k/\rho C_p$, is half of the thickness, t is the time, and Bi the Biot number, $\mathrm{Bi} = hl/k$.

A series of tests are then conducted to determine the effect of the time and distance grid size (Δx, Δt) on the output of the program. The outputs are the temperature at the surface, temperature at the center,

TABLE 1 Time and Space Size Test of the Numerical Simulation of the Frying Problem

Test No.	Δx (mm)	Δt (sec)	θ (°C)	MC* (% wb)	OC* (% wb)
1	0.100	0.0100	184.65	6.9	4.1
2	0.050	0.0025	185.08	6.5	3.1
3	0.025	0.0010	185.87	6.3	2.8
4	0.100	0.0200	186.36	6.7	4.1
5	0.100	0.0050	183.86	7.0	4.1
6	0.200	0.0100	181.41	8.1	7.4
7	0.050	0.0100	Error	Error	Error

* MC = moisture content; OC = oil content
Source: Adapted from Ref. 17.

average moisture content, and average oil content. The following tests and results, as shown in Table 1, were accomplished with a constant frying oil temperature of 190°C and frying time of 80 sec.

The temperature outputs were almost identical, but Test 2 and Test 3 required longer computation times. To minimize the time required for processing and yet give the maximum accuracy, a time step of 0.01 sec and distance step of 0.1 mm were chosen and further tests were done to test the convergence and stability of the program. For a time step of 0.01 sec, the program was still convergent if Δx changed in a certain range. However, Test 4 could hardly satisfy the accuracy requirement. The output of Test 6 was also not satisfied and Test 7 failed to converge because Eq. (23) was not satisfied.

Therefore, all computations can be performed with time step of 0.01 sec and distance step of 0.1 mm.

3 SIMULATION AND VALIDATION

The simulation of a batch of tortilla chips frying for 60 sec at 190°C was performed. The thermal, physical, and transport properties model used in the simulation are described in the following subsections.

3.1 Thermal and Physical Properties Used in the Model

The mathematical model requires the physical and thermal property data of the material to be fried and the process conditions.

The thermal conductivity, k, of the tortilla chip during frying was found to have the following correlation [19]:

For $T \leq 100°C$

$$k = 0.1085 + 0.009986\theta - 5.203 \times 10^{-6}\theta^2 \tag{24}$$

For $T > 100°C$

$$k = 0.06938 + 9.997 \times 10^{-5}\theta + 6.327 \times 10^{-8}\theta^2 \tag{25}$$

Changes in the tortilla chips specific heat (C_p) with temperature were small compared to changes with moisture content. The effect of moisture content on the tortilla chip's specific heat was calculated [19] as

$$C_p = 2.506 + 2.503M_d - 1.557M_d^2 \tag{26}$$

where M_d is moisture content in decimal dry basis (db) for $1.22 < M_d < 0.01$.

The bulk density, ρ_b, of the tortilla chips changes significantly during frying. It was found that the bulk density had a stronger relationship with moisture than with temperature or oil content [19]:

$$\rho_b = 587.287 + 1017.34M \tag{27}$$

where M is moisture content in decimal (wb) for $0.55 < M < 0.01$.

The moisture diffusion coefficient was assumed to be constant during frying but a function of different frying conditions, such as oil temperature and the chip's initial moisture content [19]:

$$D_w = 6.36 \times 10^{-9} - 1.486 \times 10^{-8}\text{IMC} + 2.253 \times 10^{-4} \exp\left(-\frac{4375}{T_{\text{abs}}}\right) \tag{28}$$

where IMC is the initial moisture content (decimal, wet basis [wb]) before frying and T_{abs} is the absolute temperature.

Oil diffusion during frying increases as the initial moisture content of the tortilla chip increases [19]:

$$D_f = 1.104 \times 10^{-8}\text{IMC}^{6.528} \tag{29}$$

The diffusion coefficients (D_f and D_w) should be a function of the x and local moisture content; however, at present, these data are not available and further work will provide the necessary information on these coefficients.

Other property data used in the program are given in Table 2. The mathematical model is then used to determine the effect of process parameters and product initial conditions on the frying process of tortilla chips.

Table 2 Parameters Used in the Simulation of a Batch of Tortilla Chips

Parameters	Value
k_d	0.00128 m/sec
h	285 W/m^2 K
k_f	1.3×10^{-6} m/sec
h_{fg}	2250 kJ/kg
h_s	1200 W/m^2 K
C_{pw}	4.2 kJ/kg K
C_{pv}	2.0 kJ/kg K
C_{pf}	2.2 kJ/kg K
S	0.0459 m^2/m^3
ϕ	0.5
A	0.00276 m^2/m^3

Source: Adapted from Ref. 17.

3.2 Methodology

The model was designed to predict temperature, moisture, and oil profile of tortilla chips during the frying process. The model can be used to analyze the effect of different process variables and initial product conditions on the final quality of frying products.

3.2.1 Tortilla Chip Preparation

All the tortilla chips were prepared at the Texas A&M Cereal Quality Pilot Plant. The preparation of tortilla chips is found elsewhere [17].

3.2.2 Moisture and Oil Content Measurements

Moisture contents at different frying times were measured. Approximately 10 pieces of each sample were ground in a household grinder. The moisture content was determined by weight loss after drying the ground sample 24 h at 103–105°C in a forced-air oven [20]. Oil contents at different frying and cooling times were determined with the petroleum ether extraction method [20] in a Soxhlet apparatus (Model HT 1043, Tecator, Sweden).

3.2.3 Temperature Measurements

The temperature history at the center of the chip (thickness of 2.6 mm was measured by inserting a thin thermocouple (Type E, 0.25 mm) in the center of the chips. The temperature of the frying oil was measured with a thermocouple (Type E, 81 mm).

3.2.4 Batch Frying

Batch frying and single-chip frying were compared by frying one chip at a time and a batch of chips (20–50) at a time using the same fryer.

3.3 Validation

Using the thermal and physical properties specified in the previous section, computer simulations were conducted to compare the mathematical model with the experimental results.

3.3.1 Temperature Changes During Frying

The temperatures at the center of the tortilla chip during frying predicted by the model and obtained experimentally are shown in Figures 10 and 11 for single and batch processes, respectively. As was expected, the temperature at the center of the chip increased to the boiling point of water, remained constant until all water vaporized (10–15 sec), then continued to rise and reached the oil temperature.

In both cases, single and batch, the model was able to predict the temperature profile behavior of the chips during the process very well.

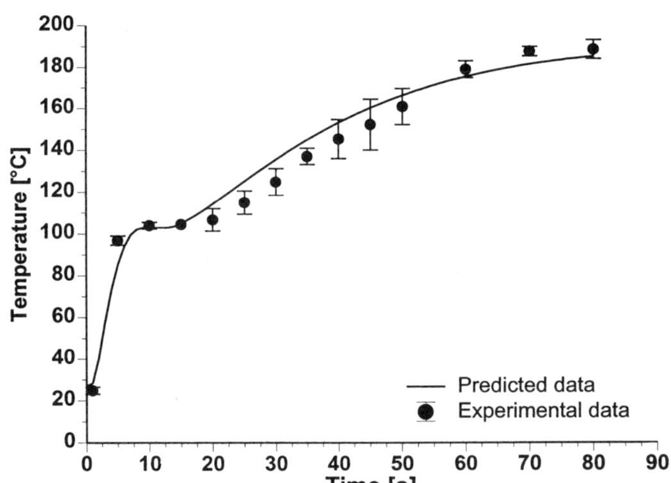

FIGURE 10 Predicted and experimental results of tortilla chip's center temperature during frying ($T = 190°C$; IMC = 0.5% wb; chip thickness = 2.6 mm; single-chip frying). (Adapted from Ref. 17.)

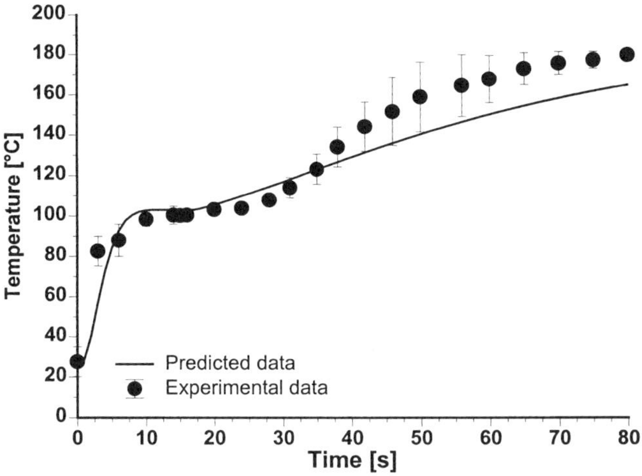

FIGURE 11 Predicted and experimental results of tortilla chip's center temperature during frying ($T = 190°C$; IMC = 0.5% wb; chip thickness = 2.6 mm; batch = 50 chips). (Adapted from Ref. 17.)

The difference, however, between the predicted and observed temperature data for both processes could have been caused by the difficulty to accurately locate and maintain the thermocouple at the center of the chips, especially during the batch frying process when the chips are continuously in motion. This factor also resulted in relatively large standard deviation (see error bar in Figures 10 and 11) in the temperature measurements.

For a single-chip frying process, the temperature of the oil is almost constant during the entire frying process. However, for batch frying, the oil temperature drops substantially and then increases gradually to the fryer setting temperature. Figure 12 shows the predicted and observed results of the frying oil temperature change during a batch frying process for tortilla chips. Good agreement was obtained between the model and the experimental results with the exception of the first 5 sec of frying. This could be caused by the lack of uniformity in the oil bath temperature before the batch of chips was placed into the fryer. As the frying proceeded, however, the chips spread uniformly in the basket so the temperature distribution became more uniform and better agreement between experimental and predicted data was obtained.

Both experimental and predicted results indicate that the temperature of the oil changes greatly during the initial seconds in batch frying process. The study of this process is very important for industrial practice.

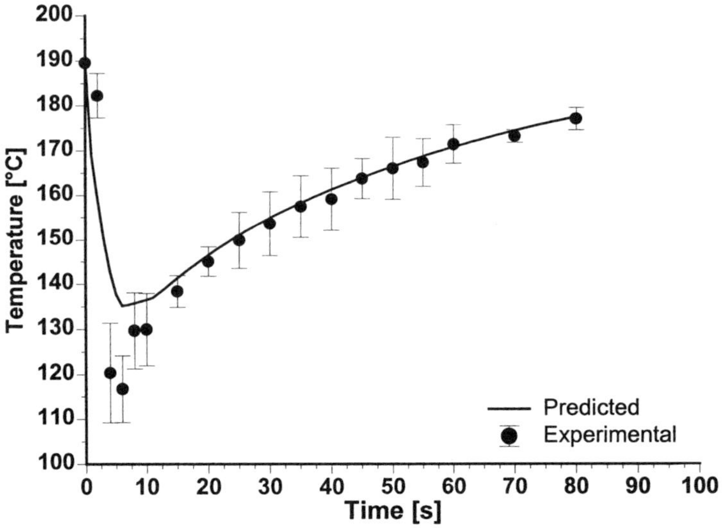

FIGURE 12 Predicted and experimental results of oil temperature change during batch frying process ($T_{fo} = 190°C$; chip thickness = 2.6 mm; IMC = 0.5% wb; batch = 50 chips). (Adapted from Ref. 17.)

3.3.2 Moisture Content Changes During Frying

The average moisture content was calculated in the computer program by numerically integrating the moisture profile at each time step. This average moisture content was then compared with the experimental data. As expected, the moisture content decreased greatly at the beginning of the frying process and then reached equilibrium at the end of the frying process (Figure 13). The predicted data were in good agreement with the experimental data.

3.3.3 Oil Content Changes During Frying

The average oil content during the frying process was determined by numerically integrating the oil profile at each time step in the computer program. Only the average oil content was calculated during the cooling process and this average value was then compared with experimental data, as shown in Figure 14. Both the experimental and the simulation results indicated that the diffusion of oil is very slow during the frying process.

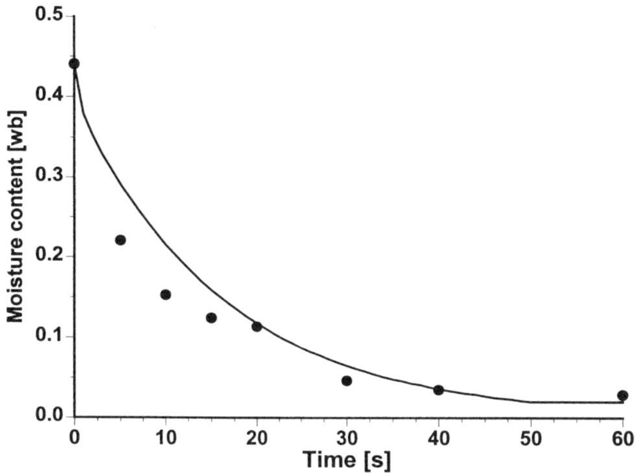

FIGURE 13 Predicted and experimental data for moisture loss of tortilla chips during frying (IMC = 0.44% wb; chip thickness = 1.6 mm, batch = 20 chips). (Adapted from Ref. 17.)

3.4 Simulation

The following sensitivity analysis was conducted to analyze the effect of different processing parameters on the frying and cooling processes of tortilla chips.

3.4.1 Frying Oil Temperature

The frying oil temperature is an important parameter in frying because the oil serves as heating medium in the process. Different oil temperatures were used in the model to analyze the effect on temperature and the moisture content of the chips.

Figure 15 shows the temperature history in a batch of 50 chips using different frying oil temperatures (150°C and 190°C). The temperature of the chips increased up to the boiling point of water at the same rate regardless of the temperature of the frying oil. It is observed that the chips' temperature increased much faster for those fried in the oil at 190°C than at 150°C.

The lower temperature gradient in the frying oil was observed when frying the chips at 150°C (Figure 15).

The average moisture content of the chips decreased faster when the oil temperature was higher, as expected (Figure 15). The higher the oil temperature, the higher the diffusion coefficient and, thus, the higher the mass transfer of the water vapor. To achieve the same final moisture content

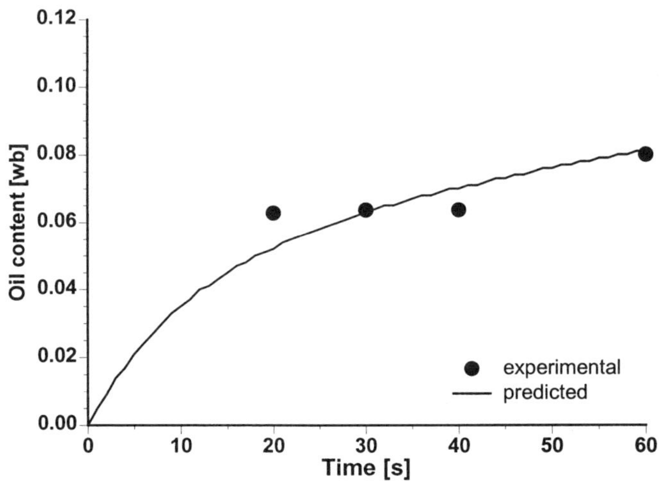

FIGURE 14 Predicted and experimental results of tortilla chip's average oil content during frying ($T = 190°C$; IMC = 0.44% wb; chip thickness = 1.6 mm). (Adapted from Ref. 17.)

(2% wb), chips fried at 190°C oil temperature only need 60 sec of frying, whereas more time was required to fry the chips in the frying oil maintained at 150°C.

The oil content was slightly affected by the oil temperature during frying (Figure 15). Less oil was absorbed during frying by the chips fried at the lower oil temperature.

3.4.2 Tortilla Chip Thickness

Figure 16 shows the effect of chip thickness on its temperature and moisture and oil content. As expected, the thinner chip reached the oil temperature faster. The plateau for the thicker chip was much longer because the thermal resistance increased significantly due to the low thermal conductivity of the frying product.

The changes in the frying oil temperature during the process (Figure 16) was more affected by the thicker chips than the thinner ones. A large temperature gradient in the frying oil was observed when frying the chips of 3 mm thickness.

The moisture removal rate was slower for the chips with a thickness of 3 mm than for the chips of 1 mm thickness (Figure 16). A longer time was required for the thicker chips to reach the equilibrium moisture content.

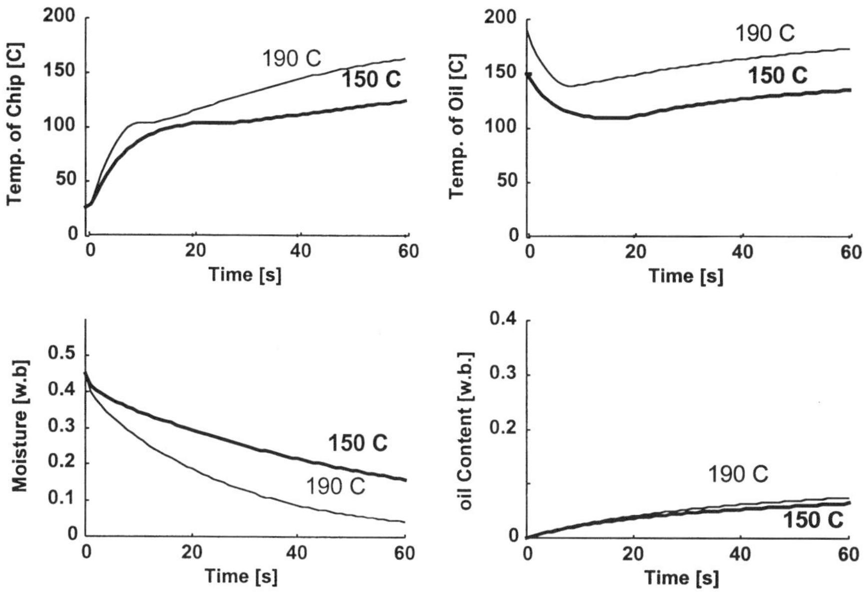

FIGURE 15 Effect of oil temperature on moisture loss, oil absorption, temperature of tortilla chips, and frying oil temperature (IMC = 0.45% wb, chip thickness = 2 mm, batch = 50 chips).

Thicker chips resulted in a fried product with lower oil content (Figure 16). A previous study suggested that a lower surface-to-mass ratio of the food decreases oil absorption.

3.4.3 Initial Moisture Content

The effect of initial moisture content on the frying process was studied by frying tortillas with initial moisture contents of 60% wb and 45% wb. Figure 16 shows the temperature at the center of the tortilla chip. The temperature of the chip with a higher initial moisture (60% wb) content was a little lower during the frying process, requiring more time to evaporate the water than the chips with lower moisture content (45% wb). The temperature profile was not greatly effected by the chip's initial moisture content during frying. The higher the initial moisture content, the higher the diffusion coefficient and, thus, the higher mass transfer of water vapor.

The oil temperature gradient was larger when frying the chips with higher initial moisture content, as expected (Figure 16).

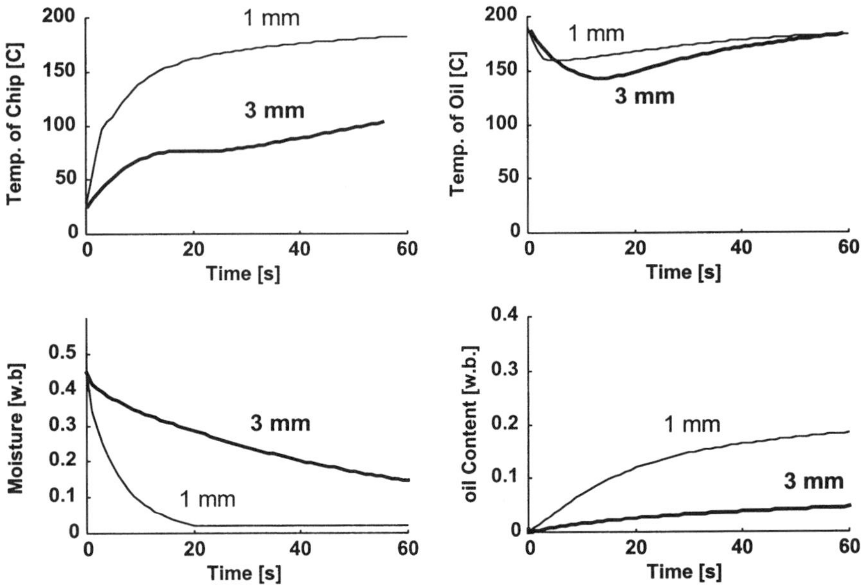

FIGURE 16 Effect of tortilla chips thickness on the moisture loss, oil absorption, temperature of tortilla chips, and frying oil temperature (IMC = 0.45% wb, $T = 190°C$, batch = 50 chips).

The moisture loss rate was faster for the tortilla chips with a higher initial moisture content (Figure 17). The equilibrium moisture content was reached in 50 sec of frying for both cases.

Chips with a higher initial moisture content have a higher final oil content (Figure 17). This is related to the effective porosity of the chip. Chips with a higher initial moisture content have more space available for oil absorption after the moisture is removed during frying.

3.4.4 Batch Frying Process

As described earlier, batch frying is quite different from single-chip frying process. It is very important to study batch frying because this is the process actually used by the food industry. For single-chip frying, the temperature of the oil is assumed to be constant, whereas for a batch process, changes in temperature of the oil depend on how many chips are fried at the same time. The temperature of the oil decreases greatly at the moment the chips are placed into the oil, then increases gradually to the setting temperature.

Figure 18 shows the center temperature of the chips during the batch frying process. The temperature of the chip increased slowly when more

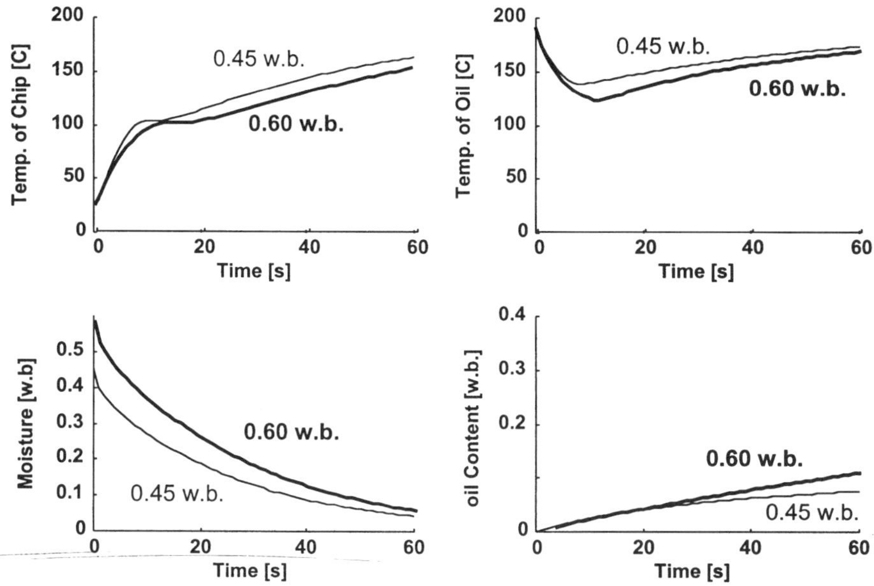

FIGURE 17 Effect of tortilla chips initial moisture content on the moisture loss, oil adsorption, temperature of tortilla chips, and frying oil temperature (chip thickness = 2 mm, $T = 190^\circ C$, batch = 50 chips).

chips were fried at the same time. The temperature of the oil dropped more dramatically (volume of oil/volume chips decreased) when more chips were fried at the same time, as shown in Figure 18.

The moisture content and oil content were not affected by the batch size in our study. Little change was observed in these variables. The most significant effect was on the frying oil and the chip's temperature.

3.5 Observations

A predictive mathematical model based on the fundamental principles of heat and mass transfer was developed to simulate the temperature, moisture content, and oil content during the frying process of food products. The model can be used to analyze processes for a single or a batch of products.

The model was simulated using MATLAB. However, any other software can be used. In our program, graphics can be generated to visualize better the simulation results on the monitor.

For a more comprehensive model of the process, which includes changes on the material structure such as pore size distribution, shrinkage,

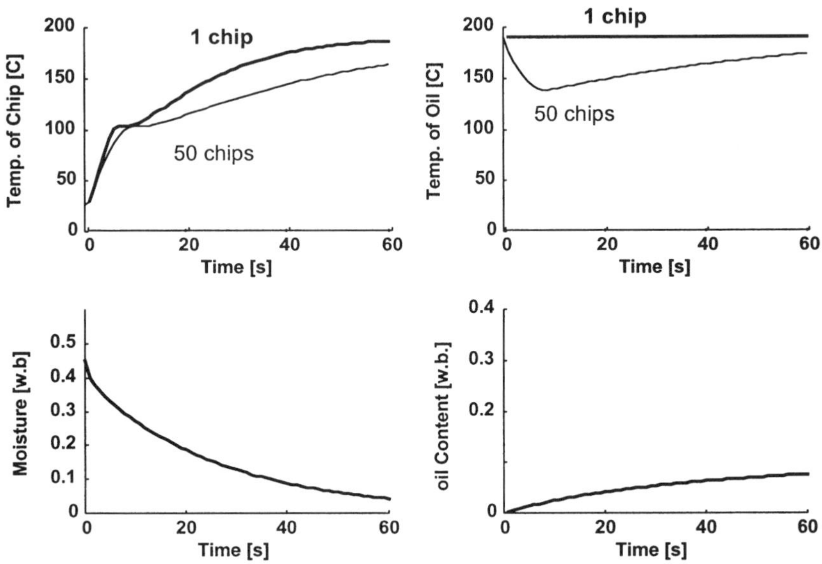

FIGURE 18 Effect of batch size on the moisture loss, oil absorption, temperature of tortilla chips, and frying oil temperature (chip thickness = 2 mm, $T = 190°C$, IMC = 0.45% wb).

and puffing, a more stable numerical solution technique would be required. Problems with stiff equations can be minimized using a finite-space/continuous-time approach or other more robust techniques.

NOMENCLATURE

A	Cross-sectional area (m^2)
Bi	Biot number
C_p	Specific heat of chip (J/kg K)
C_{pf}	Specific heat of the oil (J/kg K)
C_{pw}	Specific heat of water vapor (J/kg K)
D_f	Mass diffusivity of oil (m^2/sec)
D_w	Mass diffusivity of moisture (m^2/sec)
F	Oil content of the product in wet basis, defined as (kg oil absorbed)/(kg product) (% wb)
F	Oil content of the surrounding oil (decimal, wb)
F_e	Final oil content at surface interface between the batch and the bath oils (% wb)

Fo	Fourier number
F_0	Initial value of oil content of the product (decimal, wb)
F_{sur}	Surface oil content of the product (decimal, wb)
h	Convection heat transfer coefficient between the oil and the product (W/m^2 K)
h_g	Latent heat of vaporization (J/kg)
h_s	Convection heat transfer coefficient between frying oil and surrounding oil (W/m^2 K)
k	Thermal conductivity (W/mK)
k_d	Convection mass transfer coefficient of water vapor (m/sec)
k_f	Convection mass transfer coefficient of oil at the surface of chips (m/sec)
L	Thickness of product (m)
l	Half of the product thickness (m)
M_∞	Moisture content of the product in wet basis, defined as (kg water evaporated)/(kg product) (wb)
M	Moisture content of the surrounding oil (wb)
M_d	Moisture content of the product in dry basis, defined as (kg water evaporated)/(kg product – evaporated water (db)
M_e	Equilibrium moisture content (wb)
M_0	Initial value of moisture content (wb)
M_{sur}	Surface moisture content of the product (wb)
S	Specific surface area (m^2/m^3)
t	Time (sec)
T	Temperature of the frying oil between the product (°C)
T_{abs}	Absolute temperature of the frying oil between the product (K)
T_{fo}	Set temperature of the frying oil (°C)
T_0	Initial value of oil temperature (°C)
x	Product thickness (m)

α	Thermal diffusivity (m^2/sec)
Δt	Time step (sec)
Δt	Time step (sec)
Δx	Distance step (m)
ϕ	Porosity of the bed of product
θ	Temperature of the product (°C)
θ_0	Initial values of temperature (°C)
θ_{abs}	Absolute temperature of the product (K)
θ_{sur}	Surface temperature of the product (°C)
ρ_b	Density of the product (kg/m^3)
ρ_{oil}	Density of the oil (kg/m^3)

REFERENCES

1. SFA State of the Snack Food Industry Report. Alexandria, VA: Snack Food Association, 1997.
2. R O'Brien. Foodservice use of fat and oils. INFORM 4(8):913–921, 1993.
3. RG Moreira, ME Castell-Perez, MA Barrufet. Deep-Fat Frying: Principles and Applications. New York: Aspen Publishers, 1999.
4. G Varela, AE Bender, ID Morton. In: Frying of Foods: Principles, Changes, New Approaches. G Varela, AE Bender, ID Morton, eds. New York: VCH. 1988.
5. SL Melton, S Jafar, D Sykes, MK Tigriano. Review of stability measurements for frying oils and fried food flavor. J Am Oil Chem Soc 71:1301–1308, 1994.
6. MH Gomez, LW Rooney, RD Waniska, RL Pflugfelder. Dry corn masa flours for tortilla and snack foods. Cereal Foods World 32(5):372–377, 1987.
7. R Loewe. Role of ingredients in batter systems. Cereal Foods World 38(9):673–677, 1993.
8. M Olewnik, K Kulp. Factors influencing wheat flour performance in batter systems. Cereal World. 38(9):679–684.
9. JF Ang, WB Miller. Multiple functions of powdered cellulose as a food ingredient. Cereal Foods World 36(7):558–564, 1991.
10. BE Farkas, RP Singh. TR Rumsey. Modeling heat and mass transfer in immersion frying. I, model development. J Food Eng 29:211–226.
11. P Ateba, GS Mittal. Modelling the deep-fat frying of beef meatballs. Int J Food Sci Technol 29:429–440.
12. RG Moreira, J Palau, X Sun. Simultaneous heat and mass transfer during deep fat frying of tortilla chips. J Food Process Eng 8:307–320, 1995.
13. BE Farkas, RP Singh, TR Rumsey. Modeling heat and mass transfer in immersion frying. II, solution and verification. J Food Eng 29:227–248, 1996.
14. CK Benson, AA Caridis, LF Klein. Continuous food processing method. US Patent 5,137,740 (1992).
15. RG Moreira, X Sun, Y Chen. Factors affecting oil uptake in tortilla chips in deep fat frying. J Food Eng 31:485–498, 1997.
16. RG Moreira, MA Barrufet. A new approach to describe oil absorption in fried foods: a simulation study. J Food Eng 35:1–22, 1998.
17. Y Chen, RG Moreira. Modelling of a batch deep-fat frying process for tortilla chips. Trans Chem Eng 75(C):181–190, 1997.
18. RG Moreira, JE Palau, VE Sweat, X. Sun. Thermal and physical properties of tortilla chips as a function of frying time. J Food Process Preserv 19:175–189, 1995.
19. Y Chen. Simulation of a deep-fat frying process for tortilla chips. Master thesis. Department of Agricultural Engineering, Texas A&M University, 1996.
20. AACC. Approved Methods of the American Association of Cereal Chemists. St. Paul, MN: American Association of Cereal Chemists.

6

Mathematical Modeling of Microwave Processing of Foods: An Overview

Ashim K. Datta
Cornell University, Ithaca, New York

1 INTRODUCTION

Microwaves are part of the electromagnetic spectrum and are composed of electric and magnetic fields. In microwave heating of food, the electric fields of the microwaves interact primarily with water molecules and ions in the food material, generating heat volumetrically. Microwave heating can potentially supplement and/or replace conventional surface heating in almost any food process. Thus, the application of microwave heating can be in just about any food process involving heating—reheating, thawing, drying, pasteurization, and so on. The heat generated by the microwaves is nonuniform and therefore creates temperature gradients, which, in turn, causes diffusion, flow, and change in properties that alters the volumetric heat generation itself. This makes microwave heating and its modeling quite interdisciplinary. As shown in Figure 1, it involves electromagnetics, heat transfer, moisture transfer, and kinetics of biochemical changes. Although microwave heating involves coupling of so many different physics, to simplify things the sections in this chapter are organized in terms of a somewhat

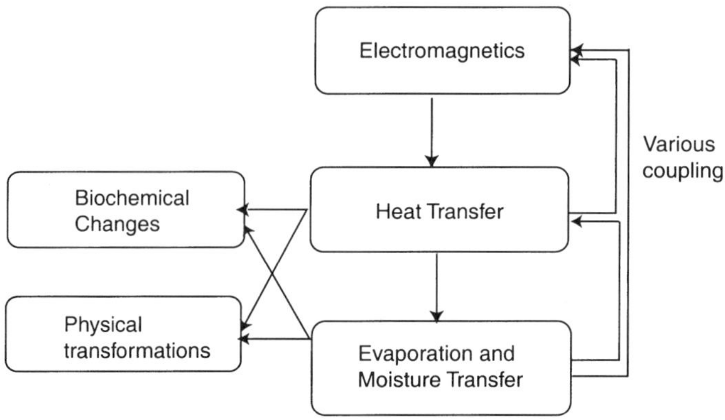

Figure 1 Schematic of various processes during microwave heating of food.

artificial division of subject matter along the lines of various engineering disciplines. Thus, they are presented in the following order: rigorous electromagnetic modeling without and with heat transfer, simplified electromagnetic modeling with application to heat transfer in solids, heat and moisture transfer in solids, and heat transfer in fluids.

Models of various complexity have been developed since microwave heating of food was established almost 50 years ago. The vast majority of these models are for solid foods without considering any internal evaporation and for semiempirical (Lambert's law) approximation of the electromagnetics. For practical food geometry, food property, oven configuration, and other factors, modeling of microwave food processing needs to be essentially numerical. Thus, detailed modeling depended to a great extent on the availability and ease of use of computing power. More detailed and sophisticated models of electromagnetics and thermal aspects have appeared mostly in the last 10 years, which are emphasized in this chapter.

This chapter is necessarily an overview because a significant amount of modeling literature now exists on the microwave processing of food. It will not be possible to discuss all of these models in complete detail, such as numerical methods, input parameters, and various results. Discussion will also concentrate on the modeling aspect and condense the physical interpretation of the results that are explained in more detail elsewhere [1,2]. Because microwave heating is used in processing ceramics, polymeric materials, and other applications, comprehensive review papers [3,4] and texts [5,6] on microwave processing cover a number of different applications. The reader is referred to these sources for obtaining more detailed understanding and more extensive bibliography on modeling of microwave processing in general. This chapter will restrict modeling of microwave processing to the context of food materials. Models ranging from very simple ones to the most complex are discussed, so that the reader has an idea of the complexities involved but may be able to choose a simpler model, depending on circumstances. Space limitation also required that discussions be more pertinent to the unique microwave heating aspect of modeling as distinct from modeling of conventional heating.

2 RIGOROUS ELECTROMAGNETIC MODELING OF MICROWAVE HEATING

A microwave heating system typically has three major components. The magnetron generates the microwave fields. The waveguide is a metallic conduit that guides the electromagnetic field from the magnetron into a metallic enclosure called a cavity (such as the food compartment in the domestic microwave oven) where the food is placed. The cavity, depending on its shape, size, and so forth forms standing waves of electromagnetic fields. When placed inside the cavity, the food experiences the electromagnetic fields that lead to its heating. The two properties that determine a material's interaction with microwaves are the dielectric constant and dielectric loss (explained later). Air in the oven can absorb very little of the microwave energy, so only the food is heated directly by the electromagnetic waves. The shape, size, and properties of the food greatly affect its spatial distribution of microwave absorption.

2.1 Governing Equations

The electromagnetic fields that are responsible for the heating of the food material, as inside a microwave oven (Figure 2), are described by the Maxwell's equations

$$\nabla \times \mathbf{E} = -\frac{\partial}{\partial t}(\mu \mathbf{H}) \tag{1}$$

$$\nabla \times \mathbf{H} = \frac{\partial}{\partial t}(\epsilon \epsilon_0 \mathbf{E}) + \epsilon''_{\text{eff}} \epsilon_0 \omega \mathbf{E} \tag{2}$$

$$\nabla \cdot (\epsilon \mathbf{E}) = 0 \tag{3}$$

$$\nabla \cdot \mathbf{H} = 0 \tag{4}$$

where **E** and **H** are the electric and magnetic field vectors, respectively. In food materials, heating is done by the electric field primarily through its

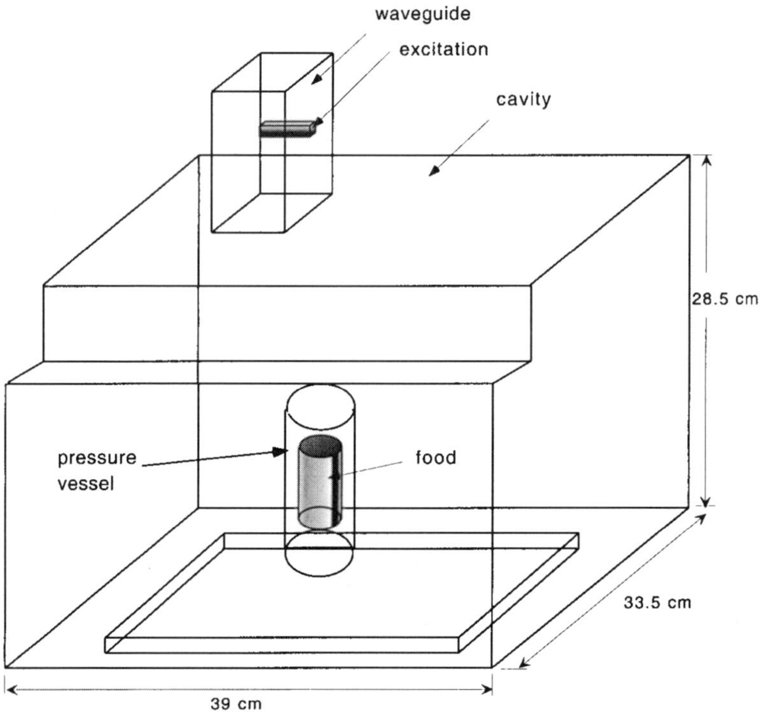

FIGURE 2 Schematic of the microwave oven used for electromagnetic calculations.

interaction with water and ions. The complex permittivity ϵ is given by

$$\epsilon = \epsilon' + j\epsilon''_{\text{eff}} \qquad (5)$$

where ϵ' is called the dielectric constant and represents the material's ability to store electromagnetic energy. The other property, ϵ''_{eff}, is the effective dielectric loss factor of materials representing the energy dissipation or heat generation. Both of the properties can be functions of locations in the food due to temperature variations. In the above equations, ϵ_0 is the permittivity of free space (8.86×10^{-12} F/m) and ω is the angular frequency of the microwaves. For a short discussion of Maxwell's equations and solutions, see Ref. 3. The Maxwell's equations are to be solved to obtain the electric field **E** as a function of position in the food and heating time. As will be discussed later, the rate of volumetric heat generation can be calculated from this electric field [Eq. (16)].

2.2 Boundary Conditions

Boundary conditions for the electromagnetic modeling of a cavity are set on the walls of the cavity, which are considered perfect conductors. The entire cavity interior is treated as a dielectric with appropriate dielectric properties of air and food in the regions that they occupy. Note that in the modeling of the entire cavity, the food–air interface does not have to be treated in any special way. In the interior of a perfect electrical conductor, the electric field is zero. This condition, together with the Maxwell's equations, lead to the boundary condition at the air–wall interface as

$$E_{t,\mathrm{air}} = 0 \tag{6}$$

$$B_{n,\mathrm{air}} = 0 \tag{7}$$

Here, the subscripts t and n stand for tangential and normal directions, respectively. These conditions are necessary to determine the solution. The conditions are translated in terms of potentials as

$$A_t = 0 \tag{8}$$

$$\psi = 0 \tag{9}$$

2.3 Input Parameters

Input parameters needed for the electromagnetic problem are the geometry of the food, inside geometry of the oven, and the dielectric properties (constant and loss) of the food material. Dielectric properties of foods can be found in a number of compilations such as Refs. 7–10. Dielectric properties are dependent on composition (moisture and salt, in particular) and temperature. Dielectric property data covering large changes in moisture levels for the same material are generally not available. Dielectric property data with temperature variation are available, but typically do not go above 100°C.

2.4 Numerical Solution Techniques

There are a number of commercial software based on various computational methods, such as the finite-difference time-domain method and the finite-element method. In the software EMAS, which is finite-element based, Eqs. (1)–(4) are transformed using a vector potential $\vec{A}$, and a scalar potential, ϕ. The final equations use a new scalar potential, ψ, where the conventional scalar potential, ϕ, is a time derivative of ψ

$$\vec{E} = -\nabla \cdot \psi - \frac{\partial \vec{A}}{\partial t} \tag{10}$$

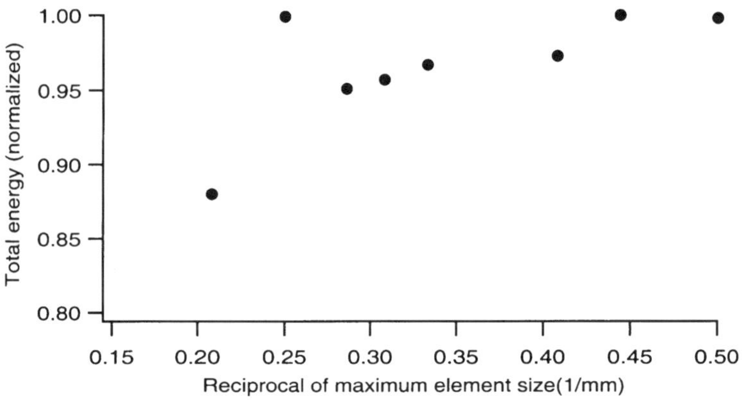

FIGURE 3 The total electromagnetic energy (normalized) calculated from the models with different mesh sizes.

Using this transformation, the final governing equations and boundary conditions are developed in terms of $\vec{A}$ and ψ. For further details, see reference manuals for the software EMAS. The convergence of the numerical results can be checked by reducing the mesh size, as shown in Figure 3.

2.5 Experimental Verification

Direct experimental verification of electric field is difficult because few, if any, sensors are available for this purpose [11]. Typically, the temperature measurements (discussed in the next section) serve as indirect validation of the electric field distributions.

2.6 Typical Results

For detailed discussion of results from electromagnetic calculations, see Refs. 12–16. Example of electromagnetic field pattern inside an oven, obtained from solving Eqs (11)–(14), is shown in Figure 4. For brevity, two important effects on shape and size of the food on heating patterns are illustrated. Figure 5 shows how microwaves can be focused inside a curved object. This is analogous to how a lens can focus light waves. Such focusing can lead to enhanced heating in the interior, causing rapid vapor generation inside a wet food and a pressure increase, sometimes leading to an explosion.

An important note about the modeling of focusing effect in microwave heating should be mentioned here. A number of studies in the literature have

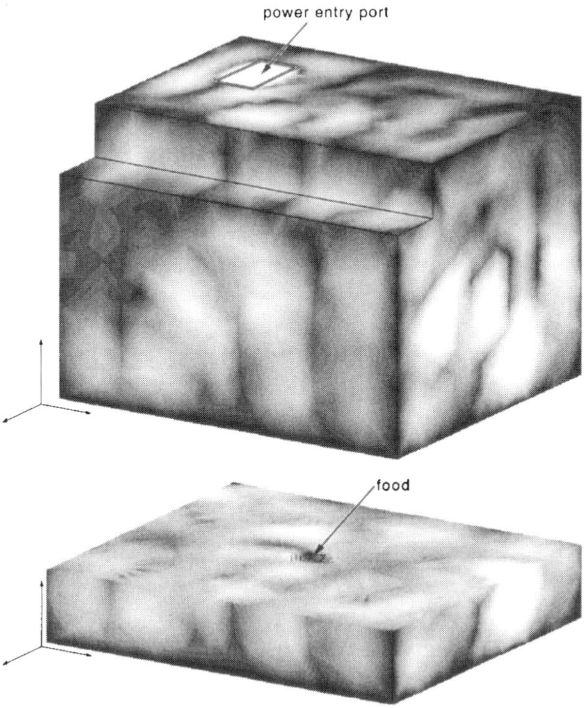

FIGURE 4 Electric field distributions inside the microwave oven shown in Figure 2 obtained from computer simulation. The color image output by the computational software has been converted to grayscale for printing purposes. Here, white signifies high values and black signifies low magnitudes of electric fields—with other shades having intermediate magnitudes of electric fields.

developed a mathematical formulation starting from exponential decay of energy from the surface and eventually "predicted" a focusing effect [17,18]. To elaborate their power formulation, suppose the exponential decay of energy flux from the surface of a cylinder is given by

$$F_r = F_0 \exp\left(-\frac{R-r}{\delta_p}\right) \tag{11}$$

where F_r is the energy flux at a radial distance r of a cylinder of total radius R and length L, and δ_p is the penetration depth. If we consider a cylindrical shell of thickness Δr between r and $r + \Delta r$, the energy absorbed in this shell

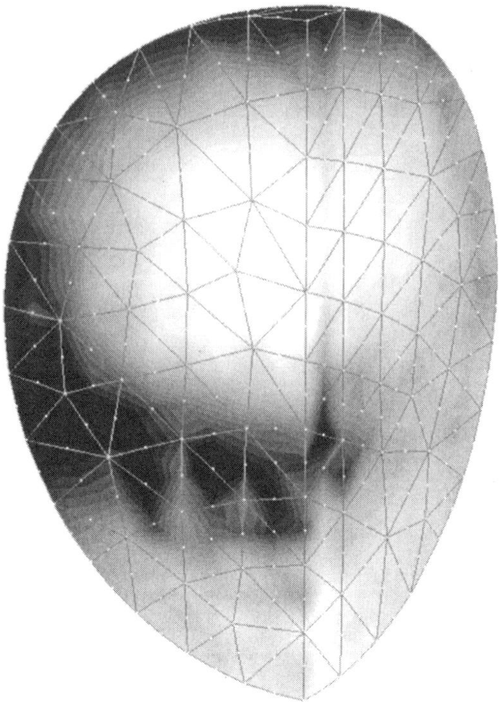

FIGURE 5 Example of focusing effect with enhanced interior heating due to a curved surface. (From Ref. 1.)

per unit volume is given by

$$Q = \frac{\overbrace{F_{r+\Delta r} 2\pi(r+\Delta r)L - F_r 2\pi r L}^{\text{Net energy absorbed}}}{\underbrace{2\pi r \Delta r L}_{\text{volume of shell}}} \tag{12}$$

$$= \frac{1}{r}\frac{\partial}{\partial r}(F_r r) \tag{13}$$

if we let $\Delta r \rightarrow 0$. The right-hand side of Eq. (13) has been used as the heat-source term in a number of literature studies. It is easy to see that this formulation is nonphysical, at least for the microwave heating process. Thus, substituting for F_r from Eq. (11) into Eq. (13) taking the derivative and rearranging, we get

$$Q = F_0\, e^{-(R/\delta_p)} \left(\underbrace{\frac{e^{r/\delta_p}}{r}}_{\text{singular at } r=0} + \frac{e^{r/\delta_p}}{\delta_p} \right) \tag{14}$$

Equation (14) shows that this formulation will be singular at $r = 0$ under all circumstances, except when $\delta_p = 0$. Previous researchers have generally avoided the inherent problems at $r = 0$ by simply avoiding the term at $r = 0$ in their numerical formulations. Thus, the idea of Eq. (12) may have some utility in qualitative explanation of focusing effect, but it is not true over the entire domain and its use in quantitative modeling is to be avoided. If focusing is an important issue for a microwave heating situation, complete electromagnetic modeling is suggested, as in Ref. 15.

Figure 6 shows that the total amount of power absorbed in a material increases with its volume and becomes asymptotic to the total power available from the system. The patterns of power deposition discussed in Figures 5 and 6 are due to the electromagnetic aspect of the heating process. Heating patterns resulting from these power depositions cannot be modeled by solving a thermal problem without consideration of the electromagnetics (or its simple approximation such as the Lambert's law).

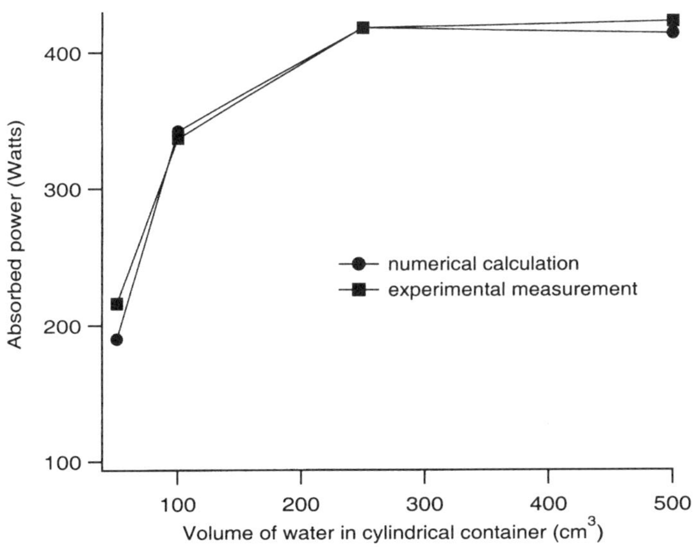

FIGURE 6 Total power absorption increases with the volume of the load.

2.7 Sensitivity Analysis and Use in Design

As the computing power becomes more easily available and software becomes more user-friendly, computer-aided design of microwave food product, process, and equipment is becoming a reality. By studying the sensitivity to various design parameters, a closer to optimum solution can be obtained by using computational software such as EMAS mentioned earlier in this section. This can reduce design time and expenses in comparison with exhaustive experimentation and prototype building. Applications of modeling to product development was studied in Refs 14–16, which investigated the effect of shape and size on total energy absorption and its spatial distribution. The application to food processing was performed in Ref. 19. The application of modeling to equipment design has been reported (see Refs 20 and 21 as examples).

3 RIGOROUS AND COUPLED ELECTROTHERMAL MODELING OF HEAT TRANSFER IN SOLIDS WITHOUT MOISTURE TRANSFER

The previous section discussed the calculation of an electric field distribution inside a food material that generates the heat. Due to the spatial variation of heat generation, the temperature starts to increase spatially nonuniformly. The increase in temperature and its spatial variation changes the dielectric properties, which, in turn, modifies the electric field distribution. Modeling of this coupled thermal–electromagnetic heating process is discussed here.

3.1 Governing Equations of Heat Transfer

To calculate temperatures, we need the governing equations for heat transfer

$$\underbrace{\rho c_p \frac{\partial T}{\partial t}}_{\substack{\text{rate of}\\ \text{temperature rise}}} = \underbrace{\nabla \cdot (k \nabla T)}_{\text{diffusion}} + \underbrace{Q(x, T)}_{\substack{\text{microwave}\\ \text{heat generation}}} \tag{15}$$

in addition to the Maxwell's equations of electromagnetics [Eqs. (1)–(4)] discussed in the previous section. The source term $Q(x, T)$ is related to the magnitude of electric field by the equation

$$Q(x, T) = \tfrac{1}{2}\omega\epsilon_0\epsilon''_{\text{eff}}E^2 \tag{16}$$

and thus couples the above heat transfer equation with the electromagnetics discussed in the previous section. As the temperature T increases spatially nonuniformly, the dielectric properties $\epsilon'(x, T)$ and $\epsilon''_{\text{eff}}(x, T)$ in Eqs. (1)–(4) change, making the electric field E in Eq. (16) vary. This coupling process is

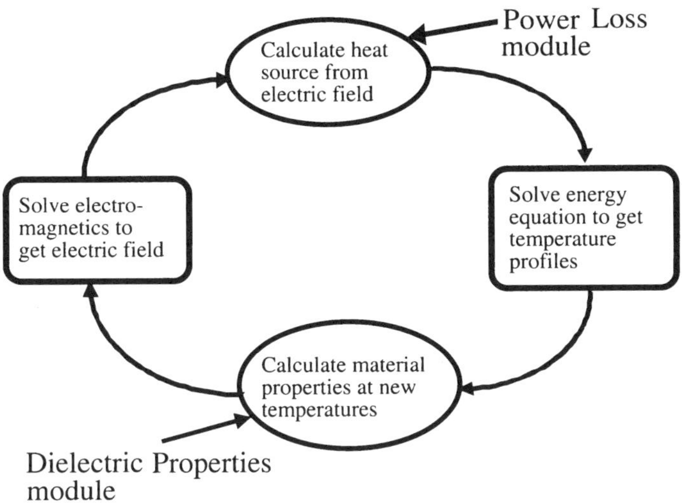

FIGURE 7 Schematic of the coupling of electromagnetic and thermal calculations.

shown schematically in Figure 7 and one software implementation of this coupling process is discussed in Section 3.4.

3.2 Boundary Conditions for Heat Transfer

The surface of the microwave-heated food typically has convection—either natural or forced. When heating a wet material, evaporative cooling at the surface can also have a strong effect on the temperature profile. Because the temperatures do not reach high values (typically lower than 100°C, the boiling point of water), radiative heat losses from the food surface to the oven wall may not be significant. However, if susceptors are used, radiative heat gain can become the dominant mechanism at the surface. These surface conditions can be included in terms of a generalized boundary condition

$$-k\frac{\partial T}{\partial n} = \underbrace{h(T - T_\infty)}_{\substack{\text{convective}\\\text{heat gain or loss}}} + \underbrace{m_w h_{fg}}_{\substack{\text{evaporative}\\\text{heat loss}}} + \underbrace{\sigma_{rad}\epsilon_{rad}T^4 - \alpha q_{rad}}_{\substack{\text{radiative}\\\text{heat gain or loss}}} \tag{17}$$

where T is the load surface temperature, n represents the normal to the surface, h is the convective heat transfer coefficient, m_w and h_{fg} are the rate of evaporation and the latent heat of vaporization of the evaporated liquid, respectively, σ_{rad} is the Stefan–Boltzmann constant, ϵ_{rad} is the radiative surface emissivity of the load, α is the surface absorptivity of the load, and q_{rad} is the incident radiant heat flow rate per unit surface area.

3.3 Input Parameters

The parameters needed for implementing the boundary condition [Eq. (17)] are hard to obtain. The literature is elusive on detailed measurements of airflow or heat transfer coefficients in microwave ovens [2]. The heat transfer coefficient calculated based on the expected small velocities approach that of natural convection, and a heat transfer coefficient of 2.6 W/m^2 K was used in Ref. 22. Even in a microwave tunnel pasteurizer, where the food slowly moves in a tunnel, air was considered relatively still and a heat transfer coefficient of 5–10 W/m^2 K was used in Ref. 19. The total moisture loss has been measured from the weight loss [23].

3.4 Numerical Solution Techniques

The coupling is implemented using the two softwares EMAS for electromagnetic calculations and NASTRAN for heat transfer. The two softwares have no built-in ways to be coupled for the thermal-electromagnetic analysis, and the only way to couple them is at the operating system level. As shown in Figure 7, the coupled solution starts with solving the electromagnetic fields in foods using EMAS. The power density data, obtained from the electromagnetic calculations, is converted into power-loss data using Eq. (16) and inserted as a module into the NASTRAN input data file for temperature calculation. Temperature distributions are calculated using NASTRAN. In order to complete the coupling, the dielectric properties of foods are modified in the EMAS input data file according to the calculated temperature data. A mean temperature for each element is used by averaging the nodal temperature values. For further details of this implementation, see Ref. 12.

3.5 Experimental Verifications

Experimental verification of calculated temperatures can be made using fiber-optic temperature sensors and infrared cameras. Fiber-optic temperature sensors provide point measurement anywhere in the food, whereas infrared cameras can provide temperature contours for an entire outside surface. Both methods are relatively expensive as compared to thermocouple measurements. More details on these measurement systems can be found in Ref. 11.

3.6 Typical Results

If the dielectric properties were not temperature sensitive, there would be no need to develop the coupling described here. It is shown [12] that such coupling is important, particularly in high-loss foods, where the dielectric properties vary more significantly with temperature. An example of the impact of the coupling process in the temperature calculation for a high-loss food is shown in Figure 8. The figure shows four quarters at four different heights of a cylinder of high-loss food (ham) at two instants during heating. When the temperature is low (after 5 sec), some hot spots are close to the center. The dielectric loss is relatively low at these temperatures and microwave energy is able to penetrate more easily. In this case, few visible maxima (hot spots) of the standing waves remain near the center. As the temperature near the surface increases at a faster rate with time, the dielectric loss near the surface also increases rapidly. Consequently, a shield develops, making the penetration of microwave power difficult and most of the power is deposited near the surface. Thus, at a later time (45 sec), hot spots from near the axis have disappeared. This trend of surface heating is expected to continue or to be enhanced at later times. Such dramatic changes in heating patterns due to the coupling of heat transfer and electromagnetics have many implications to food processing. For example, the

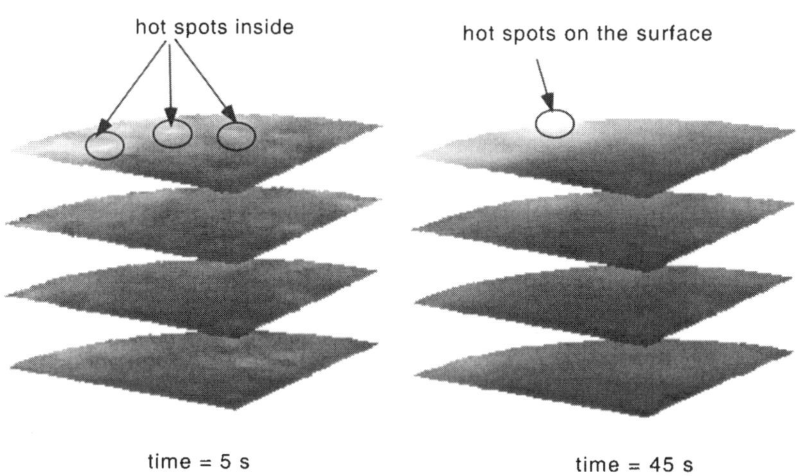

FIGURE 8 Temperature profiles (in four vertical slices along the axis of the cylinder) for high-loss foods after 5 sec (left) and 45 sec (right) of heating. For the left figure, the temperature ranges from 1.5°C to 20.9°C; for the right figure, the temperature ranges from 5.7°C to 98.5°C.

location of the cold point in a pasteurization or sterilization process can keep changing, unlike in conventional heating.

3.7 Sensitivity Analysis and Use in Design

Detailed sensitivity analysis in using a coupled electrothermal model is generally unavailable in the literature. Effect of various food properties were investigated in Ref. 12 and the application to pasteurization and sterilization was made in Ref. 13.

4 SIMPLIFIED ELECTROMAGNETIC MODEL: HEAT TRANSFER IN SOLIDS WITH NO INTERNAL MOISTURE TRANSFER

Although the rigorous electromagnetic model and its coupling with heat transfer mentioned in the previous two sections are the most comprehensive ways to model microwave processing of foods, computational resources are just becoming available to perform this routinely. Such modeling also requires interdisciplinary expertise in electrical and thermal engineering, which is often hard to locate. Most of the models to date had to use a gross simplification of the electromagnetics, called a Lambert's law model, given by

$$Q = Q_0 \exp\left(-\frac{x}{\delta}\right) \tag{18}$$

where x is the distance into the material. Lambert's law has information on the magnitude and spatial distribution of power absorption. The spatial distribution of power absorption is an exponential decay, the exact rate of decay being determined by δ, the material property called the penetration depth. The surface power absorption, Q_0, is typically found from experimental measurement of temperature rise of a known mass of material (see Section 6.3, for example).

Equation 18 is the analytical solution to the Maxwell's equation for plane waves incident on a thick lossy material. Inside a cavity, such as the domestic microwave oven, the electromagnetics is quite different and Eq. (18) is only a qualitative description of the energy absorption valid for some very restricted situations. For example, in foods that are quite lossy (large amounts of water with added salt), a rapid drop in microwaves from the surface into the material [24] can perhaps be approximated by Eq. (18). Use of Eq. (18) to predict electromagnetic effects such as focusing, as explained in Section 2.6, is clearly qualitative and can completely miss the physics [18]. Because Eq. (18) can, at times, keep the qualitative nature of the heating process and it has been used considerably in the past, the following subsections will discuss heat and mass transfer using Eq. (18) as the simplified electromagnetic model.

4.1 Solution to a Simple One-Dimensional Problem

Heating of solid foods will always involve a certain amount of evaporation in the interior of the food and at the surface. A discussion on internal evaporation is postponed to Section 5. Neglecting internal evaporation, simple semiquantitative solutions of microwave heating can be developed, as has been the majority of the models to date. For a one-dimensional (1D) slab and for constant thermophysical properties, the governing equation given by Eq. 15 can be simplified to

$$\underbrace{\rho c_p \frac{\partial T}{\partial t}}_{\text{transient}} = \underbrace{k \frac{\partial^2 T}{\partial x^2}}_{\text{diffusion}} + \underbrace{Q_0 \exp\left(-\frac{x}{\delta}\right)}_{\substack{\text{microwave}\\ \text{heat generation}}} \tag{19}$$

and can be solved analytically [25]. Some results from this study are now discussed.

4.2 Temperature Profiles

Temperature profiles calculated from the analytical solution of Eq. (19) are shown in Figure 9. We can identify three qualitatively different temperature

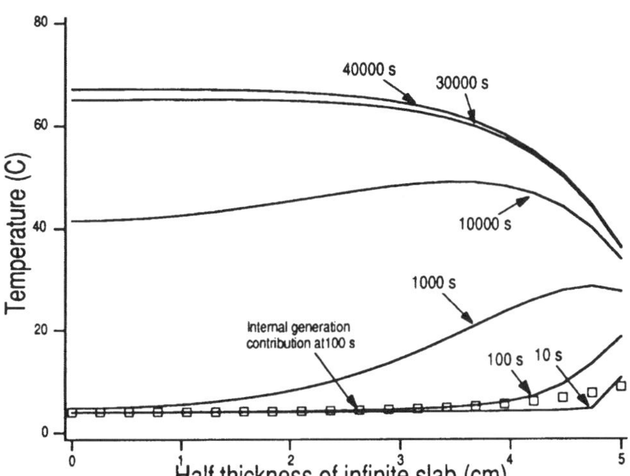

FIGURE 9 Developing temperature profiles in a thick slab during microwave heating. (From Ref. 25.)

profiles in this figure. Except for the very initial time when the temperature profiles are nearly exponential (type 1 heating profile), higher temperatures are always inside. The popular expression of microwaves heating "inside out" probably stems from the fact that the surface is always slightly colder. The type 2 profile, with the highest temperature slightly inside, is characteristic of microwave heating and is experienced most often. It occurs due to higher internal heating near the surface combined with surface cooling from convection and evaporation. Type 3 profile refers to those after very long duration of heating, with the highest temperature at the center of the slab, characteristic of steady-state conditions.

5 SIMPLIFIED ELECTROMAGNETIC MODEL: HEAT TRANSFER IN SOLIDS WITH MOISTURE TRANSFER

Because it is the moisture in the food that is primarily responsible for its heating by the microwaves, a certain amount of evaporation of this moisture is always present throughout the food material. The rate of evaporation increases with temperature. The internal evaporation, surface removal of moisture, and the nonuniform temperatures in the material lead to moisture transfer that can be important by itself or through its effect on heat transfer. A very simple semiempirical heat transfer model that includes moisture evaporation is described first, followed by a comprehensive model for moisture transport.

5.1 A Simple Heat Transfer Model Considering the Presence of Evaporation

Using a couple of major assumptions, a very simple model of moisture transport can be developed. In the first one, evaporation is ignored in the matrix until the temperature reaches 100°C. Once this temperature is reached, all of the microwave energy is assumed to be expended in the evaporation (boiling) of the liquid. The other assumption is to consider moisture transport in the solid by capillary diffusion only (i.e. ignore any resistance to vapor transport such that there is no accumulation of vapor and consequent pressure buildup that might lead to Darcy flow of liquid). This is true for high gas permeability, as shown in detailed studies [26]. The moisture content of vapor is very small compared to that of the liquid such that the moisture transport is essentially in the liquid phase. The first assumption can underestimate moisture loss, whereas the second assumption can underestimate or overestimate the moisture loss. Under these strong assumptions, the governing equations for temperature and moisture are [26]

$$\rho c_p \frac{\partial T}{\partial_t} = k\nabla^2 T + \begin{cases} Q & \text{when } T < 100^\circ\text{C} \\ 0 & \text{when } T > 100^\circ\text{C} \end{cases} \tag{20}$$

$$\rho \frac{\partial M}{\partial_t} = \begin{cases} 0 & \text{when } T < 100^\circ\text{C} \\ -\dfrac{Q}{\lambda} & \text{when } T > 100^\circ\text{C} \end{cases} \tag{21}$$

where M is the moisture content (wet basis) and λ is the latent heat of vaporization of water.

Although Lambert's law [Eq. (18)] is used here, for significant moisture loss the penetration depth δ is a function of moisture. Additionally, it is also a function of temperature. To use δ(M, T), the heat-generation term is formulated for varying penetration depth and can be written as

$$F(x) = F_0 \exp\left(-\int_0^x \frac{dx}{\delta(x)}\right) \tag{22}$$

From Eq. (22), the heat-generation term can be derived as

$$Q(x) = -\frac{dF}{dx} = \frac{F_0}{\delta(x)} \exp\left(-\int_0^x \frac{dx}{\delta(x)}\right) \tag{23}$$

Here, F_0 is the flux due to microwaves at the surface. The heat transfer boundary conditions are assumed to be convective heat loss at the top and side and insulated at the bottom. The mass transfer boundary conditions assumed no moisture loss from the top, bottom, or side. A small value of moisture capillary diffusivity is the primary reason that this assumption is valid. The initial conditions were uniform temperature and moisture content.

The total moisture loss is calculated as

$$\text{Total moisture loss at time } t = \int_{\text{vol}} \rho[M_i - M(t)]\, dv + \int_0^t J\, dt \tag{24}$$

where J is a small correction factor to include some of the evaporation until the maximum temperature ($T_{\max}$) inside reaches 100°C. It is calculated using the surface convective mass transfer equation

$$J = \begin{cases} h_v A(\rho_v - \rho_{\text{amb}}) & \text{when } T_{\max} < 100^\circ\text{C} \\ 0 & \text{when } T_{\max} > 100^\circ\text{C} \end{cases} \tag{25}$$

where h_v is the mass transfer coefficient, A is the top surface area, ρ_v is the concentration of water vapor next to the solid surface at saturation, and ρ_{amb} is the vapor density in the surrounding air, which is assumed zero.

The nonuniformity of heating at any instant is quantified using the standard deviation with respect to the mass average temperature at that instant; that is,

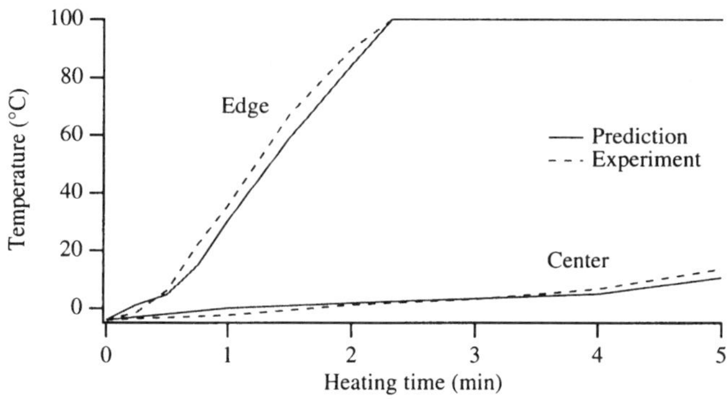

FIGURE 10 Experimental data on temperature profile at two locations in the partially frozen food used to predict the surface flux value for frozen food.

$$\sigma = \sqrt{\int_{\mathrm{vol}} \left[T - \left(\int_{\mathrm{vol}} T \, dv_f \right)^2 \right]^2 dv_f} \tag{26}$$

where v_f is the volume fraction of material at temperature T.

The governing equations together with the boundary and initial conditions were solved numerically using the commercial finite-element code FIDAP. An axisymmetric geometry is used. The strong nonlinearity in the governing equations due to the apparent specific-heat formulation of the phase-change problem required many iterations to reach a solution. Furthermore, close to 100°C, the moisture transfer equation is solved together with the energy equation, which significantly reduces the time increment and increases the CPU time.

Experimentally measured temperature changes in the food itself are used to calculate the surface microwave flux. An average surface flux value for the entire heating period, obtained by trial and error, is the value that best matches the temperature profile as shown in Figure 10. Note that in this process, moisture modeling remains independent.

Spatial distribution of moisture is shown in Figure 11a. The moisture distribution can be easily explained from the spatial distribution of temperature at the same time, as shown in Figure 11b. The top edge is heated the fastest and the center is heated the slowest. The bottom edge heats at a rate in between the top and the center. Thus, the moisture loss is highest at the top edge, lowest in the center, and in between everywhere else. After heating of about 5 min, almost 30% of the moisture is lost near the top edge.

As an example of sensitivity analysis, heating profiles for a varying aspect ratio are shown in Figure 12. Here, the surface flux for a different aspect ratio (flatter shape) is modified using experimentally obtained rela-

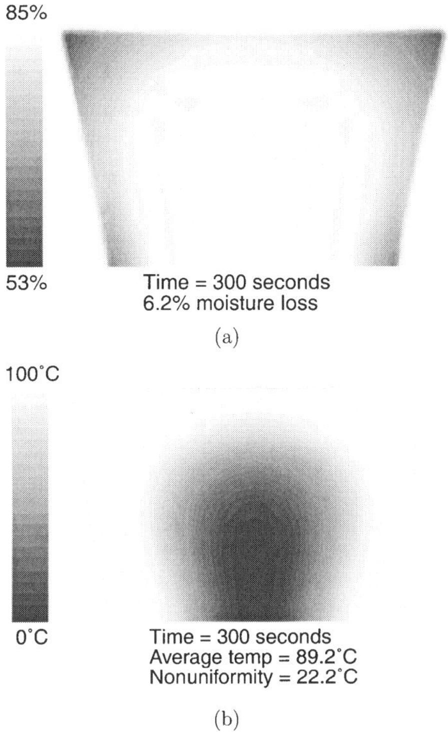

FIGURE 11 Moisture (a) and temperature (b) contours in partially frozen food after 5 min of heating. The surface flux values used are obtained by matching temperature profiles from Figure 10.

tionships [26]. The increased surface area makes heating more uniform. Thus, Figure 12 shows that at the same average temperature, moisture loss is lower, which is attributed to more uniform heating. Although the surface flux is lower with increased surface area (decreasing 9%), the total energy absorbed is higher, leading to a faster rise in average temperature. Thus, flatter shapes heat faster and lose less moisture due to more uniform heating. In a separate study that used a similar model [27,28], the sensitivity of thawing time to microwave power level (Figure 13) was developed for the design of the microwave thawing processes.

5.2 Governing Equations for a Multiphase Porous Media Model

A comprehensive model for moisture transport is now described [29,30] for microwave heating that can also be combined with infrared heating. The

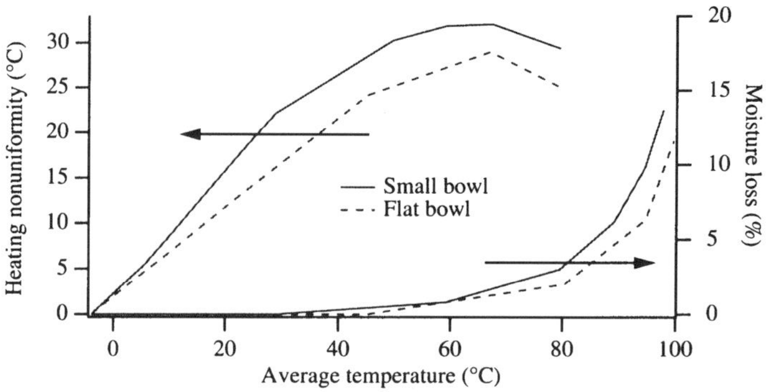

FIGURE 12 Comparison of heating nonuniformity and moisture loss for different aspect ratios (partially frozen foods).

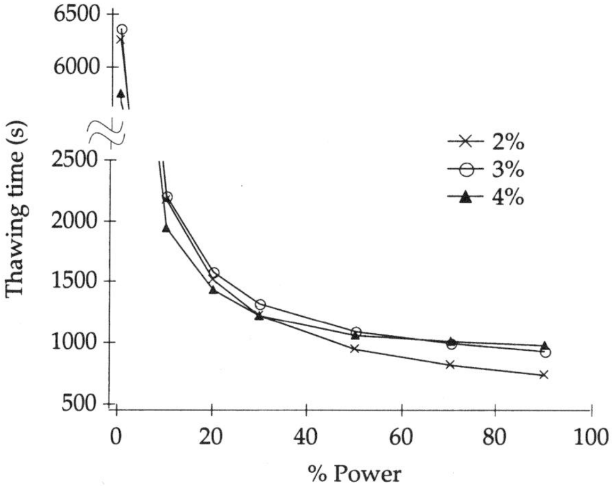

FIGURE 13 Thawing time for a tylose block, as a function of microwave oven power levels. Thawing with conventional heat (0% power) at the same surrounding temperature of 26°C and $h = 20\,W/m^{2\circ}C$ is added for comparison. (From Ref. 27.)

food is treated as a porous media to describe multiphase transport of liquid water, vapor, and air. Such a multiphase porous media model can accommodate a wide range of moist solid foods such as potato, meat, and cookie dough. More importantly, such a model can include explicitly the pressure-driven flow, which can be a major driving force in microwave heating. The major assumptions of this model include the following:

1. The solid, liquid, and gas phases are continuous.
2. Local thermal equilibrium exists between the phases.
3. Vapor pressure is a function of saturation and temperature.
4. Liquid transport results from convective flow due to the gradient in the total gas pressure and from capillary flow due to the gradient of the capillary force, which is a strong function of moisture content.
5. Vapor and air transport are driven by convective flow due to the gradient in the total gas pressure and diffusion due to the vapor concentration gradient.
6. The contribution of convection to energy transport can be ignored.
7. The geometry does not change during the heating and overall shrinkage is ignored. An equivalent porosity is defined as the fraction of the total volume occupied by the liquid water, vapor, and gas. This equivalent porosity is assumed constant during the heating process and is used to calculate the concentration of each phase—liquid water, vapor, and air. Different food systems are accommodated by using different initial equivalent porosity values. The effect of structure change during heating manifests as changed gas porosity and related transport properties. For example, loss of water increases gas porosity and, therefore, the intrinsic permeability.
8. Moisture removal from the surface consists of two parts: Vapor diffuses to the boundary and is convected away from the surface area occupied by the gas fraction on the boundary. Liquid water evaporates at the boundary and is convected away from the surface area occupied by the liquid fraction on the boundary. For a high-moisture food or when significant water is pushed to the surface from inside due to pressure driven flow, surface evaporation of liquid dominates. For a drier surface, vapor diffusion dominates.

A schematic diagram of the model (1D slab) is shown in Figure 14. The microwaves are assumed to be incident on the left side (open boundary), where a flow of heat and moisture from the porous medium is convected away. The right side (closed boundary) is insulated and the heat and moisture fluxes are zero. On the open boundary, the convective transport of energy and water vapor takes place. Liquid water reaching the surface from inside is considered to be evaporated first and then convected away.

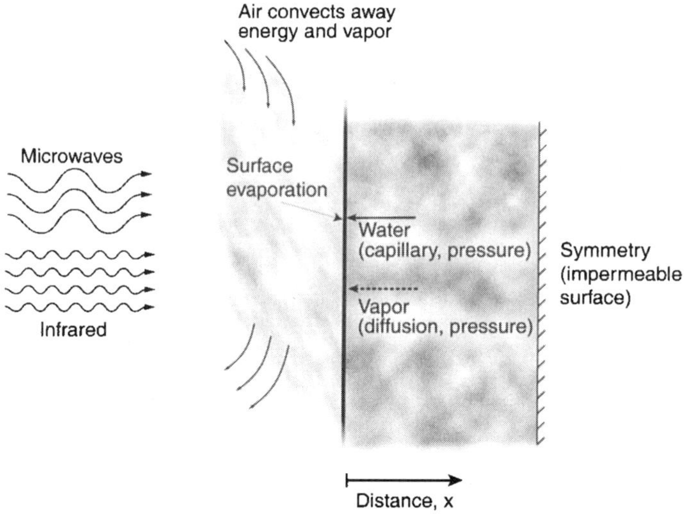

FIGURE 14 Schematic diagram of the heating process showing transport of liquid water and vapor to the surface from inside, evaporation of liquid water at the surface, and convective transport of energy and vapor at the surface.

The equivalent porosity and saturations are used to describe the concentrations, c, of various phases:

$$c_v = \frac{p_v S_g \phi M_v}{RT} \tag{27}$$

$$c_a = \frac{p_a S_g \phi M_a}{RT} \tag{28}$$

$$c_w = \rho_w \phi S_w \tag{29}$$

where p is partial pressure and S is saturation. The liquid saturation, S_w, and the equivalent porosity, ϕ, are related to the moisture content by

$$M = \frac{\rho_w \phi S_w}{(1-\phi)\rho_s} \tag{30}$$

The conservation equations for vapor, liquid water, air, and energy in the porous medium are written respectively as

$$\frac{\partial c_v}{\partial t} + \nabla \cdot (\vec{n}_v) = \dot{I} \tag{31}$$

$$\frac{\partial c_w}{\partial t} + \nabla \cdot (\vec{n}_w) = -\dot{I} \tag{32}$$

$$\frac{\partial c_a}{\partial t} + \nabla \cdot (\vec{n}_a) = 0 \tag{33}$$

$$(\rho c_p)_{\text{eff}} \frac{\partial T}{\partial t} = \nabla \cdot (k_{\text{eff}} \nabla T) - \lambda \dot{I} + \dot{q}_{\text{mic}} + \dot{q}_{\text{inf}} \tag{34}$$

Here, $\dot{I}$ is the rate of internal evaporation and $\dot{q}_{\text{mic}}$ and $\dot{q}_{\text{inf}}$ are rate of heat absorption from microwave and infrared, respectively. The total flux of vapor, $\vec{n}_v$, and that of air, $\vec{n}_a$, are composed of convective (Darcy) flow [31] and diffusion [32], respectively:

$$\vec{n}_v = -\rho_v \frac{k_g}{\mu_g} \nabla P - \frac{C_g^2}{\rho_g} M_a M_v D_{\text{eff},g} \nabla x_v \tag{35}$$

$$\vec{n}_a = -\rho_a \frac{k_g}{\mu_g} \nabla P - \frac{C_g^2}{\rho_g} M_a M_v D_{\text{eff},g} \nabla x_a \tag{36}$$

where P is the total pressure. The total liquid flux is composed of convective (Darcy) flow that can be further expanded in terms of total pressure and capillary pressure p_c, the latter being a function of saturation, S_w:

$$\vec{n}_w = -\rho_w \frac{k_w}{\mu_w} \nabla (P - p_c) \tag{37}$$

$$= -\rho_w \frac{k_w}{\mu_w} \nabla P - D_w \rho_w \phi \nabla S_w \tag{38}$$

The coefficient D_w is equal to

$$D_w = -\frac{k_w}{\phi \mu_w} \frac{\partial p_c}{\partial S_w} \tag{39}$$

The temperature dependence of capillary pressure, p_c, and the gravitational effects are ignored in Eq. (38).

The microwave power flux is assumed as an exponential decay with a varying penetration depth [Eq. (23)]. The infrared power flux, when used, is also modeled as an exponential decay [30] which leads to an additional volumetric heat-source term similar to Eq. (23) where δ_{inf} replaces the microwave penetration depth. Thus, if the infrared penetration depth is large, infrared energy is absorbed quite similar to microwave energy.

5.3 Boundary and Initial Conditions

The initial conditions are given by uniform initial temperature T_i, uniform total pressure equal to ambient pressure P_{amb}, and uniform liquid water saturation equal to S_{w_i}. The boundary conditions are quite complex, particularly when significant pressure develops inside, leading to liquid "pumping" at the surface. These are discussed in detail in Ref. 29. At the symmetry line, no heat or mass exchange takes place. At the open (left side) boundary, there is an exchange of energy, vapor, liquid, and air. Two situations can occur, depending on whether the liquid from the interior crosses the boundary with or without a phase change. In the first case, the water flux (liquid and vapor) reaching the boundary from the interior is fully evaporated and convected away as vapor to the ambient. Regardless of the volumetric evaporation rate inside, any remaining liquid flux arriving at the surface is evaporated at the open surface. The second case occurs when the internal pressure pushes more liquid to the open boundary than can be evaporated there and the surface of the porous medium becomes fully saturated. Even a small pressure gradient in the liquid toward the surface can cause liquid to flow across the open boundary without a phase change—a process referred to as "pumping." The pumped water is not fully evaporated at the boundary (it can leave the surface as liquid).

5.4 Input Parameters

A large number of input parameters are required for the multiphase porous media model, most of which are hard to find. They are discussed in detail in Refs. 29 and 30.

5.5 Numerical Solution Techniques

The governing equations are solved using a finite-difference method with central space differencing for the diffusion terms, Crank–Nicolson time differencing, and upwind differencing of the convection terms when they are included. A uniform grid system is used with 41 nodes over a total material thickness of 1 cm. The heat-source term is evaluated with values from the previous time step; this kept the computations stable. A rather small time step is used. Convergence of the numerical results is verified through mesh convergence, going up to 101 nodes in the 1-cm thickness. Energy and mass conservations of the numerical model were also checked at different times during heating. Further details of the numerical methods, including the finite-difference equations, are available in Ref. 33.

5.6 Experimental and Other Verifications

To check the accuracy, calculations from the present model were compared with previous numerical predictions for convective drying of clay brick [34]. The temperature and water saturation were close and showed similar trends.

Measured moisture loss versus heating rate followed the same trend as the model predictions. One obvious possibility for more comprehensive experimentation of moisture and temperature profiles would be to use nuclear magnetic resonance imaging (MRI). This has only recently been accomplished [35].

5.7 Temperature, Moisture, and Pressure Profiles

Temperature, moisture, and pressure profiles in microwave heating are shown in Figures 15 and 16. Higher rates of internal evaporation and resulting pressure generation modify the moisture transport considerably. Also, the moisture removal capacity of the air is drastically reduced due to its cold temperature (air is not directly heated by the microwaves). During the initial time, when sufficient pressure from evaporation has not built up, pressure-driven moisture flow is insignificant and moisture content drops similar to that in a surface convective-heating situation. As the internal temperature approaches 100°C, evaporation increases and pressure starts to build. Even small amounts of pressure in a low-moisture material (Figure 15) can cause enough moisture to reach the surface, exceeding its moisture removal capacity. This causes moisture accumulation near the surface, a characteristic of microwave heating resulting in soggy foods that were crispy before heating started. Moisture profiles after any significant pressure built up are fundamentally different from profiles due to capillary (or other) diffusion mechanisms in a surface convective-heating situation. In surface convective heating, moisture moves from a higher concentration to a lower concentration and the profile would decrease from high values in the inside region to low values near the surface.

During microwave heating of a high-moisture food, pressure rises much quicker and reaches a much higher value (Figure 16c). At higher pressures, enough moisture is pushed to the surface such that the surface is fully saturated and cannot retain water. After this, liquid is pumped across the open boundary without undergoing a phase change. Such pumping causes a large drop in internal moisture at about 3 min (Figure 16d). This can cause excessive moisture loss in microwave heating of a high-moisture material.

5.8 Sensitivity Analysis and Use in Design

The addition of infrared increases the surface temperature, as shown in Figure 17. This, in turn, increases surface evaporation. The infrared power level is varied by changing the infrared surface flux. The effect of infrared power level on the surface moisture and temperature is shown in Figure 17 for zero infrared penetration depth. At a lower infrared power level of 2000 W/m^2, the

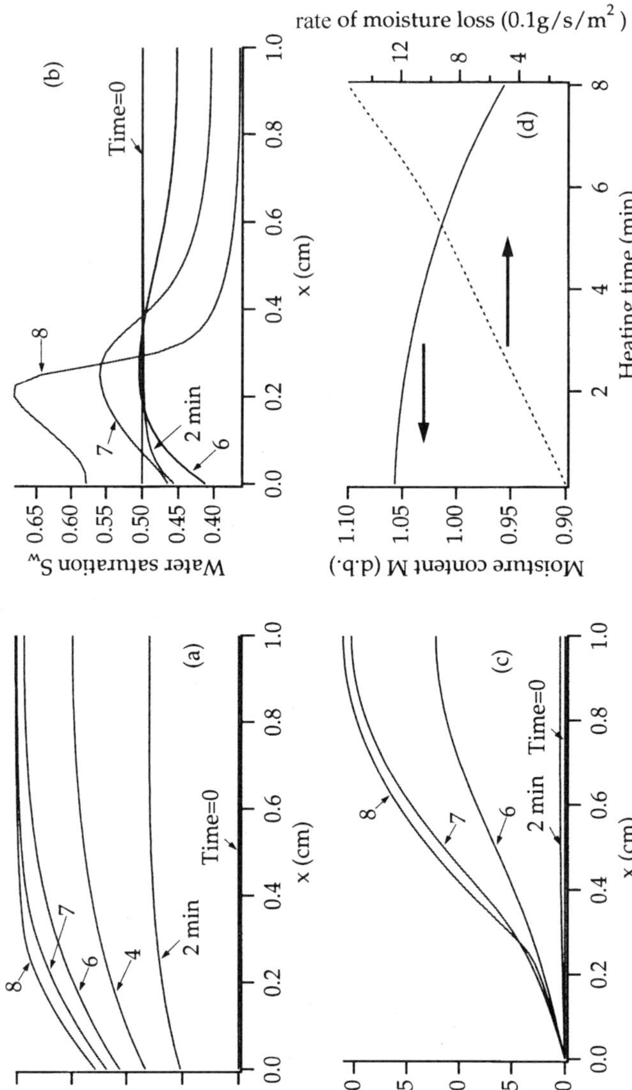

FIGURE 15 Temperature, water saturation, and pressure profiles and moisture loss in microwave heating of a low-moisture potato. (From Ref. 36, with permission.)

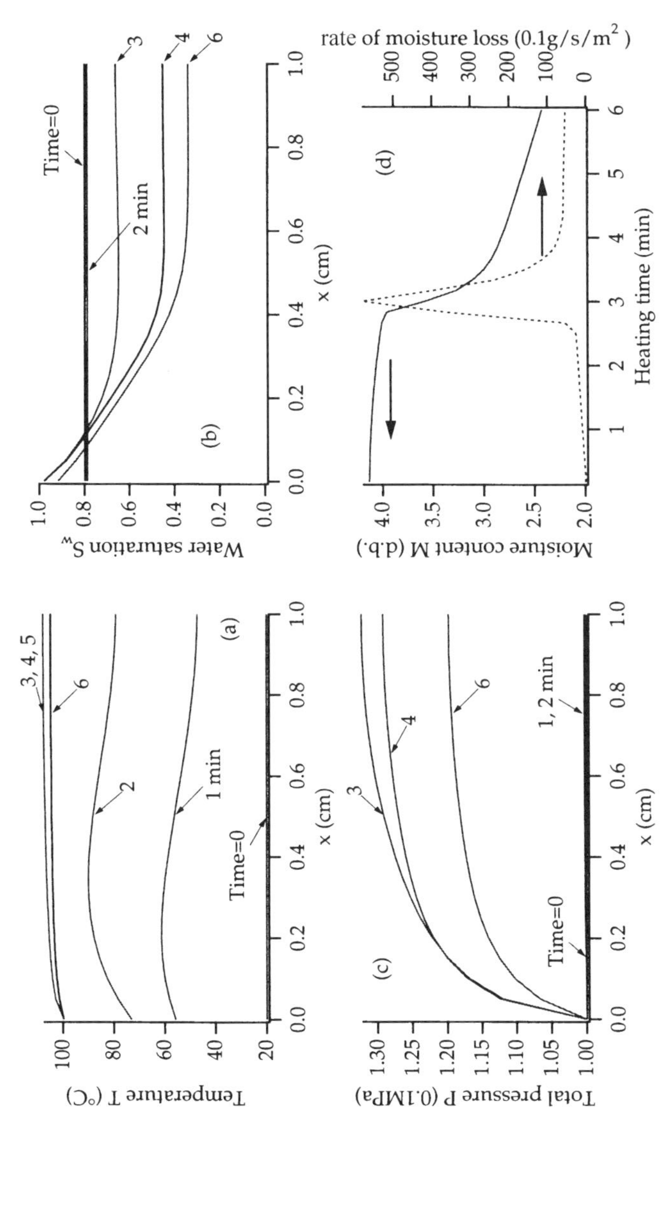

FIGURE 16 Temperature, water saturation, and pressure profiles and the loss of moisture in rapid microwave heating of a high-moisture food. Increased moisture loss after 3 min is due to a "pumping effect" whereby liquid water leaves the boundary without being evaporated. (From Ref. 36, with permission.)

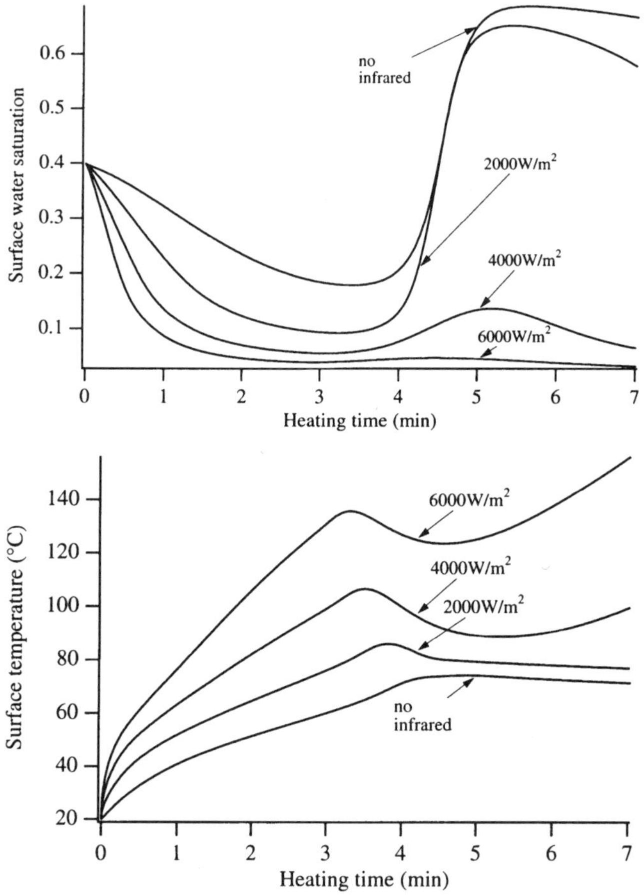

FIGURE 17 Effect of infrared power levels at zero penetration depth on surface water saturation and temperature.

moisture levels are only slightly lower compared to the microwave-only heating situation shown in the same figure. At a higher infrared power level of 4000 W/m^2, the surface moisture can reduce significantly and is drier than the initial value. With an increased power level, the surface moisture can be reduced to a very low value, perhaps leading to a dried crust.

The effect of adding hot air on the surface moisture and temperature is shown in Figure 18. Compared to the addition of infrared (Figure 17), the hot air at temperatures of 149°C, 177°C, and 204°C is unable to reduce the surface moisture significantly, although some reductions in surface moisture

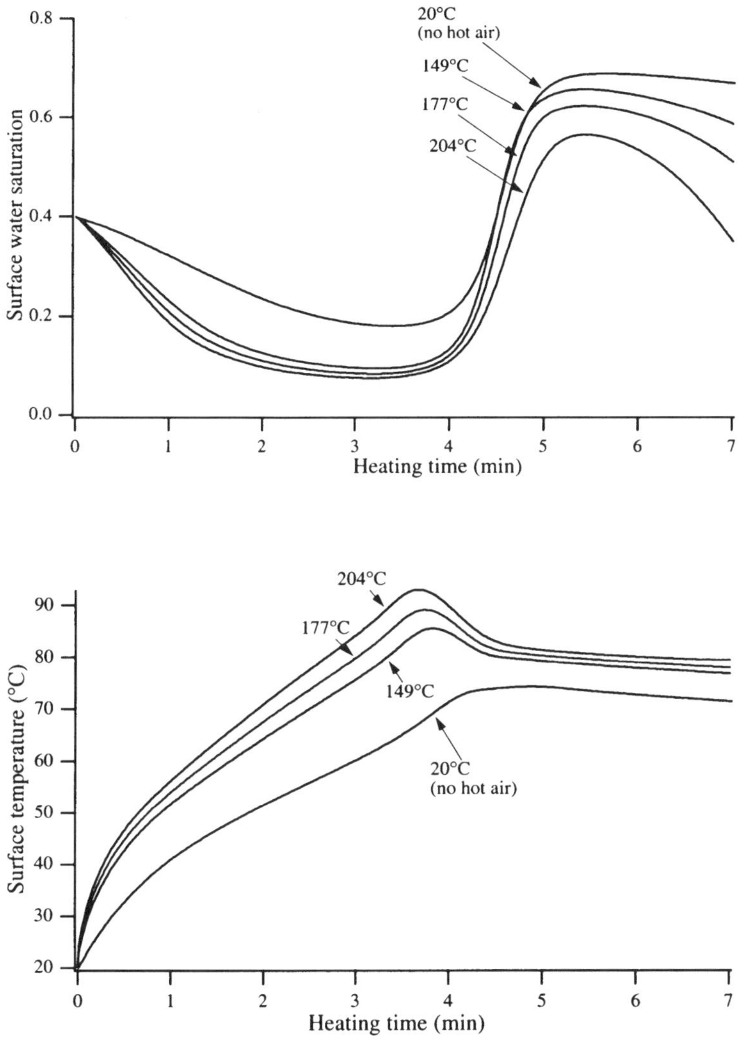

FIGURE 18 Effect of hot-air temperatures on surface saturation and temperature.

do take place. The surface temperature also increases somewhat, for the air temperature range of 149–204°C. It appears that the addition of hot air is not as effective in reducing the surface moisture, likely due to a lower heat flux for hot air as compared to that of infrared. Realistic increases in air temperature cannot increase the surface heat flux to values comparable to those of infrared.

6 SIMPLIFIED ELECTROMAGNETIC MODEL: BATCH HEATING OF LIQUIDS

The most common situation of microwave heating of liquids is in a container, without any agitation. If the container is large enough, the use of Lambert's law to simplify the electromagnetics may be a reasonable assumption. The decay in energy from the surface to the interior would lead to a hotter liquid near the surface than in the interior. This would cause buoyancy-driven flow and mixing of the liquid.

6.1 Governing Equations

For microwave heating of containerized, nonagitated liquid foods, the governing equations are the coupled energy and the Navier–Stokes equations, with the energy equation having the spatially varying microwave source term. In cylindrical coordinates, these equations are given by

$$\frac{\partial T}{\partial t} + v\frac{\partial T}{\partial r} + u\frac{\partial T}{\partial z} = \frac{k}{\rho c_p}\left[\frac{1}{r}\frac{\partial}{\partial r}\left(r\frac{\partial T}{\partial r}\right) + \frac{\partial^2 T}{\partial z^2}\right] + Q_0 \exp\left(-\frac{R-r}{\delta_p}\right) \tag{40}$$

$$\rho\left(\frac{\partial u}{\partial t} + v\frac{\partial u}{\partial r} + u\frac{\partial u}{\partial z}\right) = -\frac{\partial p}{\partial z} + \mu\left[\frac{1}{r}\frac{\partial}{\partial r}\left(r\frac{\partial T}{\partial r}\right) + \frac{\partial^2 u}{\partial z^2}\right] + \rho g \tag{41}$$

$$\left(\frac{\partial v}{\partial t} + v\frac{\partial v}{\partial r} + u\frac{\partial v}{\partial z}\right) = -\frac{\partial p}{\partial r} + \mu\left[\frac{\partial}{\partial r}\left(\frac{1}{r}\frac{\partial (rv)}{\partial r}\right) + \frac{\partial^2 v}{\partial z^2}\right] \tag{42}$$

Density is treated constant, except in the term containing gravity, g [Eq. (42)], where the following equation is used:

$$\beta = -\frac{1}{\rho}\frac{\partial \rho}{\partial T} \tag{43}$$

6.2 Boundary Conditions

For unagitated natural convection heating, the boundary condition for the thermal problem is considered insulated at all boundaries. For the velocities, the boundary conditions used are no-slip at all the walls. Symmetry at the centerline is used, reducing the computations. Symmetry boundary conditions (zero gradients) for velocity and temperature are used at the centerline.

6.3 Input Parameters

Estimation of the surface heating rate, Q_0, is one of the most difficult aspects of microwave heating. For this study [37], temperatures measured at different grid points were used to calculate the mass average temperature rise and this was equated to the temperature rise that would be obtained assuming Eq. (18):

$$\int_0^V Q\,dv = \frac{d}{dt}(mc_pT) \tag{44}$$

which was implemented using the grid point measured temperature data as

$$\sum_i \int_i^{i+1} Q_0 \exp\left(-\frac{R-r_i}{\delta_i}\right) dv\,\Delta t = \sum_i m_i c_{pi}\,\Delta T_i \tag{45}$$

where i stands for the locations where temperatures were measured. An interesting aspect of this study was that the estimate of Q_0 varied during heating, as shown in Figure 19. A major reason for this variation was attributed to the decreased conversion efficiency of the magnetron with temperature, leading to less power output from the magnetron. The temperature of the magnetron increases with time, and, consequently, the power output drops until it steadies (in about 30 sec) for a steady magnetron temperature.

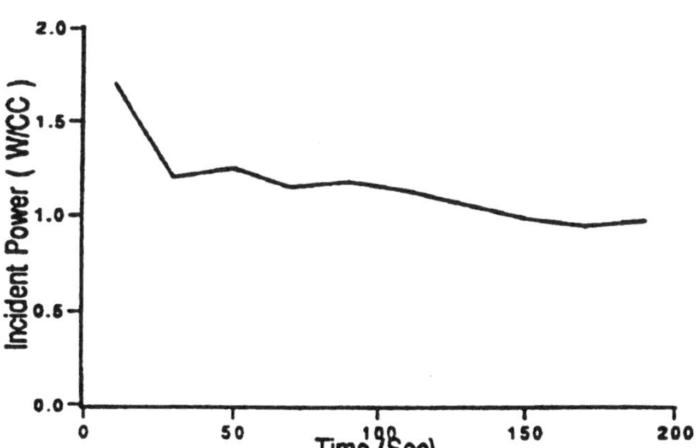

FIGURE 19 Variation of Q_0 at different times during heating, estimated from Eq. (45).

6.4 Numerical Solution Techniques

Equations (40)–(42) are solved using a finite-difference technique. A simple algebraic method was used to generate a nonuniform grid system that had finer grids near the wall where boundary layers are expected. To facilitate numerical solution in several ways, the new variables vorticity (ω) and stream function (ψ) are defined as

$$\omega = \frac{\partial v}{\partial z} - \frac{\partial u}{\partial r} \tag{46}$$

$$u = \frac{1}{r}\frac{\partial \psi}{\partial r} \tag{47}$$

$$v = -\frac{1}{r}\frac{\partial \psi}{\partial z} \tag{48}$$

Complete details of the solution method can be seen in Ref. 38.

6.5 Experimental and Other Verifications

Tap water was heated in polypropylene microwaveable containers of diameters 11.2 cm and 17 cm, respectively. The top and the bottom of the container were covered with aluminium foil at the liquid surfaces. Three temperature sensors were used in repeated experiments to obtain temperatures at all locations. Further details of the experiments are provided in Ref. 37.

6.6 Temperature and Velocity Profiles

Spatially nonuniform heating of the microwaves leads to spatial variations in temperature and, therefore, density. This leads to buoyancy driven recirculating flows, as shown in Figure 20 [37,39]. The flow of liquid leads to mixing and generally some reduction of nonuniformity, as shown in the radial profiles in Figure 21. Because the hot liquid rises while colder liquid sinks, the liquid becomes thermally stratified, as can be seen in the axial temperature profiles of Figure 21. Note that the direction of circulation would depend on electric field variations inside the container. If focusing effects are present, as might be true in a small-diameter container, the flow pattern can be reversed (i.e., liquid can rise at the center) [40].

6.7 Sensitivity Analysis and Use in Design

The extent of flow and mixing depends on thermal as well as dielectric properties. The increase in viscosity reduces the natural convection and associated mixing, making temperature profiles more nonuniform and the

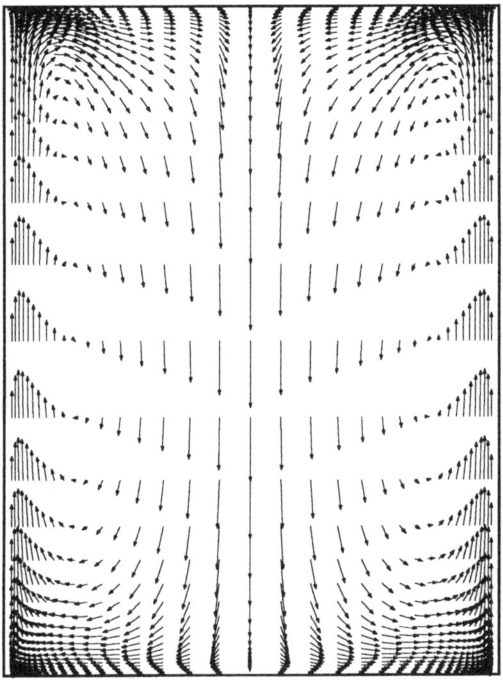

FIGURE 20 Flow patterns during microwave heating of water in a cylindrical container in the absence of any focusing effect.

core liquid colder [40]. The increase in the dielectric loss also causes more energy to be deposited near the surface, making the core region colder [40]. Selective electrical shielding of top, bottom, or sides of the container can be used to modify the flow and temperature profiles [41].

7 SIMPLIFIED ELECTROMAGNETIC MODEL: CONTINUOUS HEATING OF LIQUIDS

As a concept, continuous microwave heating of flowing liquids has been discussed over the years, often in the context of the continuous pasteurization of liquids [42–46]. Commercial use of such continuous microwave heating is not known. Successful radio-frequency heating for continuous flow through a tube was reported by Houben et al. [47], with applications for sausage emulsions. Heat transfer during microwave heating of liquids flowing through a microwave transparent cylindrical tube was modeled in Ref. 43, which is now discussed.

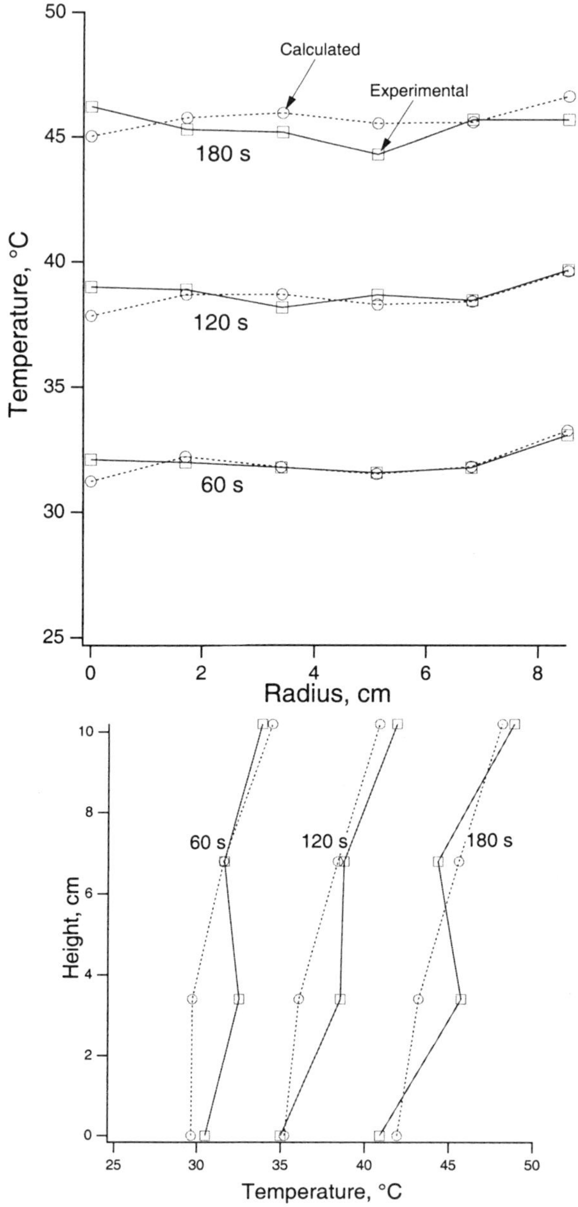

FIGURE 21 Radial and axial temperature patterns during microwave heating of water in a cylindrical container in the absence of any focusing effect.

7.1 Governing Equations and Boundary Conditions

The modeling of continuous microwave heating in a tube was performed in Ref. 43 using the Lambert's law approximation for the microwave field for flow through a circular pipe. The energy equation for steady, developing laminar flow is given by

$$u\frac{\partial T}{\partial z} = \alpha\left[\frac{1}{r}\frac{\partial}{\partial r}\left(\frac{\partial T}{\partial r}\right)\right] + Q_0\ \exp\left(-\frac{R-r}{\delta_p}\right) \tag{49}$$

where z is the axial direction. The governing momentum equation is given by

$$0 = -\frac{\partial p}{\partial z} + \mu\left[\frac{1}{r}\frac{\partial}{\partial r}\left(\frac{\partial u}{\partial r}\right)\right] \tag{50}$$

The thermal times, F_0, representing biochemical changes such as destruction of nutrients or bacteria is calculated from [43]

$$\begin{aligned} F_0 &= \int_0^L 10^{(T-T_R)/Z}\,dz \\ &= \int_0^L \frac{10^{(T-T_R)/Z}}{dz/dt}\,dz \\ &= \int_0^L \frac{10^{(T-T_R)/Z}}{u(r)}\,dz \end{aligned} \tag{51}$$

The tube surface was considered to be thermally insulated. A nonslip boundary condition was used for the velocities at the wall.

7.2 Numerical Solution Techniques

The finite-element package FIDAP was used to solve the governing energy and momentum equations [Eqs. (49) and (50)] for an axisymmetric heating situation. A graded mesh was used with more nodes near the walls of the tube to resolve the larger variation of temperature and velocities in that region. Although mesh convergence for this problem was accomplished, it was not reported. Also, due to lack of any available experimental data or setup, the computed results could not be compared with experimental data.

7.3 Temperature and Velocity Profiles

The velocity profile is unchanged from conventional heating because strong coupling to change fluid properties with temperature was not implemented.

7.4 Sensitivity Analysis and Use in Design

Nonuniformity of temperatures and temperature–time histories are described in terms of thermal time distributions that provided the volume fractions associated with any temperature value or thermal time, F_0. Details of thermal-time distributions can be seen in Ref. 43. Figure 22 shows that as the microwave power is increased, the nonuniformity of heating increases significantly more. At higher rates of microwave heating, larger differences in heating rates from the surface to the center occur. This increases the spread in temperature values and, therefore, the spread in thermal-time distributions, as shown in Figure 22. To achieve the same lower value of F_0, the average F_0 increases at higher heating rates, giving lower processed quality due to increased destruction of nutrients. To improve the quality, more uniform heating can be achieved at slower heating rates. This conclusion is analogous to the conventional case, where lower surface temperatures provide more uniformity.

For certain ranges of parameters, microwaves can lead to a higher spread in thermal times (signifying more nonuniform heat treatment) as compared to conventional heating, depending on process parameters. Figure 23a shows a microwave-heated liquid having worse uniformity than conventional heating. Only when the microwave heating rate was reduced to a rather low value (Figure 23b) did the microwave heating show a smaller spread in thermal time (less nonuniformity) than conventional heating. This conclusion is true for microwave heating in general and

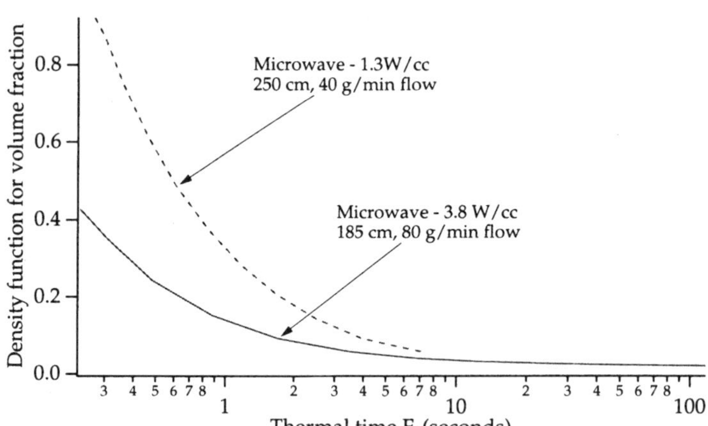

FIGURE 22 Nonuniformity of heating increases with microwave power (as represented by higher W/cc value at the surface).

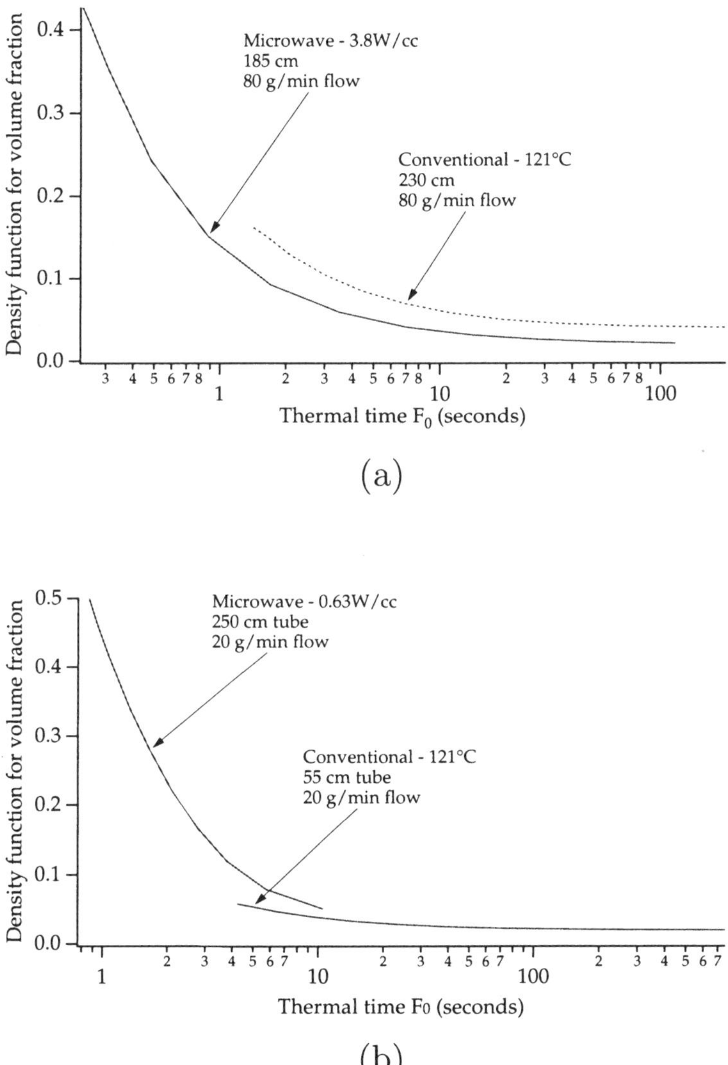

FIGURE 23 Microwave heating can be less (a) or more (b) uniform, depending on heating rates and other process parameters.

contradicts the popular belief that microwaves are always more uniform than conventional heating.

8 SUMMARY

Modeling of microwave processing of food in the literature vary considerably over their complexities. Although more comprehensive models are desirable, computational complexities and lack of available data on input parameters to the model make this difficult to achieve. Due to the coupled nature of microwave heating, although software are available to simulate separately some of the physics (electromagnetics, heat transfer, etc.), they are generally less useful for microwave food processing as they are not interlinked. Microwave heating is also volumetric and quite intensive. Volumetric evaporation and related pressure development are typically not implemented in the commercial software of today. The intensive heating also makes modeling of microwave heating of food computationally challenging. As future software include these missing aspects, we should see more use of modeling in design of microwave product, process, and equipment in the food industry.

ACKNOWLEDGMENTS

Most of the work cited here with the author as one of the coauthors was performed by his former graduate students Hanny Prosetya, Jerry Liu, Steven Lobo, Haitao Ni, Montip Chamchong, and Hua Zhang. This manuscript would not exist without their work.

REFERENCES

1. H Zhang, AK Datta. Electromagnetics of microwave oven heating: Magnitude and uniformity of energy absorption in an oven. In AK Datta, S Anantheswaran, eds. Handbook of Microwave Food Technology. New York: Marcel Dekker, Inc., in production.
2. AK Datta. Fundamentals of heat and moisture transport for microwaveable food product and process development. In: AK Datta, S Anantheswaran, eds. Handbook of Microwave Food Technology. New York: Marcel Dekker, Inc., in production.
3. C Saltiel, AK Datta. Heat and mass transfer in microwave processing. Adv Heat Transfer 33:1–94, 1999.
4. KG Ayapa. Modelling transport processes during microwave heating: A review. Rev Chem Eng 13(2):1–68, 1997.
5. G Roussy, J Pearce. Foundations and Industrial Applications of Microwaves and Radio Frequency Fields. New York: John Wiley & Sons, 1995.

6. AC Metaxas. Industrial Microwave Heating. London: Peter Peregrinus, 1983.
7. S Nelson. Electrical properties of agricultural products—A critical review. Trans ASAE 16(2):384–400, 1973.
8. M Kent. Electrical and Dielectric Properties of Food Materials. Hornchurch, England: Science and Technology Publishers, 1987.
9. E. Sun, AK Datta, S Lobo. Composition-based prediction of dielectric properties of foods. J Microwave Power Electromagn Energy, 30(4):205–212, 1995.
10. S Nelson, AK Datta. Dielectric properties of food materials and electric field interactions. In: AK Datta, S. Anantheswaran, eds. Handbook of Microwave Food Technology. New York: Marcel Dekker, Inc., in production.
11. AK Datta, H Berek, D Little. Measurement of electric field, temperature, heating uniformity, temperature-time history and moisture profiles in microwave heating of foods. In: AK Datta, S. Anantheswaran, eds. Handbook of Microwave Food Technology. New York: Marcel Dekker, Inc., in production.
12. H Zhang, AK Datta. Coupled electromagnetic and thermal modeling of microwave oven heating of foods. J Microwave Power Electromagn Energy 35(2):71–85, 2000.
13. H. Zhang, AK Datta. Experimental and numerical investigation of microwave sterilization of solid foods. Am Inst Chem Eng J, in production.
14. H Zhang, AK Datta. Heating concentrations of microwaves in spherical foods. Part one: In plane waves. J Microwave Power Electromagn Energy, in production.
15. H Zhang, AK Datta. Heating concentrations of microwaves in spherical and cylindrical foods Part two: In a cavity. J Microwave Power Electromagn Energy, in production.
16. H. Zhang, AK Datta. Microwave power absorption in single and multicompartment foods. J Microwave Power Electromagn Energy, in production.
17. L Zhou, VM Puri, RC Anantheswaran, G Yeh. Finite element modeling of heat and mass transfer in food materials during microwave heating—model development and validation. J Food Eng 25:509–529, 1995.
18. D-S Chen, RK Singh, K Haghighi, PE Nelson. Finite element analysis of temperature distribution in microwaved particulate foods. J Food Eng 18:351–368, 1993.
19. D Burfoot, CJ Railton, AM Foster, R Reavell. Modeling the pasteurisation of prepared meal with microwave at 896 MHz. J Food Eng 30:117–133, 1996.
20. M Sundburg, PO Risman, T Ohlsson. Analysis and design of industrial microwave ovens using the finite-difference time-domain method. J Microwave Power Electromagn Energy 31(3):142–157, 1996.
21. G Bellanca, S Botti, P Bassi, G Falciasecca. Sensitivity of fdtd simulations to small mesh modifications in microwave oven design. In Microwave and High Frequency Heating '97, 1997 pp. 60–63.
24. SR Lobo. Characterization of spatial non-uniformity in microwave reheating of high loss foods. Master's thesis, Cornell University, 1988.
23. CK Wei, HT Davis, EA Davis, J Gordon. Heat and mass transfer in water-laden sandstone: Microwave heating. AIChE J 31(5):842–848, 1985.

24. F Peyre, AK Datta, CE Seyler. Influence of the dielectric property on microwave oven heating patterns: Application to food materials. J Microwave Power Electromagn Energy 32(1):3–15, 1997.
25. JJ Dolande, AK Datta. Temperature profiles in microwave heating of solids: A systematic study. J Microwave Power Electromagn Energy 28(2):58–87, 1993.
26. H Ni, AK Datta, R Parmeswar. Moisture loss as related to heating uniformity in microwave processing of solid foods. J Food Process Eng 22(5):367–382, 1999.
27. M Chamchong, AK Datta. Thawing of foods in a microwave oven: I. Effect of power levels and power cycling. J Microwave Power Electromagn Energy 34(1):9–21, 1999.
28. M Chamchong, AK Datta. Thawing of foods in a microwave oven: Effect of load geometry and dielectric properties. J Microwave Power Electromagn Energy 34(1):22–32, 1999.
29. H Ni, AK Datta, KE Torrance. Moisture transport in intensive microwave heating of wet materials: A multiphase porous media model. Int J Heat Mass Transfer 42(8):1501–1512, 1999.
30. AK Datta, H Ni. Infrared and hot air additions to microwave heating of foods for control of surface moisture. J Food Eng, 1999.
31. J. Bear. Dynamics of Fluids in Porous Media. New York: American Elsevier, 1972.
32. RB Bird, WE Stewart, EN Lightfoot. Transport Phenomena. New York: John Wiley & Sons, 1960.
33. H Ni. Multiphase moisture transport in porous media under intensive microwave heating. PhD thesis, Cornell University, 1997.
34. SB Nasrallah, P Perre. Detailed study of a model of heat and mass transfer during convective drying of porous media. Int J Heat Mass Transfer 31(5):957–967, 1988.
35. KP Nott, LD Hall, JR Bows, M Hale, ML Patrick. Three-dimensional mri mapping of microwave induced heating patterns. Int J Food Sci Technol 34: 305–315, 1999.
36. H Ni, AK Datta, KE Torrance. Moisture transport in intensive microwave heating of wet materials: A multiphase porous media model. Int J Heat Mass Transfer, 1998.
37. AK Datta, H Prosetya, W Hu. Mathematical modeling of batch heating of liquids in a microwave cavity. J Microwave Power Electromagn Energy 27(1):38–48, 1992.
38. AK Datta, AA Teixeira. Numerical modeling of natural convection heating in canned liquid foods. Trans ASAE. 30(5):1542–1551, 1987.
39. H Prosetya, AK Datta. Batch microwave heating of liquids: an experimental study. J Microwave Power Electromagn Energy 26(3):215–226, 1991.
40. RC Anantheswaran, L Liu. Effect of viscosity and salt concentration on microwave heating of model non-newtonian liquid foods in a cylindrical container. J Microwave Power Electromagn Energy 29(2):119–126, 1994.

41. RC Anantheswaran, L Liu. Effect of electrical shielding on time–temperature distribution and flow profiles in water in a cylindrical container during microwave heating. J Microwave Power Electromagn Energy 26(3):156–159, 1991.
42. Y Emami, T Ikeda. System and method for sterilization of food material. UK Patent Application GB2 098 040, 1982.
43. AK Datta, J Liu. Thermal time distributions in microwave and conventional heating. Trans Inst Chem Eng 70(C), 83–90, 1992.
44. HS Ramaswamy, S Tajchakavit. Continuous-flow microwave heating of orange juice. ASAE Paper 93-3588, 1993, pp. 1–16.
45. T. Ohlsson. In-flow microwave heating of pumpable foods. In International Congress on Food and Engineering, 1993, pp 1–7.
46. JJR Thomas, EM Nelson, RJ Kares, R. Stringfield. Temperature distribution in a flowing fluid heated in a microwave resonant cavity. Mater Res Soc Symp Proc 430:565–569, 1996.
47. J Houben, L Schoenmakers, E van Putten, P van Roon, B Krol. Radiofrequency pasteurization of sausage emulsions as a continuous process. J Microwave Power Electromagn Energy 26(4):202–205, 1991.

7

Infrared Heating of Biological Materials

Oladiran O. Fasina
U.S. Department of Agriculture–Agricultural Research Service and North Carolina State University, Raleigh, North Carolina

Robert Thomas Tyler
University of Saskatchewan, Saskatoon, Canada

1 INTRODUCTION

Infrared refers broadly to that portion of the electromagnetic spectrum starting at the deep red (the point at which light just begins to become visible, hence the name infrared) and extending to the microwave radar region. As shown in Figure 1, the relative position of infrared region of the electromagnetic spectrum is in the wavelength range of 0.75 to 1000 μm. Infrared waves are described as short, medium, or long wave. Short infrared waves (or near infrared) are closest to visible light. Because much of this energy is light, it is easily reflected. Short infrared waves occupy the region of the electromagnetic spectrum in the wavelength between 0.75 and 3.0 μm. The long infrared (or far infrared) waves, spanning the wavelength region of 25–1000 μm, are readily absorbed by most materials as heat. The medium waves (or middle infrared) occupy the region between the short infrared and long infrared regions [1].

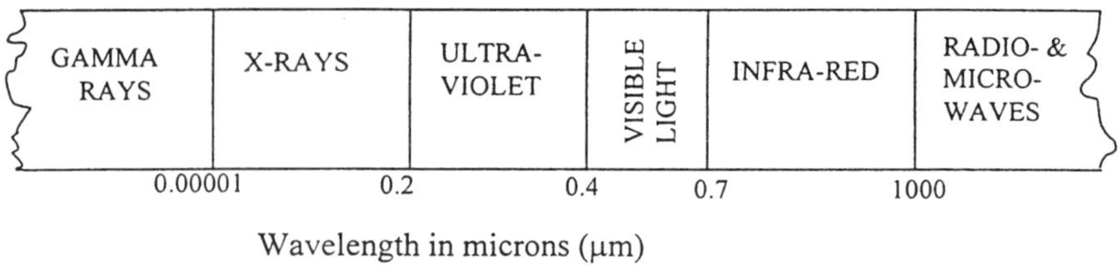

FIGURE 1 The electromagnetic spectrum.

When radiant electromagnetic energy impinges upon a food surface, it may induce changes in the electronic, vibrational, or rotational states of atoms and molecules [2]. The types of mechanisms for energy absorption are determined by the wavelength range of the incident energy. Changes in the electronic state correspond to wavelengths in the range between 0.2 and 0.7 μm (ultraviolet and visible rays); changes in the vibrational state correspond to wavelengths in the range 2.5–100 μm (part of infrared region); and changes in the rotational state correspond to wavelengths above 100 μm (microwaves). Infrared radiation causes molecular vibration changes; hence, heating occurs when biological materials are exposed to infrared radiation.

The fundamental relationships of infrared energy are established by three basic laws [Eqs. (1)–(3)] that determine the distribution and quantity of infrared energy. The laws are written for a blackbody, which assumes that a surface will absorb all and reflect none of the radiation falling on it. An ideal "blackbody" is a surface that absorbs and in turn radiates all the energy incident upon it.

Stefan–Boltzmann Law

$$W = \sigma T_{\mathrm{ab}}^4 \tag{1}$$

Wien's Displacement Law

$$\lambda_{\max} = \frac{2897.6}{T_{\mathrm{ab}}} \tag{2}$$

Planck's Equation

$$E = \frac{C_1 \lambda^{-5}}{\exp(C_2/\lambda T) - 1} \tag{3}$$

where the first (C_1) and second (C_2) radiation constants have values of $3.742 \times 10^8\ \mathrm{W\,\mu m^4/m^2}$ and $1.469\ \mathrm{\mu m\,K}$, respectively. The symbols are defined in Section 6 nomenclature.

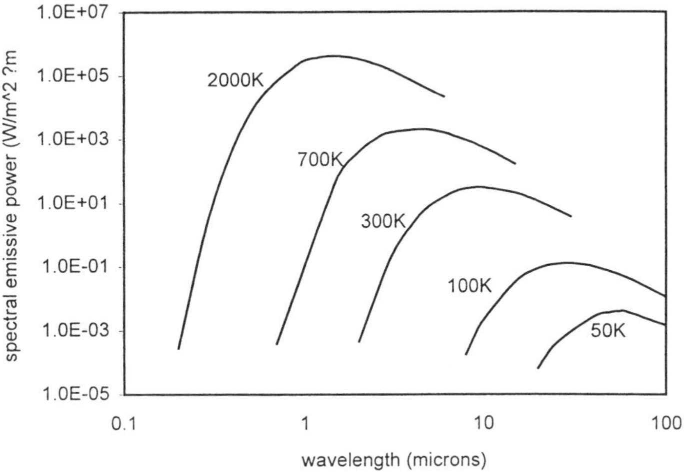

Figure 2 Spectral blackbody emissive power.

The Stefan–Boltzmann law and Planck's equation provide the means for determining the intensity and spectral distribution of the emission from a radiator. The Stefan–Boltzmann law shows that the intensity of heat at the surface of a body exposed to infrared radiation is proportional to the fourth power of temperature. The Wein's displacement law states that the peak wavelength (λ_{max}) varies inversely with the absolute temperature of the radiating object. Figure 2 shows the emissive power curves of blackbody radiation versus wavelengths at different temperatures. The hotter the object, the shorter the wavelength of infrared radiation. The total emitted energy is the integral or the area under the curve shown in Figure 2 and can be calculated using the Stefan–Boltzmann relationship.

Because no object is a perfect emitter (i.e., a blackbody), real substances are characterized by an efficacy of radiant emission called emittance or emissivity (ε). The radiant energy flux for real objects is then calculated from

$$W = \varepsilon \sigma T_{\mathrm{ab}}^4 \tag{4}$$

Emissivity varies with wavelength and temperature. The common practice in engineering, however, is to represent the emissivity for the entire band of the electromagnetic spectrum involved rather than for a particular wavelength. Emissivity data for biological materials is scarce. Typical values of emissivity for agricultural crops vary from 0.7 to 0.9 [3,4]. Where there are no data for a particular application, an emissivity value of 0.9 is often used.

2 INFRARED HEAT GENERATION

Most generators of infrared energy are either electrically heated or gas fired. The electrical generators of infrared radiation include quartz lamp, tungsten arc lamp, xenon arc lamp, nonsheathed radiator, and resistance element (Table 1). For electrically heated radiators, infrared radiation is obtained by passing an electric current through an element [6]. Quartz, tungsten, and xenon lamps generally have maximum radiation at wavelength less than 1.3 μm. They are therefore referred to as light (short-wave) radiators. These lamps emit at temperatures of 1773–2073 K [7]. Resistance elements and gas-type generators are generally dark (long-wave) radiators because they have maximum radiation in the invisible infrared rays (> 1.3 μm). Gas-fired generators are made of perforated plate (metal or refractory) that is heated by gas flames in one of the surfaces, thereby causing the plate to rise in temperature and emits radiant energy [8]. The characteristics of commercially used infrared heat sources are compared in Table 1 [5].

3 APPLICATIONS TO BIOLOGICAL MATERIALS

Sun drying is the oldest method that has been used to dry agricultural products. Because most of the radiant energy of the sun is in the infrared region, infrared energy is indirectly the oldest and most traditional energy source for agricultural applications. Despite the historical nature of infrared energy in food preservation, the use of infrared radiation is mostly at the industrial level, such as in drying of coatings (powders, paints, inks, adhesives, films), in hazardous heating (space heating for oil and gas and petrochemical industries) and in electronics and metal processing applications.

Interest in the use of infrared heating in food processing has increased in the past few years due to recent developments in the design of infrared heaters that offer rapid and economical methods for production of food products with high organoleptic and nutritional value. The most significant advantage of infrared drying when used for drying is the reduction in drying time. Other advantages of infrared heating include the following [2,8]:

(a) High efficiency to convert electrical energy into heat when electric heaters are used.
(b) Efficient heat transfer to the food reduces processing time and energy costs.

Table 1 Characteristics of Commercially Used Infrared Heat Sources

Infrared source		Source temperature: Usual range (K)	Source temperature: Max (K)	Peak wavelength (μm)	Power (kW/m^2)
Electrically heated radiators					
Nonsheathed radiators	Sylite	1,750–1,800	2,200	1.65	Up to 80
	Graphite	2,300–2,800	3,500	1.2	Up to 1200
	Metallic–filament tungsten	1,900–2,200	2,700	1.2	$(1–1.4) \times 10^5$
	Metallic–molybednum	1,600–2,000	2,000	0.9	$(1–2) \times 10^5$
Sheathed radiators	Light bulbs	1,900–2,500	2,500	1.3	Up to 20
	Quarta lamp	1,900–2,500	2,800	1.0	30–400
	Plate radiators	700–1,200	1,200	4.0–9.0	4–14
	Xenon arc lamp	5,000–10,000	10,000	0.8–1.1	Up to 50
	Tungsten arc lamp	3,200–4,000	7,000	0.72	Up to 1400
Gas-heated					
Flame	Direct flame (Bunsen, Teclu, or Mecker burner)	500–1,600	1,800	2.8–4.3	20–30
	Indirect flame—ceramic element	600–800	1,500	4.0	50–60
	Indirect flame—metallic element	300–900	1,000	3.6	20–30
Flameless	Heated porous plate with internal burning	350–850	1,200	4.0	40–90
	Heated porous plate with external burning	1,000–1,700	2,000	1.5–2.0	160–2400

Source: Ref. 5.

(c) The air surrounding the equipment is maintained at ambient level.
(d) Infrared heaters are less expensive when compared to dielectric and microwave sources and they have longer service life and low maintenance.
(e) Surface irregularities on foods have insignificant effect on infrared heating—uniform heating of product is easily achieved.

Some of the disadvantages of infrared heating are (a) proper scaling up of heaters from laboratory model to full-plant model, and (b) infrared heating is essentially a surface heating method and is therefore best for thin materials.

Infrared heat is generally applied to biological materials in order to achieve thermal effects such as controlling insect infestation in stored product, inactivation of toxic and antimicrobial factors and degradative enzymes, reduction of microbial counts, enhancement of the dehulling of legume grains, and starch gelatinization in starch bearing materials [7,9–10]. The determination of appropriate equations to describe a process requires an understanding of the physical, chemical, and microbiological changes that occur when the process is applied to biological materials. In this section, some of the applications of infrared heating in food and agricultural industries are discussed. Examples of the various changes that occur in infrared heated foods are also presented.

3.1 Applications Involving Insect Disinfestation

Kirkpatrick [11] showed a 99% death rate of *Sitophilus oryzae* and a 93% death rate of *Rhyzopertha dominica* when insect-infested wheat samples were exposed to infrared radiation. The temperature of the wheat samples increased to 48.6°C. In another study, Kirkpatrick et al. [12] found that the natural infestations of stored wheat by the weevil *S. oryzae*, the grain borer *R. dominica*, *Crypolestes pusillus Schonh*, and *Tribolinum castaneum* were controlled by raising sample temperature to 55°C. Despite these encouraging results, there is no evidence that infrared heating is used commercially to disinfect food and agricultural materials. This is probably due to the limited use of infrared heating in the food industry at the time these studies were conducted. Due to the energy crisis of the 1970s, it was less expensive for food manufacturers then to use chemicals for food preservation.

3.2 Applications Involving Legume and Oil-Bearing Materials

Most of the commercial use of infrared heat processing in the food industry involves the inactivation of antinutritional factors in legume seeds (mostly soybeans) and enzymes that cause product degradation and development of rancidity. Several studies carried out by researchers at the Agricultural University in Wageningen, the Netherlands [13–16], showed that infrared heating can be used to improve the nutritive value of soybeans. The researchers, in addition to results from other published studies [17–19], found that infrared radiation can be used to inactivate lipoxygenase enzyme (that causes oxidative rancidity), reduce the trypsin inhibitor and other antinutritional factor levels, and increase the binding, emulsion power, water holding capacity, and shelf life of full-fat soybean flour. This has generally led to a longer shelf life of the product. Perhaps the most important conclusion made by the Wageningen researchers is that infrared treatment of soybeans offers the possibility for reducing energy requirements and production costs in comparison to the conventional steam-heating method used for soybean processing. The researchers showed that soybeans can be infrared heated to surface temperatures of 125–133°C for 60 sec. Steam heating is usually carried out at temperatures of 110–125°C for 20–30 min [20–22]. In addition, steam-heated samples have to be dried after treatment thus increasing processing cost.

When cocoa beans were infrared heated prior to dehulling, there was a significant improvement in winnowing performance during the separation of nib or beans from the shell. The shells became lighter due to expansion and are thus more effectively removed during air separation [20–22]. In addition, bacteria and contamination levels were reduced by 95%. The effect of infrared heating on the microbial counts of cocoa nibs is shown in Table 2 [9]. Infrared heating of the nibs was carried out for 10 sec under a ceramic plate heated to 970°C.

Cenkowski and Sosulski [23] investigated the effect of infrared heating on the physical and cooking properties of lentils. They found that cooking time was shortened from 30 mm for the controlled seeds to 15 mm for lentils adjusted to 25.8% moisture content and infrared heated to 55°C. Infrared

TABLE 2 Effect of Infrared Treatment of Cocoa Nibs on Microbial Counts

	Before infrared treatment (counts/g)	After infrared treatment (counts/g)
Total count	5×10^6	2×10^5
Enterobacteria	10^4	10
Yeasts	8×10^4	$<10^2$
Molds	6×10^4	$<10^2$

Source: Ref. 9.

heating was effective in the gelatinization and solubilization of the starch. When maize germ was infrared heated at different temperatures (98–118°C), the degree of starch gelatinization and water absorption increased and the protein dispersibility and enzyme activity decreased with higher infrared temperatures [24]. The processed wheat germ samples were able to keep for 12 months before any appreciable change in rancidity (free fatty acid [FFA] and peroxide value) occurred.

Fasina et al. [25] investigated the effect of infrared heat treatment on the physical, chemical, mechanical, and functional characteristics of five legume seeds (kidney beans, green peas, black beans, lentil, and pinto beans). Within a duration of 15 sec or less, the seeds were heated to a surface temperature of 140°C. Significant changes (Table 3) in the properties of the seeds in terms of increased volume, lower rupture point and toughness, higher water uptake, and higher leaching losses (when seeds were soaked in water) were obtained in comparison to unprocessed seeds. The changes in the physical and mechanical properties of the seeds were attributed to seed cracking during infrared heating. The authors also found the functionality of the flour (pasting characteristics and protein solubility)

TABLE 3 Water Uptake and Loss of Solubles When Raw and Heated Legume Seeds Were Soaked in Water

Legume sample	Absorbed moisture (g/100 g seed)	LL[a] (g/100 g seed)
Kidney beans		
Raw	133.6	1.1
Processed	137.5	10.3
Green peas		
Raw	119.8	5.5
Processed	118.2	11.0
Black beans		
Raw	130.7	1.7
Processed	144.3	11.1
Lentil		
Raw	136.4	2.5
Processed	140.3	11.5
Pinto bean		
Raw	130.8	3.3
Processed	141.7	11.7

[a] Leaching losses
Source: Ref. 25.

Table 4 Pasting Characteristics of Flours from Raw and Infrared-Heated Legume Seeds

Legume sample	Peak height (BU[a])	95°C height (BU)	15 min height (BU)	50°C height (BU)
Kidney bean				
Raw	180	10	60	180
Processed	220	20	120	220
Green pea				
Raw	30	10	20	30
Processed	120	10	100	120
Black bean				
Raw	50	10	20	50
Processed	280	40	120	280
Lentil				
Raw	540	200	360	520
Processed	420	300	410	420
Pinto bean				
Raw	100	20	100	80
Processed	460	60	220	400

[a] BU-Brabender unit.
Source: Ref. 25.

from infrared-heated seeds were superior to those of flour from unprocessed seeds. Table 4 shows the improvement in pasting characteristics of flour from infrared-heated legume seeds; Figure 3 shows the change in protein solubility of the legume seeds due to infrared heating.

3.3 Applications Involving Cereal Grains

The most common reason for applying infrared heating to cereal grains is to gelatinize the starch in them [27–32]. The processed grains are often used in the production of ready-to-eat breakfast cereals, as cereal adjunct in the brewing industry, or as livestock feed. However, there is no substantial evidence showing that significant improvement is obtained when infrared heated cereal grains are fed to livestock [33–35]. Other changes that are achieved when cereal grains are infrared heated include reduction in protein solubility [26, 36], inactivation of peroxidase enzyme in oat flakes [37], reduction of microbial counts on edible wheat bran [38], and reduction of tannin content in sorghum [39]. Figures 4–7 and Tables 5–7 show the changes in the physical and functional characteristics of hull-less and pearled barley infrared heated at different temperatures and moisture content [32].

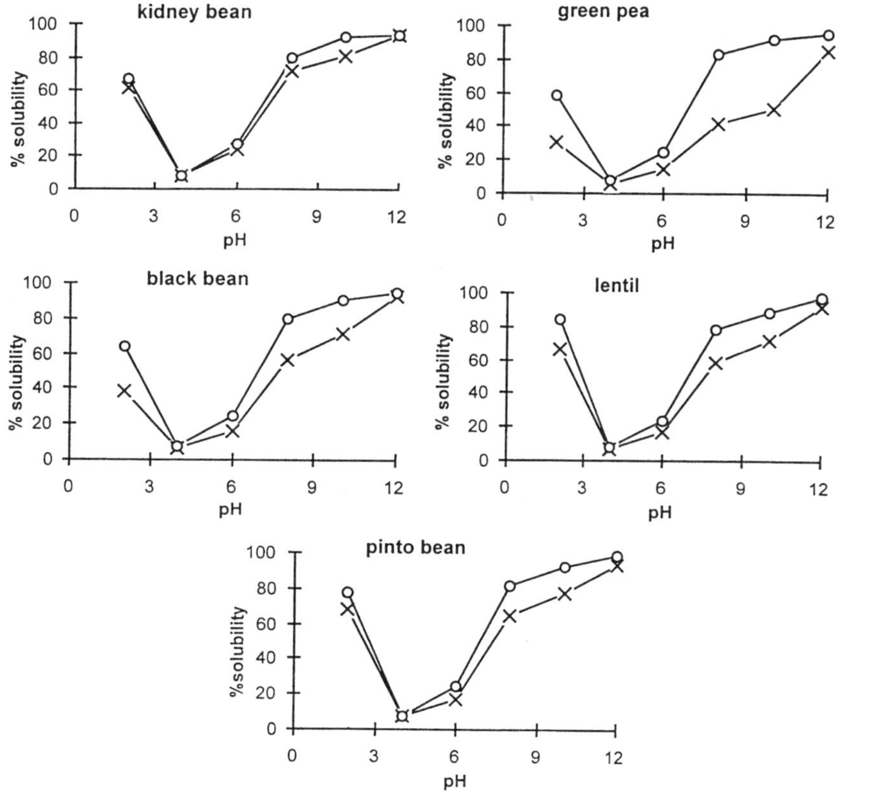

FIGURE 3 Protein insolubility(percent) of legume seeds infrared heated at 140°C ○: unprocessed; ×: processed. (From Ref. 26.)

3.4 Other Applications

In the last decade, there has been increased diversity in the use of infrared radiation in the food industry. Some of the studies that have been reported in the literature include pasteurization of packaged foods [1], heating of oysters and eggs [2], thawing of frozen foods [9], drying of various agricultural materials [40,41], surface sterilization of vegetables [42], enhancement of value of tofu and soy milk [43], and browning of foods [44]. Infrared lamps are now integral parts of buffet tables and cafeteria lines because they are used to keep food hot and to give food a rich red color while under the lamps [28]. Infrared lamps are also used in poultry barns as a heat source for brooding chicks and turkey poults.

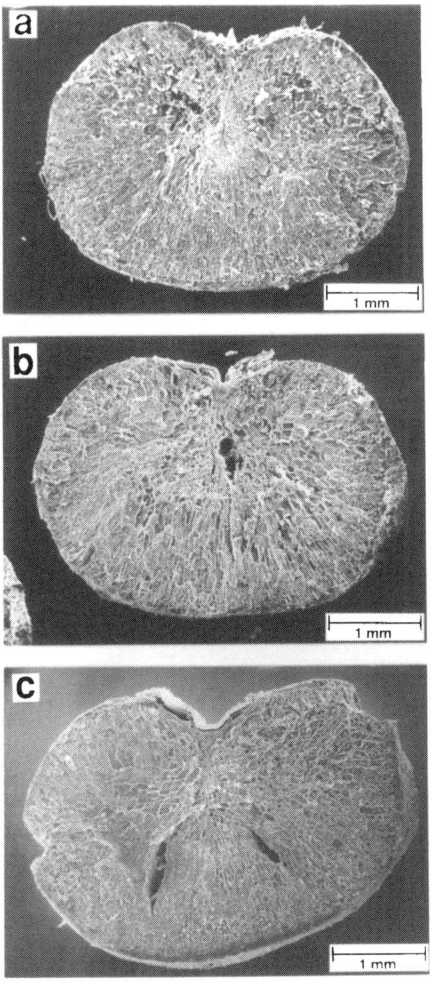

Figure 4 Scanning electron micrographs of cross sections of hull-less barley showing kernel expansion due to micronization: (a) unprocessed; (b) initial moisture content of 13.3%, temperature of 150°C; (c) initial moisture content of 26.5%, temperature of 115°C. (From Ref. 32.)

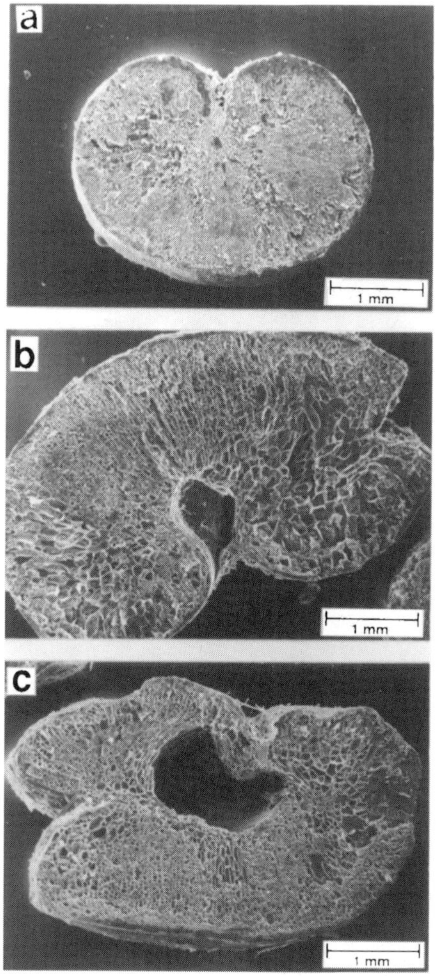

Figure 5 Scanning electron micrographs of cross sections of pearled barley showing kernel expansion due to micronization: (a) unprocessed; (b) initial moisture content of 12.2%, temperature of 150°C; (c) initial moisture content of 25.9%, temperature of 115°C. (From Ref. 32.)

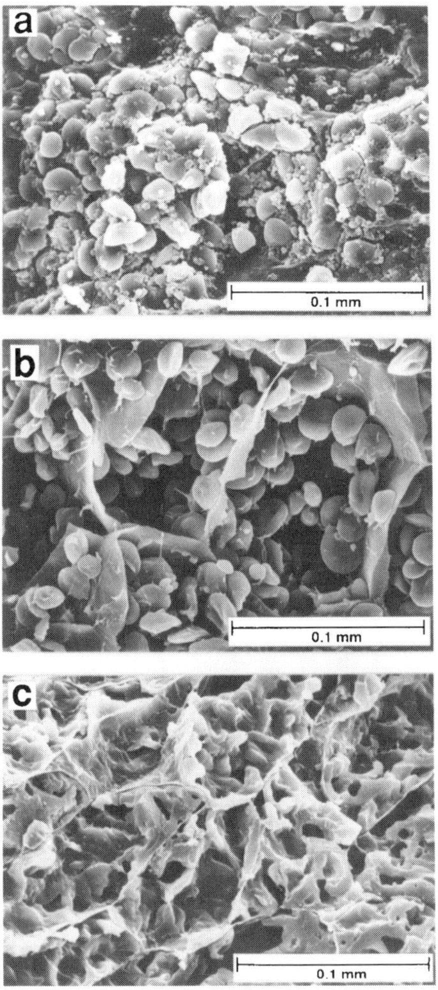

FIGURE 6 Scanning electron micrographs of cross sections of hull-less barley showing changes in microstructure due to micronization: (a) unprocessed; (b) initial moisture content of 13.3%, temperature of 150°C; (c) initial moisture content of 26.5%, temperature of 115°C. (From Ref. 32.)

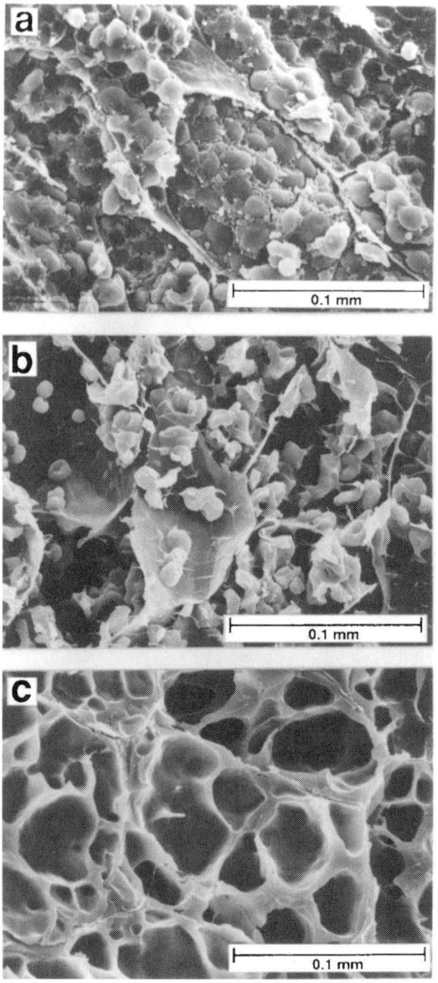

FIGURE 7 Scanning electron micrographs of cross sections of pearled barley showing changes in microstructure due to micronization: (a) unprocessed; (b) initial moisture content of 12.2%, temperature of 150°C; (c) initial moisture content of 25.9%, temperature of 115°C. (From Ref. 32.)

TABLE 5 Percent Amount of Starch Gelatinized[a] in Hull-less and Pearled Barley Samples During Infrared Heating

Kernel surface temp. (°C)	Hull-less		Pearled	
	Initial m.c.[b] (%, w.b.[c])	Gelatinized starch (%)	Initial m.c. (%, w.b.)	Gelatinized starch (%)
115		1.5		0.1
135	13.3[d]	11.2	12.2[d]	2.9
150		47.4		14.0
105		14.3		1.9
115	19.2	34.3	19.3	20.5
135		85.3		44.8
105	26.5	84.1	25.9	26.5
115		93.8		67.3

[a] Maximum deviation of 0.8 from mean values.
[b] m.c. = moisture content.
[c] w.b. = wet
[d] Unconditioned.
Source: Ref. 32.

TABLE 6 Thermal Parameters for Infrared-Heated Hull-less and Pearled Barley Flours Showing the Effect of Kernel Surface Temperature

Temp. (°C)	Hull-less T_p (°C)	ΔH (J/g)	Pearled T_p (°C)	ΔH (J/g)
Raw	67.6	8.02	65.0	7.58
115	67.3	7.80	65.1	7.47
135	67.3	7.30	65.1	7.23
150	67.3	4.85	66.1	6.93

Note: Kernels had initial moisture contents of 12.2%.
Source: Ref. 25.

Even though this is not the focus of this chapter, we should mention that a major use of infrared radiation in the food industry is in the area of analytical measurements. Modern analytical instrumentations that are based on infrared energy measurements are being used for rapid and routine analyses that are needed in the formulation, quality control and development of food products.

TABLE 7 Water Hydration Capacities (WHC)[a] of Flours from Infrared-Heated Hull-less and Pearled Barley

Kernel surface temp. (ωC)	Hull-less		Pearled	
	Initial m.c.[b] (%, w.b)	WHC (%)	Initial m.c. (%, w.b.)	WHC (%)
Unprocessed		272.2		241.8
115		319.7		261.9
135	13.3[d]	326.8	12.2[d]	300.8
150		429.3		288.8
105		330.8		276.4
115	19.2	347.9	19.3	329.4
135		642.4		434.9
105	26.5	760.1	25.9	356.1
115		857.1		497.6

[a] Maximum deviation of 9.3 from mean values.
[b] m.c. = moisture 8 content.
[c] w.b. = wet basis.
[d] unconditioned.
Source: Ref. 32.

4 MODELING THE INFRARED HEATING OF BIOLOGICAL MATERIALS

4.1 Model Development

As mentioned in the previous sections, infrared heating is becoming an important source of heat treatment in the food industry because of advantages such as simplicity of equipment, fast transient response, significant energy savings, and easy accommodation with convective, conductive, and microwave heating [1]. Factors such as type of raw material (e.g., cereal grain, oil seed, legume), infrared burner temperature, height of infrared burner from the grain bed, residency time of grain under infrared heat, and the initial moisture content of the sample will affect final product temperature and moisture content. The optimum combination of these factors can be obtained by complete experimentation or by the use of computer modeling and simulation with subsequent verification of the models by experimental results. The latter is preferable based on cost and time considerations [45].

Despite the wide application of infrared heating in the food and feed industries, few studies have been reported on transfer of heat and mass in agricultural crops subjected to this heating method. Zuilichem et al. [9] presented the following set of heat transfer equations to describe the infrared heating of agricultural seeds:

$$\rho c_g \frac{\partial T}{\partial t} = k\left(\frac{\partial^2 T}{\partial r^2} + \frac{2}{r}\frac{\partial T}{\partial r}\right). \tag{5}$$

Equation (5) was solved by applying the Fourier integral and Fourier series to obtain

$$T - T_e = \frac{2h(T_0 - T_e)}{T} \sum_{h=1}^{\infty} \exp(-a\alpha_n^2) \times \frac{A^2\alpha_n^2 + (Ah-1)^2}{\alpha_n^2[A^2\alpha_n^2 + Ah(Ah-1)]} \sin A\alpha_n \sin r\alpha_n \tag{6}$$

where $\alpha_n (n = 1, 2, 3, \ldots)$ are the roots of the equation

$$A\alpha_n \operatorname{cotg}(A\alpha_n) + (Ah - 1) = 0 \tag{7}$$

$$h = \frac{\sigma F(T_e^4 - T_{\text{av}}^4)}{k(T_e - T_s)} \tag{8}$$

$$T_{\text{av}} = T_i + \frac{\sigma T_e^4}{\rho c_g d} t \tag{9}$$

In developing Eqs. (5)–(9), the authors neglected mass transfer and convection of applied heat to the ambient air surroundings. Parameter F in Eq. (8) was defined as the configuration factor. If this is the case, the calculation of the radiation constant h will be in error because the emissivity of the product is not included in the equation. The appropriate equation to use in the calculation of h (the radiative heat transfer coefficient) will be presented later in this section. Unfortunately, the predictions the authors obtained from these Eqs. (5)–(8) were not validated with experimental results.

There is substantial evidence in the literature indicating that agricultural products experience an increase in temperature and lose moisture when exposed to infrared radiation. Zheng et al. [26] showed that unconditioned legume seeds (initial moisture content 9.5–10%) lose about 2 percentage point of moisture when infrared-heated to a surface temperature of 115°C. When lentil seeds with initial moisture contents of 22.5%, 25.8%, and 38.6% were infrared-heated, the measured final seed temperature were inversely related to initial moisture content [23]. After about 60 sec of infrared heating, the moisture contents of the seeds were respectively reduced to 11%, 13%, and 28%. The corresponding final seed temperature were measured to be 110°C, 125°C, and 130°C. Stephenson and McKee [46] investigated the possibility of using infrared heating for the accelerated drying of oats and barley. The moisture content of oats and barley when infrared-heated to surface temperatures of 100°C and 110°C reduced from 20% to 12.1% for oats and from 17.2% to 11.8% for barley.

As mentioned in Section 1, when an infrared wave impinges on a surface, the molecules at the product surface vibrate at a frequency of 8.8×10^7 to 1.7×10^8 MHz (corresponding to wavelength of 1.8–3.4 μm that is typically used in the food industry). The friction of the intermolecular structure of the product will cause a temperature rise, hence the heating and moisture loss at the surface [47]. Heat passes into the food by conduction and moisture migration to the product surface occurs. Kuang et al. [48] studied the transfer of heat and mass during the use of infrared heating to dry paper. The set of equations used for mass transfer were based on capillary moisture movement. Ratti and Mujumdar [7] presented a set of equations to describe the infrared drying of foodstuff. A lumped moisture model (i.e., no resistance to internal moisture movement) was used to describe the mass transfer component. Numerous studies have shown that the transfer of mass in agricultural crops is mainly a diffusion process [49–51]. In diffusion processes, the mass transfer models take into account the internal and external resistances of the single grain particle to moisture movement [52].

Irudayaraj et al. [53] modeled the drying of cereal grains using two models. The first model assumed that diffusion of moisture occur through vapor and liquid phases within the grain kernel. Luikov's coupled system of partial differential equations [54] for heat and mass transfer in porous media was used to describe the diffusion process. The second model used by the authors assumed that moisture diffuses to the outer boundaries of the kernel in liquid form and that evaporation takes place only at the surface of the grain (conduction model). Simulation results showed that temperature predictions by the Luikov's model approached the equilibrium temperature faster than the conduction model. The author, however, found that the overall grain kernel temperature and moisture predictions from the conduction model were better than that of the model.

The conduction model was used by Fasina et al. [55] to model temperature changes and moisture loss in barley grains subjected to infrared heating. The equations which take into account the internal and external resistances to transport processes are given by the following:

Mass transfer

$$\frac{\partial M}{\partial t} = \frac{1}{r^2} \frac{\partial}{\partial r} D_m r^2 \frac{\partial M}{\partial r} \tag{10}$$

Heat transfer

$$\rho c_g \frac{\partial T}{\partial t} = \frac{1}{r^2}\frac{\partial}{\partial r} kr^2 \frac{\partial T}{\partial r} \tag{11}$$

The initial and boundary conditions to Eqs. (10) and (11) are given as

$$M = M_i \qquad \text{at } t = 0,\ 0 < r \leq R \tag{12}$$

$$T = T_i \qquad \text{at } t = 0,\ 0 < r \leq R \tag{13}$$

$$-D_m \frac{dM}{dr}\bigg|_s = h_m(M_s - M_{\text{eq}}) \qquad \text{at } t > 0,\ r = R \tag{14}$$

$$-kA_g \frac{\partial T}{\partial r}\bigg|_s = q_r + h_c A_g(\theta - T_s) + \rho V h_{\text{fg}} \frac{\partial M}{\partial t} \qquad \text{at } t > 0,\ r = R \tag{15}$$

It is assumed that (a) agricultural materials are opaque to radiation and, therefore, impinged radiation is converted to heat at the surface of the material [7], (b) the flow of material in infrared equipment is in a thin or single-kernel layer, and (c) the kernel is spherical in shape.

The variable q_r in Eq. 15 is the heat radiated to the kernel from (a) the infrared heater directly above the grain bed and (b) the two side plates that enclose the space between the infrared heater and the grain bed. Equations (16) and (17) describe the contribution of each radiative term to the kernel heat and mass transfer phenomena are given below [56,57]; they state that the total resistance to radiation exchange between a surface (emitter or side plate) and the barley grain is comprised of the two surface resistances [the first and third terms of the denominator in Eqs. (16) and (17)] and the geometrical resistance (F_{ge}).

For the infrared heater,

$$q_{r2} = \sigma A_g(T_p^4 - T_s^4)\left(\frac{1-\varepsilon_g}{\varepsilon_g} + \frac{1}{F_{\text{ge}}} + \frac{1-\varepsilon_e}{\varepsilon_e(A_e/A_g)}\right)^{-1} \tag{16}$$

For the two side plates,

$$q_{r2} = 2\sigma A_g(T_p^4 - T_s^4)\left(\frac{1-\varepsilon_g}{\varepsilon_g} + \frac{1}{F_{\text{gp}}} + \frac{1-\varepsilon_p}{\varepsilon_p(A_p/A_g)}\right)^{-1} \tag{17}$$

F_{ge} and F_{gp} are the view or configuration factors between the grain and the heater and plate surfaces, respectively, and were obtained from the following relation for a small sphere (barley) and a rectangular plane (infrared emitter) [58]:

$$F_{gp} \text{ or } F_{ge} = \tan^{-1}\left(\frac{x(y - \cos\theta)}{\sqrt{(1 + x^2 + y^2 - 2y\cos\theta)}}\right) + \tan^{-1}\left(\frac{x\cos\theta}{\sqrt{(1 + x^2)}}\right) \tag{18}$$

where $x = b/c$ and $y = a/c$. The constants a and b are the width and length of the rectangular plane, respectively. The constant c is the distance from the sphere to the rectangular plane.

4.2 Numerical Simulation

4.2.1 Mass Transfer

The mass transfer equation [Eq. (10)] can be rewritten as

$$\frac{\partial M}{\partial t} = D_m\left(\frac{\partial^2 M}{\partial r^2} + \frac{1}{r}\frac{\partial M}{\partial r}\right) + \left(\frac{\partial M}{\partial r}\right)^2 \frac{\partial D}{\partial M} \tag{19}$$

Using the central finite difference scheme (nomenclature given in Figure 8),

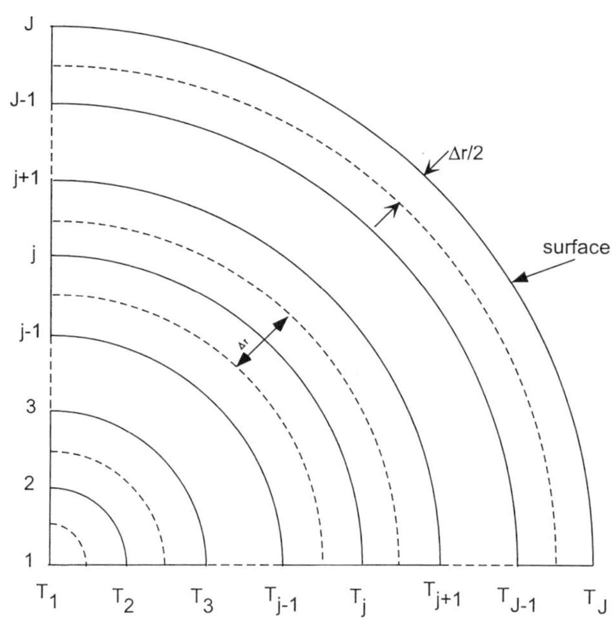

FIGURE 8 Nomenclature used for finite-difference formulation in spherical geometry.

we obtain,

$$M_{j+1}^{n+1} = M_j^n + \Delta t\left[D\left(a + \frac{b}{r}\right) + D'b^2\right] \tag{20}$$

where

$$a = \frac{M_{j-1}^n - 2M_j^n + M_{j+1}^n}{(\Delta r)^2} \tag{21}$$

$$b = \frac{M_{j+1}^n - M_{j-1}^n}{2\Delta r} \tag{22}$$

and

$$D' = \frac{\partial D}{\partial M} \tag{23}$$

When $r = 0$, Eq. (19) has a term with zero denominator. This makes the term $(1/r)(\partial M/\partial r)$ indeterminate. With the assumption that $\partial M/\partial r$ is a continuous function, l'Hospital's rule was therefore applied to obtain

$$\lim_{r\to 0} \frac{1}{r}\frac{\partial M}{\partial r} = \left.\frac{\partial^2 M}{\partial r^2}\right|_{r=0} \tag{24}$$

Substituting Eq. (24) into Eq. (19) with $\partial M/\partial r = 0$ at the center yields

$$\frac{\partial M}{\partial t} = 3D\,\frac{\partial^2 M}{\partial r^2} \tag{25}$$

At the center, $M_{j+1} = M_{j-1}$ and Eq. (25) takes the form

$$M_{j+1} = 6\Delta t D\,\frac{M_2^n - M_1^n}{\Delta r^2} \tag{26}$$

Simpsons's rule was used to obtain the average moisture concentration (M_{av}) using the relation

$$M_{\text{av}} = \frac{1}{(n-1)^3}\sum[(j-2)^2 M_{j-1} + 4(j-1)^2 M_j + j^2 M_{j+1}]$$

$$\text{for } j = 2, 4, 6, \ldots, n-2, n \tag{27}$$

A time step of 0.5 s and a size increase equal to one-tenth of the radius of the barley grain were used as step increase in the simulation program.

4.2.2 Heat Transfer

The heat transfer equation [Eq. (11)] was solved using the thermal resistance concept [59,60]. For a node d contained by the volume element ΔV_d, the steady-state energy balance equation can be written as

$$\sum_{j=1}^{n} \frac{T_i - T_d}{R_{jd}} = \rho c_g \Delta V_d \frac{dT}{dt} \tag{28}$$

The net heat flow to a point d from its surrounding nodes (denoted by j) is equal to the change in internal energy of the mass associated with the point. Applying Eq. (28) to any node j to $j-1$ gives

$$\frac{T_{j-1} - T_j}{B_{j-1,j}} + \frac{T_{j+1} - T_j}{B_{j+1,j}} = \rho c_g \Delta V_d \frac{dT}{dt} \tag{29}$$

where

$$B_{j-1,j} = \frac{\Delta r}{4\pi(j\Delta r - \Delta r/2)^2 k} = \frac{1}{4\pi j^2 \Delta r(1 - 1.2j)^2 k} \tag{30}$$

$$B_{j+1,j} = \frac{\Delta r}{4\pi(j\Delta r_{\Delta} r/2)^2 k} = \frac{1}{4\pi j^2 \Delta r(1 + 1/2j)^2 k} \tag{31}$$

and

$$\Delta V_j = 4\pi(j\Delta r)^2 \Delta r \tag{32}$$

Substituting Eqs. (30)–(32) into Eq. (29), we obtain the following:

$$(1-u)^2 T_{j-1}^n - 2vT_j^n + (1+u)^2 T_{j+1}^n = \kappa(T_j^{n+1} - T_j^n) \tag{33}$$

where

$$u = \frac{1}{2j}$$

$$v = 1 + \frac{1}{4j^2}$$

$$\kappa = \frac{\rho c_g (\Delta r)^2}{k \Delta t}$$

At the grain surface, the heat required to evaporate moisture from the surface is subtracted from the total heat flow into the surface node J. Applying Eq. (28) to Eq. (15) we obtain

$$\frac{T_{J-1} - T_J}{B_{J-1,J}} + h_c A_g(y - T_s) + q_r + \rho V h_{fg} \frac{\partial M}{\partial t} = \Delta V_J \rho c_g \frac{dT_J}{dt} \tag{34}$$

Substituting for $B_{J-1,J}$, A, V, and ΔV_J, Eq. (34) becomes

$$g_1 T_{j-1}^n + g_2 T_j^n + g_3 = \frac{\kappa}{2}(T_j^{n+1} - T_j^n) \tag{35}$$

where

$$g_1 = \left(1 - \frac{1}{2J}\right)^2$$

$$g_2 = -g_1 - \Delta r\left(\frac{h_r - h_c}{k}\right)$$

$$g_2 = \Delta r\left(\frac{h_r}{k}\right)T_e - \Delta r\left(\frac{h_c}{k}\right)T_a + g_4 h_{fg}$$

$$h_r = \frac{q_r}{A_g(T_e - T_s)}$$

$$g_4 = \frac{J(\Delta r)^2 \rho}{3k}\frac{\partial M}{\partial t}$$

At the center where $r = 0$ and applying Eq. (28) yields

$$6(T_2^n - T_1^n) = \kappa(T_1^{n+1} - T_1^n) \tag{36}$$

The nodal temperatures required for numerical solution at time steps n and $n+1$ are therefore given by Eqs. (33), (35), and (36). The Crank–Nicolson method was used to numerically solve the equations. This method is a modification of the implicit method of the finite-difference solution in that stable solutions are obtained even though there are no restrictions on the time step used in the numerical algorithm [61]. Applying this method to Eqs. (33), (35) and (36) yields the following:

$$(1-u)^2 T_{j-1}^{n+1} - 2vT_j^{n+1} + (1+u)^2 T_{j+1}^{n+1} + (1-u)^2 T_{j-1}^n - 2vT_j^n + (1+u)^2 T_{j+1}^n = 2\kappa(T_j^{n+1} - T_j^n) \tag{37}$$

$$g_1 T_{j-1}^{n+1} + g_2 T_j^{n+1} + g_1 T_{j-1}^n + g_2 T_j^n + 2g_3 = \kappa(T_j^{n+1} - T_j^n) \tag{38}$$

$$T_2^{n+1} - T_1^{n+1} + T_2^n - T_1^n = \frac{\kappa}{3}(T_1^{n+1} - T_1^n) \tag{39}$$

Rearranging Eqs (37)–(39) to bring all the times for time $n+1$ to the left-hand side and those for time n to the right hand side, we obtain

$$\left(\frac{(1-u)^2}{2}\right)T_{j-1}^{n+1} + (-\nu - \kappa)T_j^{n+1} + \left(\frac{(1+u)^2}{2}\right)T_{j+1}^{n+1} = \left(-\frac{(1-u)^2}{2}\right)T_{j-1}^n + (\nu - \kappa)T_j^n + \left(-\frac{(1+u)}{2}\right)T_{j+1}^n \tag{40}$$

$$\frac{g_1}{2}T_{j-1}^{n+1} + \left(\frac{g_2 - \kappa}{2}\right)T_j^{n+1} = \frac{g_1}{2}T_{j-1}^{n} + \left(-\frac{g_2 - \kappa}{2}\right)T_j^{n} - g_3 \qquad (41)$$

$$(-3 - \kappa)T_1^{n+1} + 3T_2^{n+1} = (-3 - \kappa)T_1^{n} - 3T_2^{n} \qquad (42)$$

Equations (40)–(42) can be written in a matrix form as follows:

$$[A]\{T\}^{n+1} = [B]\{T\}^{n} + \{C\} \qquad (43)$$

$[A]$ and $[B]$ are square banded matrices whose elements are the constants of Eqs. (40)–(42). The Gauss elimination procedure can be used to solve Eq. (43) to obtain a new set of temperature vector at any time step t.

4.3 Simulation Results

Figures 9 and 10 show the closeness of predicted to experimental data for the surface temperature and average moisture content of barley kernels subjected to infrared heating. Predicted temperature and moisture at the

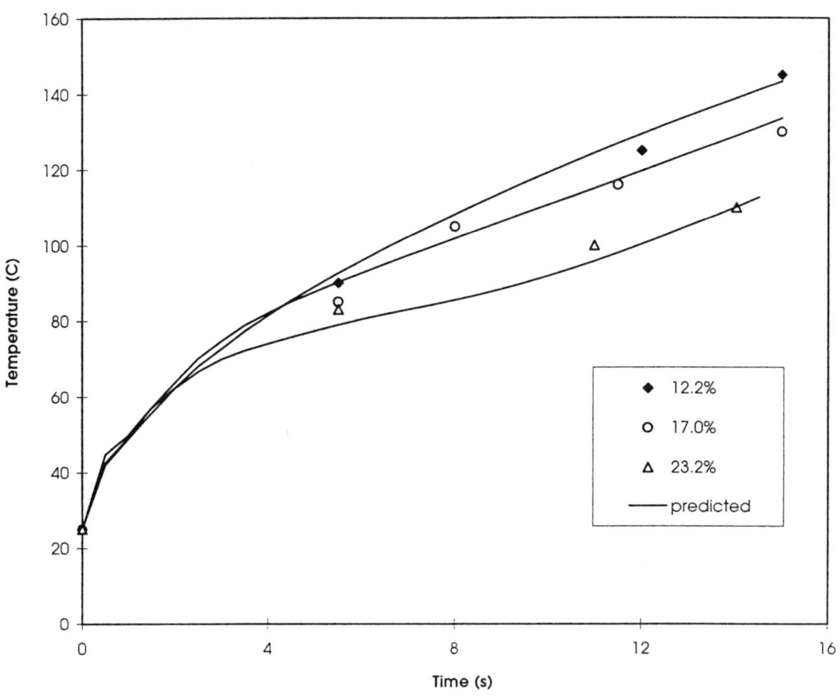

FIGURE 9 Comparison of predicted and experimental surface temperatures of infrared; heated hull-less barley at different initial moisture contents. (From Ref. 53.)

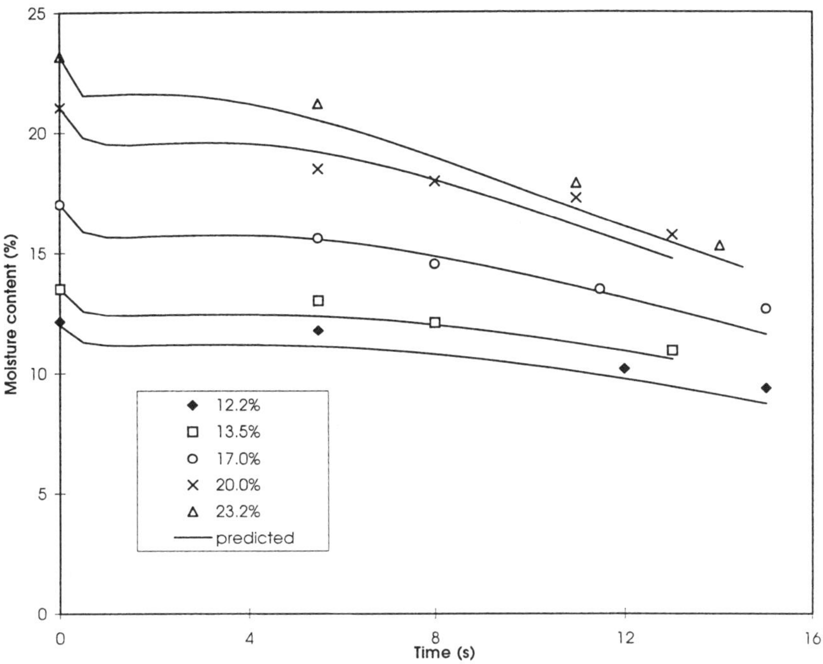

FIGURE 10 Comparison of predicted and experimental average moisture contents of infrared-heated hull-less barley at different initial moisture contents. (From Ref. 53.)

grain surface and center as affected by the initial moisture content are given in Figures 11 and 12. Exposure to infrared heat resulted in an immediate increase in kernel surface temperature. Surface and center temperatures were inversely related to moisture content due to the evaporative cooling effect. The difference in temperature between the surface and center of the kernels varied between 20°C and 45°C during the 15 sec of infrared exposure, supporting the applicability of infrared radiation to applications such as microbial decontamination and dehulling, which involves thermal and moisture treatment.

In order to compare the efficiency of heating of barley grains with infrared heating in comparison with hot-air (conventional) heating, we eliminated the radiative heat term (q_r) from Eq (15) and used hot-air temperature (θ) of 180°C. The value of the convective heat transfer coefficient was initially set at 30 W/m^2 K because this is numerically equal to the value of the radiative heat transfer coefficient that was calculated from Eq. (35) Figures 13 and 14 show that the time required for the grains to attain a

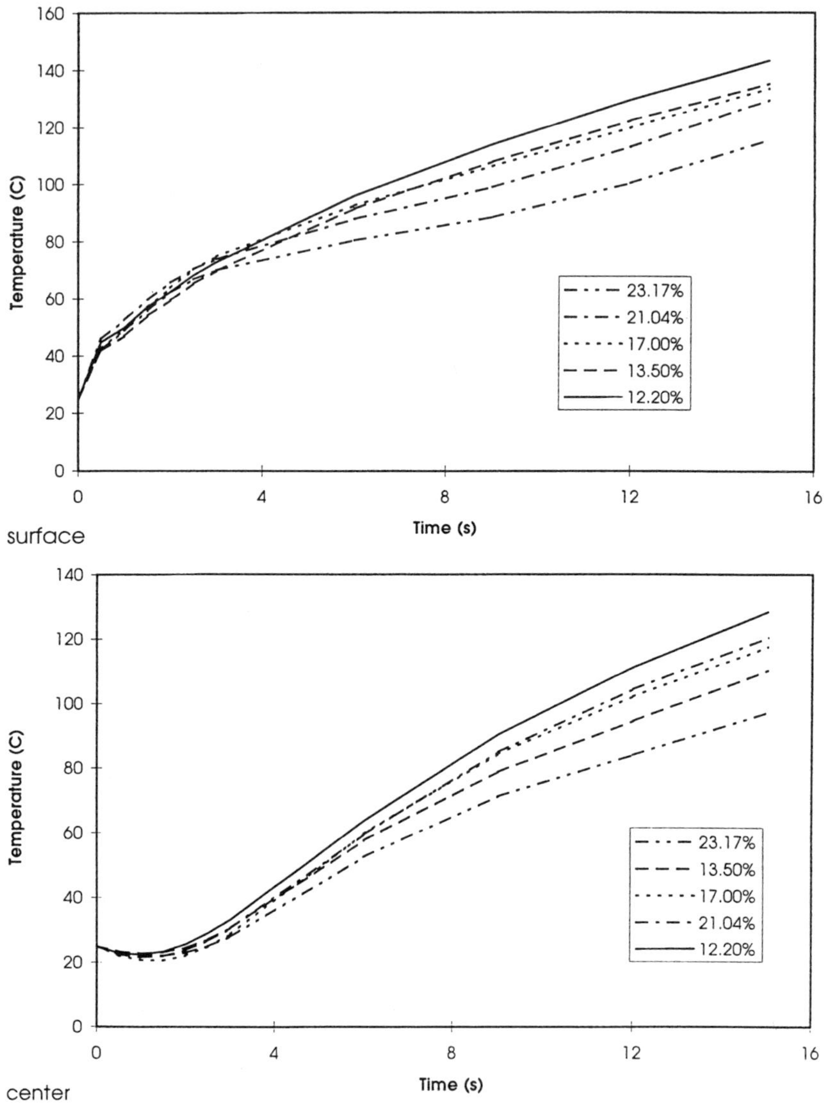

FIGURE 11 Effect of initial moisture content on predicted surface and center temperatures of hull-less barley. (From Ref. 53.)

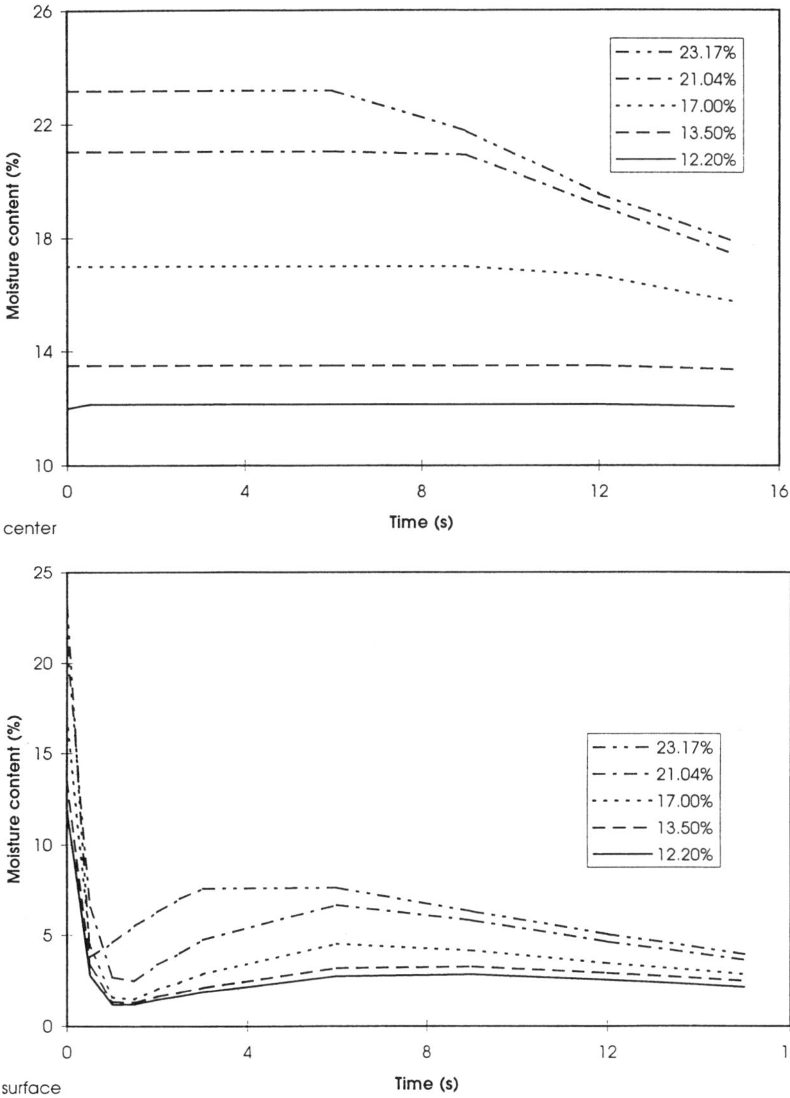

FIGURE 12 Effect of initial moisture content on predicted surface and center moisture contents of hull-less barley (From Ref. 53.)

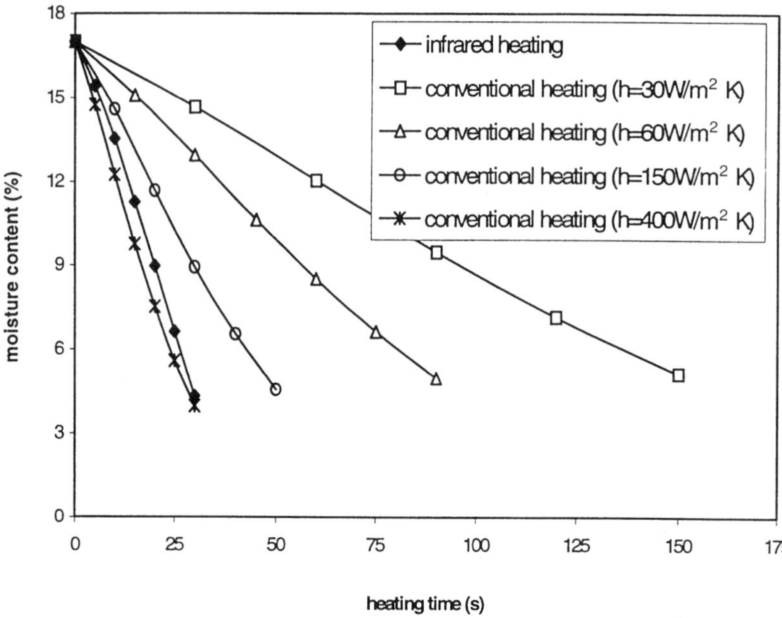

FIGURE 13 Average moisture content in barley grains when infrared heated (using emitter temperature of 850°C) in comparison with conventional heated grains using air temperature of 180°C.

surface temperature of about 170°C with hot-air heating is about 10 times of that required when the grains were infrared heated (30 sec). Similarly, the grains have to be hot-air heated for 150 sec before the average moisture content can be reduced to about 5 percentage points in comparison to 30 s required for infrared heating. Increasing the values of the convective heat transfer coefficient reduces the time required to attain the desired surface temperature of 170°C average moisture content of 5%. The value of the convective heat transfer coefficient required to achieve the desired temperature and moisture content was of 400 W/m^2 K. This is much higher than the h_c range of 50–200 W/m^2 K that is typically used in hot-air application [62]. This supports the claim of infrared heater manufacturers that energy is transferred much faster by infrared radiation. Dostile et al. [63] compared the an infrared dryer to that of convection dryer used for the drying of sheating panels and acoustic tiles. The hot air used for drying in the convection dryer was at a temperature of 170°C. Experimental results showed that it took 20 min to remove about 4 kg/m^2 of moisture from sheating panels using an infrared dryer in comparison to

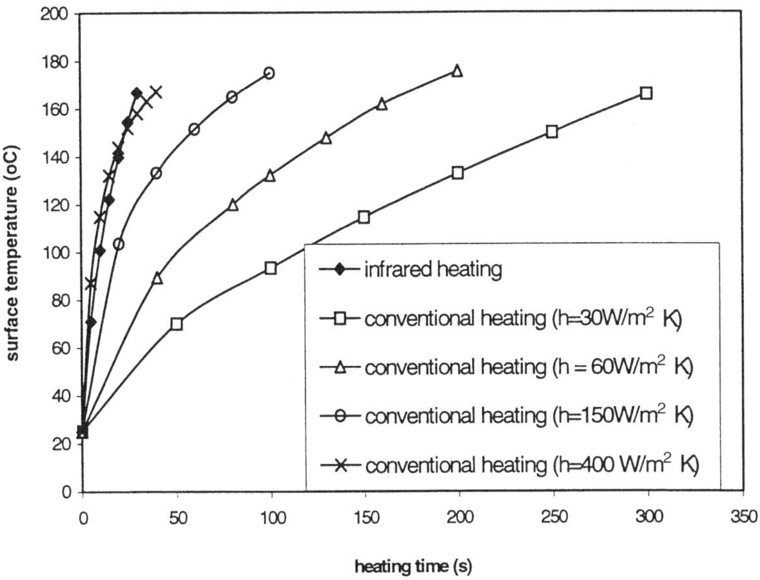

FIGURE 14 Surface temperature of barley grains when infrared heated (using emitter temperature of 850°C) in comparison with conventional heated grains using an air temperature of 180°C.

about 55 min for the conventional dryer. The difference in time required to remove 5 kg/m^2 of moisture from the acoustic tiles was also about 35 min (60 min for the infrared dryer and the 95 min for the convectional dryer).

Figures 15 and 16 show the sensitivity of the infrared system to initial grain moisture content, infrared burner distance from grain bed, and infrared heater temperature. It can be summarized as follows.

1. Surface temperatures of grain were significantly increased with increases in burner temperature and with decreases in burner height.
2. At the end of 15 sec of exposure, kernel surface and center temperatures were respectively increased by approximately 115% and 110% for every 50°C increase in burner temperature and for every 0.04-m decrease in burner height from the grain bed.
3. The burner temperature and burner height did not significantly affect the rate and quantity of moisture change in barley subjected to infrared heating.

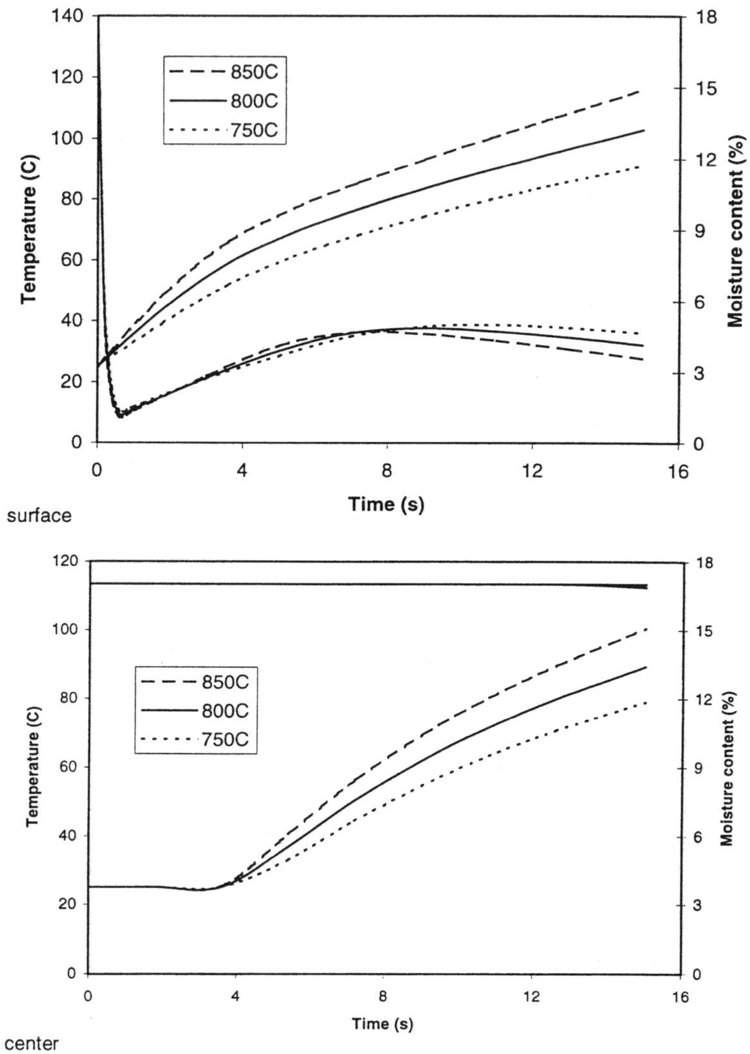

FIGURE 15 Effect of burner height on moisture content and temperature of hull-less barley. Burner temperature = 850°C; initial moisture content = 17% wet basis (wb) (From Ref. 53.)

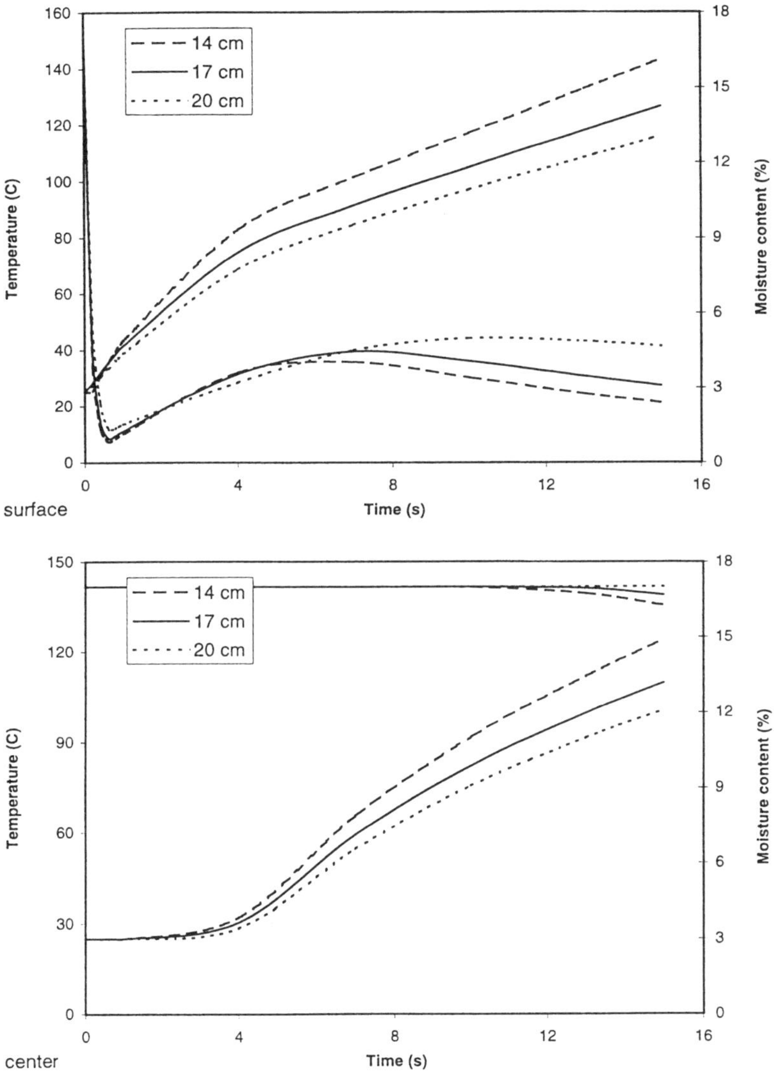

FIGURE 16 Effect of burner temperature on moisture content and temperature of hull-less barley. Height of burner = 20 cm; initial moisture content = 17% wb. (From Ref. 53.)

5 CLOSING REMARKS

In this chapter, we have shown the possibilities for the use of infrared heating in the processing of biological materials. In addition to using infrared heating as a means of moisture removal, the method can be used to alter the functional, chemical, and physical properties of cereal grains, legumes, and oilseeds, to reduce microbial load on surfaces of any material, and for rapid and routine determination of composition of food products. Even though the rapid rise in surface temperature has limited the possible application of infrared heating in the meat, milk, fruit, and vegetables industries, there is still room for the unique/novel use of this method in these industries such as in the drying/preservation of meat and fish especially due to its high heating rate in comparison to the conventional method of using hot air.

With increasing emphasis on biotechnology and production of transgenic seeds, infrared heating may be an alternative and quick way to conventional drying in order to achieve the purposes mentioned in the previous paragraph. However, further research is needed on the penetration of infrared radiation into biological material and the possible influence of infrared heating on the thermophysical properties of biological materials. The possibility of incorporating a heat-source term into Eq. (4) due to infrared penetration in biological materials should also be investigated. In addition, the models presented in this chapter need to be verified for other biological materials and for various configurations of infrared heating systems.

NOMENCLATURE

A	Surface area (m^2)
c	Specific heat (J/kg K)
D_m	Moisture diffusivity (m^2/sec)
E	Radiant energy emitted per unit area per unit wavelength interval ($W/m^2 \mu m$)
F	Configuration factor
h_c	Convective heat transfer coefficient (W/m^2 K)
h_{fg}	Latent heat of vaporization (J/kg)
h_m	Mass transfer coefficient (m/sec)
k	Thermal conductivity (W/m K)
M	Moisture concentration (%, dry basis)
q_r	Radiative heat component (W)
R	Particle distance (m)
r	Radial distance (m)
T	Temperature (K)

t	Time (sec)
V	Volume (m^3)
W	Radiant energy emitted per unit area (W/m^2)

Greek Letters

ρ	Density (kg/m^3)
σ	Stephan–Boltzman constant (5.6697×10^{-8} $W/m^2\ k^4$)
ε	Emissivity
Δ	Change in parameter
λ_{max}	Wavelength for peak emission intensity (μm)
θ	Emitter temperature (K)

Subscripts

e	Infrared emitter
eq	Equilibrium
g	Grain
i	Initial
p	Side plate
s	Surface

REFERENCES

1. C Sandu. Infrared radiative drying in food engineering: A process analysis. Biotechnol Prog 2(30):109–119, 1986.
2. N Sakai, T Hanzawa. Applications and advances in far-infrared heating in Japan. Trends Food Sci Technol 5:357–362, 1994.
3. SG Il'yasov, VV Krasnikov. Physical Principles of Infrared Irradiation of Foodstuffs. Revised, Augmented and Updated Edition. New York: Hemisphere, 1991, p 397.
4. EA Arinze, GJ Schoenau, FW Bigsby. Determination of solar energy absorption and thermal radiative properties of some agricultural products. Trans ASAE 30(1):259–267, 1987.
5. C Strumillo, T Kudra Drying: Principles, Applications and Design. New York: Gordon and Breach Science, 1986, p 448.
6. B Halistrom, C Skjoldebrand, C Tragard. Heat Transfer and Food Products, New York: Elsevier Applied Science, 1988, p 263.
7. AS Ginzburg. Application of Infrared Radiation in Food Processing. London: Leonard Hill Books Strand, 1969, p 413.
8. C Ratti, AS Mujumdar. Infrared drying. In: AS Mujumdar, ed. Handbook of Industrial Drying, 2nd Edition, Revised and Expanded. New York: Marcel Dekker, Inc, 1995, pp 567–588.

9. DJ Zuilichem, K van Riet, W Stolp. An overview of new infrared radiation processes for agricultural products. In: LM Maguer, P Jelens, ed. Food Engineering and Process Applications, Transport Phenomena, Vol. 1, New York: Elsevier Applied Science, 1985, pp 595–610.
10. W Jones. A place in the line for the micronizer. Special Report, Micronizing Company (UK) Ltd., Framlinghem Suffolk, UK, 1992, pp 1–3.
11. RL Kirkpatrick. Infrared radiation for control of less grain borers and rice weevils in buck wheat. J Kans Entomol Soc 48:1549–1551, 1975.
12. RL Kirkpatrick, JH Bower, EW Tilton. Gamma, infrared and microwave radiation combinations for control of Rhyzopertha dominica in wheat. J Stored Prod Res 9:19–24, 1973.
13. M Kouzeh-Kanani, DJ van Zuilichem, JP Roozen, W Pilnik. A modified procedure for low temperature infrared radiation of soybeans. I. Improvement of nutritive quality of full-fat flour. Lebensm Wiss U Technol 14:242–244, 1981.
14. M Kouzeh-Kanani, DJ van Zuilichem, JP Roozen, W Pilnik. A modified procedure for low temperature infrared radiation of soybeans II. Inactivation of lipoxygenase and keeping quality of full-fat flour. Lebensm Wiss U Technol 15:139–142, 1982.
15. M Kouzeh-Kanani, DJ van Zuilichem, JP Roozen, W Pilnik. A modified procedure for low temperature infrared radiation of soybeans III—Pretreatment of whole beans in relation to oil quality and yield. Lebensm Wiss U Technol 17:39–41, 1984.
16. M Kouzeh-Kanani, DJ van Zuilichem, JP Roozen, W Pilnik, JR van Delden, W Stolp. Infrared processing of soybeans. Qual. Plant Foods Human Nutr 33:139–143, 1983.
17. Anon. Micronised soya flour produced commercially. South Afr Food Rev 2(2):22–23, 1975.
18. DJ van Zuilichem, AFB van der Poel. Effect of HTST treatment of Pisum Sativum on the inactivation of antinutritional factors. In: J Huismam, AFB van der Poel, JF Liener, eds. Recent Advances of Research on Antinutritional Factors in Legume Seeds. Wageningen, The Netherlands: Pudoc, 1989, pp. 263–267.
19. I. Bozovic. Testing the suitability of methods of evaluating the quality of processed soybean. Arhiv Za Pojoprivredne Nauke 52(187):255–270, 1991.
20. Anon. New infrared machine for cocoa processing. Confect Product 47(6):308–309, 1981.
21. M van Liere. Process for treating raw soybeans. US Patent 4810513, 1989.
22. RL Anderson. Effects of steaming on soybean proteins and trypsin inhibitors. J Am Oil Chem Soc 69(12):1170–1176, 1992.
23. S Cenkowski, FW Sosulski. Physical and cooking properties of micronized lentils. J Food Process Eng 20(3):249–264, 1997.
24. M Kouzeh-Kanani, DJ van Zuilichem, JP Roozen, W Pilnik. Infrared processing of maize germ. Lebensm Wiss U Technol 17:237–239, 1984.
25. OO Fasina, WD Ziehl, RT Tyler, MD Pickard. Adaptation of micronization to the development of functional food ingredients from waxy barley and pulses,

including the adaptation and testing of a small scale, gas fired micronization unit. Project No. 32232 p. Report to submitted to the National Research Council of Canada, 1997 pp 95.
26. GH Zeng, OO Fasina, FW Sosulski, RT Tyler, Nitrogen solubility of cereals and legumes subjected to micronization. J Agric Food Chem 46(10):4150–4157, 1998.
27. DE Blendford. Potential applications of micronizing in food processing. Confect Manuf Market 16(4):3–5, 7, 1979.
28. I Rosenthal. Electromagnetic Radiations in Food Science. New York: Springer-Verlag, 1992, p. 244.
29. BA Rusnak, CL Chou, LW Rooney. Effect of micronizing on kernel characteristics of sorghum varieties with different endosperm type. J Food Sci 45(6): 1529–1532, 1980.
30. JA Collier. The application of recent technical advances to commercial production of brewery materials. Brewer 59:507–510, 1973.
31. JA Collier. Trends in UK usage of brewing adjuncts. Brewing Distilling Int 16(3):15–17, 1986.
32. OO Fasina, RT Tyler, MD Pickard, GH Zheng. Infrared heating of hulless and pearled barley. J Food Process Preserv 23:135–151, 1999.
33. TU Lawrence. Some effects on the growth and composition on the carcass of the bacon pig of feeding micronized or ground maize or barley based diets to give three different digestible energy intakes. Livestock Prod Sci 4(4):343–353, 1977.
34. TU Lawrence. An evaluation of the micronization process for preparing cereals for the growing pigs. II. Effects on growth rate, food conversion efficiency, and carcass characteristics. Anim Product 16(2):109–116, 1973.
35. JC Aimone, DG Wagner. Micronized wheat. I. Influence of feedlot performance digestibility, VFA and lactose levels in cattle. J Anim Sci 44(6):1088–1095, 1977.
36. SY Shiau, SP Yang. Effect of micronizing temperature on the nutritive value of sorghum. J Food Sci 47(3):965–968, 1982.
37. D Meyer, H Zwingelber, AW El-Baya, Experimental production of oat flakes with the micronizer. Getreide Mehl Brot 36(10):259–263, 1982.
38. G Spicher, H Zwingelberg. The micronizer—Equipment for reduction of the microflora in wheat bran. Getreide Mehl Brot 35(11):296–299, 1987.
39. CW Glennie, KH Daiber, RJN Taylor. Reducing the tannin content in sorghum grain by heat. South Afr Food Rev 9(3):51–55, 1982.
40. TM Afzal, T Abe. Thin layer infrared radiation drying of rough rice. J Agric. Eng Res 67(4): 289–297, 1997.
41. WR Lein, WR Fu. Small fish dehydration by far infrared heating. Food Sci Taiwan 24(3):348–356, 1997.
42. JS Townsend, S Cenkowski, M Friesen-Fischer. The thermal effects of high intensity infrared radiation on fresh lettuce leaves. Proceedings of an International Conference of Harvest and Postharvest Technologies for Fresh Fruits and Vegetables, Guanajuato, Mexico, 1995, pp 268–275.
43. R Metussin, I Alli, S Kermasha. Micronizaton effects of composition and properties of tofu. J Food Sci 57(2):418–422, 1992.

44. C Friedrich. Browning of foods. New applications for IR techniques. Process 1089:40, 1993.
45. JF Metzger, WE Muir. Computer model of two-dimensional conduction and forced convection in stored grain. Can Agric Eng 25(1):199–225, 1983.
46. KQ Stephenson, GW McKee. Accelerated drying of seeds with infrared radiation. Trans ASAE 229–231, 1964.
47. CW Hall. Theory of infrared drying. Trans ASAE 5(1):14–16, 1962.
48. H Kuang, J Thibault, BPA Grandjean. Study of heat and mass transfer during IR drying of paper. Drying Technol 12(3):545–575, 1994.
49. M Fortes, MR Okos. A non-equilibrium thermodynamics approach to transport phenomena in capillary porous media. Trans ASAE 24:760–80, 1981.
50. S Sokhansanj Improved heat and mass transfer models to predict grain quality. Drying Technol 5:511–525, 1987.
51. OO Fasina, S Sokhansanj. Estimation of moisture diffusivity coefficient and thermal properties of alfalfa pellets. J Agric Engng Res 63:333–344, 1996.
52. M Parti. Selection of mathematical models for drying grains in thin layers. J Agric Eng Res 54:339–353, 1993.
53. J Irudayaraj, K Haghighi, RL Stroshine. Finite element analysis of drying with application to cereal grain. J Agric Eng Res 53:209–229, 1992.
54. AV Luikov. Heat and Mass Transfer in Capillary Porous Bodies. Oxford: Pergamon Press, 1966, p 523.
55. OO Fasina, RT Tyler, MD Pickard. Modeling the infrared heating of agricultural crops. Drying Technol 16(9&10):2065–2082, 1998.
56. FP Incropera, DP DeWitt Fundamentals of Heat and Mass Transfer, 4th ed. New York: John Wiley & Sons, 1996, p 886.
57. MN Ozisik. Radiative Transfer and Interactions with Conduction and Convection. New York: John Wiley & Sons, 1973, p 575.
58. LD Albright. Environmental Control for Animals and Plants. St. Joseph, MI: American Soc of Agricultural Engineers, 1990.
59. MN Ozisik. Heat Transfer—A Basic Approach. New York: McGraw Hill, 1985.
60. S Sokhansanj, DM Bruce. Finite difference solutions of heat conduction and moisture diffusion equations in single kernel grain drying. Divisional Note 1351, National Institute of Agricultural Engineering, Silsoe, Bedford, UK, 1986.
61. PK Chandra, RP Singh. Applied Numerical Methods for Food and Agricultural Engineers. Boca Raton, FL: CRC Press, 1994.
62. RP Singh, DR Heldman. Introduction to Food Engineering, 2nd ed. New York: Academic, 1993.
63. M Dostie, JN Seguin, D Maure, QA Ton-That, R Chatigny. Preliminary measurements on the drying of thick porous materials by combination of intermittent infrared and continuous convection heating. In: AS Mujumdar, MA Roques, eds. Drying '89. New York: Hemisphere, 1989, pp. 513–519.

8

Modeling Electrical Resistance ("Ohmic") Heating of Foods

Peter J. Fryer and Laurence J. Davies
University of Birmingham, Birmingham, United Kingdom

1 BACKGROUND

1.1 Conventional Processes for Heating Foods

Using heating to reduce the bacterial level of foods is widespread [1]. Thermal processing aims to reduce contamination so that the food will not cause a health hazard during its shelf life. The classical method of thermal preservation is canning, in which food is placed in the package, which is then sealed and sterilized. Each part of the material must be processed to a level prescribed to ensure product safety. Heat transfer to cans is slow, however. The commonest method of providing heat is the use of condensing steam. Heat must then be transferred within the can, either by thermal conduction (if the food is a solid or a very viscous liquid) or by convection if the food is a less viscous liquid. This can lead to overcooking of some or all of the material. As a result, canned food has a taste and texture significantly different than unprocessed foods.

Several sets of reactions occur when a food is heated: those which reduce the level of bacterial contamination, those which result in losses in product quality, in terms of nutrition, taste, and texture, and some that improve taste and texture. It is possible to take advantage of the kinetics of these processes to improve product quality, by changing the temperature of sterilization. The reactions which reduce the number of bacterial spores have a higher activation energy than those which lower product quality; for example, Holdsworth [1] gives activation energies for the destruction of *Clostridium botulinum* spores as about 300 kJ/mol, whereas those of the loss of enzyme activity are ~120 kJ/mol. This implies that to maximize quality for a given level of sterility, it is best to process at as high a temperature as possible. Continuous ultrahigh temperature (UHT) or high-temperature short-time (HTST) processes exploit this to produce food of a higher quality than canning. In these processes, food is rapidly heated to 140°C, held there for a short period, and then rapidly cooled [1]. At 140°C, food can be sterilized in a few seconds, rather than the several minutes needed at canning temperatures. This requires rapid heating rates to minimize time spent at high temperatures and to minimize the quality loss, and it is best done in a continuous process. Processes incorporate three sections:

1. A *heating* section in which the product is first heated to the required temperature
2. A *holding* section in which it is held at the high temperature long enough to ensure sterility
3. A *cooling* section prior to packaging

Processing single-phase liquid foods this way is straightforward, because they can be heated and cooled rapidly. It is possible to process low-viscosity liquids using forced convective heat transfer in plate or tubular heat exchangers, in which the food is contacted indirectly with hot fluid. In this equipment, high heating and cooling rates (> 1°C/sec) are possible. This type of process can efficiently process liquids such as milks, fruit juices, soups, and sauces. Higher heating rates can be obtained by direct contact with steam. For high-viscosity products, such as creams and viscous sauces, which are shear sensitive and may foul heavily, direct steam injection can be used; steam infusion, in which the fluid to be heated is passed through a steam chamber, is also possible. Both infusion and injection commonly reduce the fluid temperature via evaporative flash after a holding section.

These techniques cannot readily be applied to foods which contain particles because of the slowness of thermal conduction into the solid. Conduction heating requires a temperature driving force between particles and liquid, and for particles larger than about 2 mm, rapid heating is not feasible. If HTST methods were used to process a solid liquid mixture containing very large particles, a very overprocessed liquid phase would result in the time required to make the particles sterile. It is possible to use conduction/convection methods to process foods with a high solids fraction (e.g., in scraped surface exchangers), but low heating rates and long hold times are needed, giving poor product quality. The aim of this chapter is to describe one possible solution to this problem, in which heat is directly generated within the material.

1.2 Heat Generation: Electrical Resistance Heating

The need to conduct heat is the limiting factor in the thermal processing of particles and, thus, for solid–liquid mixtures. Volumetric heat-generation techniques can solve this problem. Various techniques are available which use electric fields. In microwave or radio-frequency (RF) heating [2], a high-frequency electric field excites the water molecules within the material, whereas in resistance heating ("ohmic heating"), the passage of electrical current results in heating throughout the material. The process is more energy efficient than microwave heating, because nearly all of the electrical energy goes to heat the food. It requires the passage of electric current through the material; electrodes that make good contact with the food are required, unlike microwave heating, which needs no physical contact.

Like many of the "new" sterilization techniques, electrical heating is not novel [3]. An electrical pasteurization process was successful in the United States in the 1930s, and applications have been found in areas such as fish processing [4]. For a process to be commercially acceptable, a number of process and safety criteria must be satisfied:

- Electrical design to avoid electrolysis and product contamination
- Effective control of food heating and flow rates
- Efficient aseptic packaging techniques for a two-phase mixture
- Overall cost-effectiveness

A number of early processes did not meet these criteria; for example, by using a dc supply (which leads to electrolysis) or by using packages containing expensive electrode material. The absence of efficient packing plant also limited success. The resurgence of interest in the process in the last 15 years is due to the development of a commercial system by APV Baker [5–8], which incorporates solutions to the three problems described above. The APV Baker ohmic heater was originally developed by EA Technology at Capenhurst, UK. In 1984, APV Baker secured a licence for the heating system and then developed a commercial process. Commercial-scale systems are available with power outputs of 75 and 300 kW, corresponding to product capacities of ~750 and 3000 kg/h. A schematic process

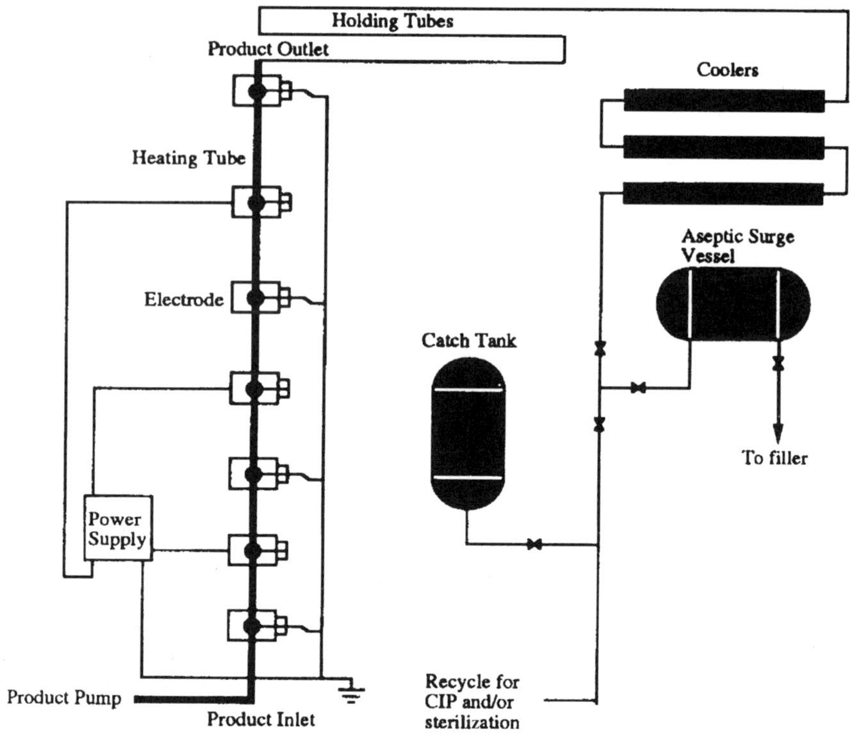

FIGURE 1 Schematic process flowsheet for the APV Baker "ohmic heater." (From Ref. 8.)

flowsheet is given in Figure 1. The system incorporates heat, hold, and cool sections. Food passes from a product pump into a vertical or near-vertical pipe containing a series of electrodes between which current flows. Sufficient pressure is maintained to ensure that the material does not boil; this can be up to 4 bar for sterilization at 140°C. Figure 1 shows a seven-electrode machine, in which each of the live electrodes is surrounded by two which are earthed; four-electrode machines have also been produced [6], in which the outer two are earthed.

Each electrode housing is machined from polytetrafluoroethylene (PTFE) and contains a cylindrical cantilever electrode (supported only at one end) across the tube. Tubes are made of different lengths to ensure similar impedances between each pair of electrodes; they thus increase in length from inlet to outlet, reflecting the increase in food electrical conductivity with temperature. From the heater, the material passes to a holding tube, which may be very short. Cooling may be provided by water flowing on the outside of the cooling tube, or precooled liquid can be pumped into the food to increase the heating rate [8].

The process was originally designed to sterilize foods of high solids fraction, up to about 60%, at heating rates in the region of 1°C/sec. Large particles, up to 25 mm in diameter, can be processed, and the technique has found commercial application in the United Kingdom, the United States, Europe, and Japan. It has been developed for a number of different applications:

- Aseptic processing of high-added-value ready-prepared meals for storage and distribution at ambient temperatures
- Pasteurization of particulate food products (e.g., fruit pieces)
- Preheating of food products prior to in-can sterilization
- The hygienic production of high-added value ready-prepared meals for storage and distribution at chilled temperatures.

Rapid pasteurization of particles gives very high product quality and has proved very successful. These applications have been reviewed elsewhere by Fryer [9], together with a discussion of the nutritional aspects of the process.

It is important to prove that an ohmic heater carries out the required time–temperature indicators or treatment. Process efficiency can be measured by passing alginate beads impregnated with spores through the heater [10,11] and measuring the loss of activity. This technique displays the effect of the three-step process (heating, holding, and cooling) rather than the behavior of the heater alone. To understand what happens during heating, it is necessary to construct a model for the process, both for designing the process and for proving that it is acceptable to regulatory bodies. As discussed below, models are needed both to predict how the plant behaves (which needs a simple model relating power and temperature), to identify local heating behavior around individual parts of the system, and to explain the limitations of the process.

1.3 Modeling Ohmic Heating

Ohmic and conventional processes differ significantly. The process is also significantly different from microwave heating, which has a finite penetration depth of energy into any food [2]; electrical heating has no such limitation.

Ohmic heating offers the possibility of rapidly heating solid–liquid mixtures on a commercial scale. The process does, however, lead to problems for formulation designers and food engineers. Some of these are conceptual; many of the situations possible in ohmic heating are not generally encountered in food processing. It is necessary to develop simple models to explain how the process operates. The majority of practical problems arise because of the need to ensure product and process safety, together with high product quality. It is easy to produce a safe product that is so overprocessed that it is not possible to sell it.

The need is to understand the process enough to exploit the benefits which ohmic heating can provide. Models can give that sort of understanding. Numerical models of electrical heating have concentrated on a set of problems on different length and timescales:

- The relationship between current and voltage (i.e., the electrical conductivity of a food mixture)
- The types of temperature patterns found in solid–liquid mixtures
- Predicting the output properties of material undergoing commercial processing

This requires different types of model:

- Simple ones which can describe a process as a basis for control
- More complex ones which can study the local thermal and electric fields around particles in a mixture

2 THE BASIS OF OHMIC HEATING PROCESSES

2.1 Governing Thermal Equations

To design a food formulation to exploit the advantages of electrical heating requires understanding of the process and the factors which affect product sterility and quality. The process is controlled by the rate of heat generation, which is governed by a number of factors, the most important of which is the electrical conductivity of the food material. However, temperature is also affected by the way food flows through the heater and, thus, the residence time within the plant. To design a food and process requires that the factors which affect the heating rate of the material and the time that it spends in the heater be understood. A number of models for the conventional heating of two-phase mixtures have been given [12–15]. These models cannot be applied to ohmic heating because the mechanism of heating is different; new types of model are needed.

At the frequencies at which electrical heating is carried out, heat generation is ohmic (i.e., given by the familiar I^2R, where R is the resistance and I the current). Preliminary work [16] demonstrated that this equation is appropriate (i.e., that [unlike in microwave heating] it is not necessary to solve Maxwell's equations). A more useful form of Ohm's law considers voltage gradients, where the heating rate at point i is given as

$$Q_i = \kappa_i E^2 = \kappa_i(\text{grad } V \cdot \text{grad } V) \tag{1}$$

where E is the voltage gradient and κ_i is the electrical conductivity (S/m) at point i. To calculate the distribution of heat generation requires a known voltage distribution. The voltage field within a system in which the electrical conductivity varies with position is found by solving Laplace's equation:

$$\nabla \cdot (\kappa \nabla V) = 0 \tag{2}$$

throughout the material with appropriate boundary conditions, such as constant voltage on the electrodes. The distribution of electrical conductivities thus controls the voltage distribution and critically affects the heating rates of the different phases. Local voltage gradients may well be different to global ones because of local conductivity changes. Laplace's equation can only be solved analytically in very simplistic situations, such as for an isolated sphere or cylinder of constant physical properties in a uniform field [16]. For example, for a sphere of radius R_p in an infinite parallel field of strength E_0, Eq. (2) can be rewritten in spherical coordinates and solved by separation of variables to give

inside the sphere

$$V_S = V_0 - \left(\frac{3\kappa_L}{2\kappa_L + \kappa_S}\right) E_0 r \cos\theta \tag{3}$$

outside

$$V_L = V_0 - \left[1 + \left(\frac{R_p}{r}\right)^3\right]\left(\frac{\kappa_L - \kappa_S}{2\kappa_L + \kappa_s}\right) E_0 r \cos\theta \tag{4}$$

where the subscript L denotes the liquid phase and S denotes the solids. The presence of the particle distorts the electric field; current diverts around an insulator or into a conducting particle. These equations can be manipulated to give the relative heat-generation rates in the two phases in estimating the relative temperature changes. Neglecting thermal conduction, the ratio of the heat generation in the solid to that at a point in the liquid where the particle has no effect on the field is given by substitution of Eqs. (3) and (4) into Eq. (1) as

$$\frac{Q_S}{Q_L} = \frac{9\kappa_S\kappa_L}{(\kappa_S + 2\kappa_L)^2} \tag{5}$$

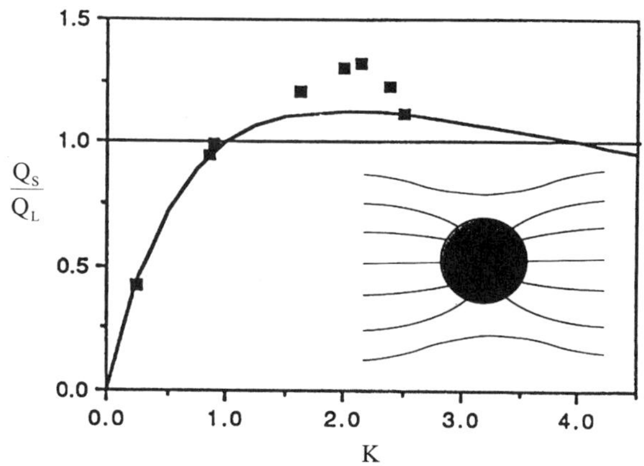

FIGURE 2 A plot of the electric field (inset) for $1 < K < 4$, and Eq. (5), showing complex heating effects. (From Ref. 9.)

The relative heating rate is thus a function of both solid and liquid electrical conductivities. Complex heating effects are found, as a function of the ratio of the solid to liquid conductivity, $K = \kappa_S/\kappa_L$ [16,17].

Figure 2 sketches the shape of the electric field for the case of a solid with a higher electrical conductivity than the liquid, together with a plot of Eq. (5) for Q_S/Q_L in terms of K. Some experimental data has been superimposed on the plot, without any attempt to distinguish between the physical properties of the two phases. The shape of the theory curve matches the experimental results, and the maxima also occur at the same point. This shows that Laplace's equation and the Ohm's law analysis are appropriate.

The form of the response to the electric field will depend strongly on particle shape and its orientation to the electric field (i.e., the distribution of the electrical conductivity in the system). Figure 3 demonstrates this clearly using experimental results; the same particle will *overheat* the fluid if placed parallel to the electric field or *underheat* if placed at right angles to the field. Particles with a smaller aspect ratio, such as cubes, will have less heating-rate variation with orientation [18], but some effect will always be seen around sharp edges.

The rate of heat generation is further related to the local temperature change by the thermal properties of the solid. The factors that affect heating are as follows:

1. The thermal capacity of the material, the product of density ρ, and specific heat c_p. It is convenient to define G, the inherent heating

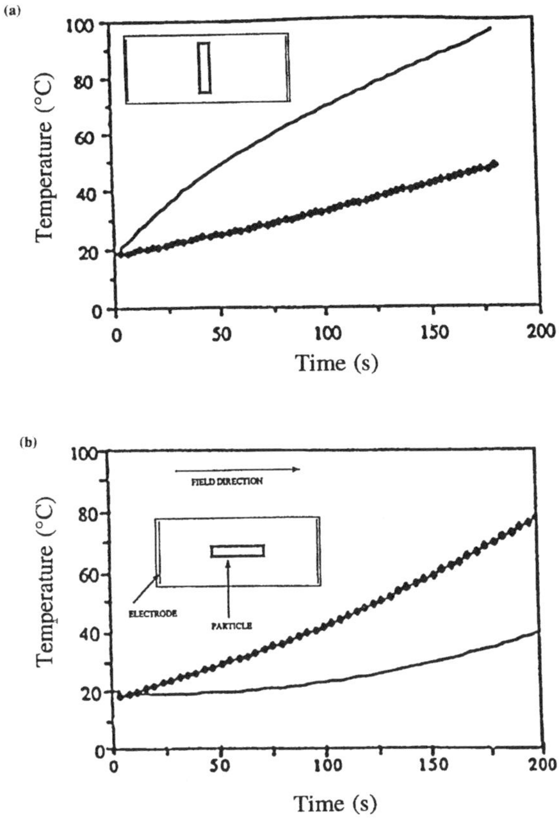

FIGURE 3 Overheating and underheating of a particle (●), in liquid (—), in different orientations to the electric field (shown in inset.) (From Ref. 16.)

rate of the material in the absence of thermal conduction and convention, as

$$G = \frac{dT}{dt} = \frac{Q}{\rho c_p} = \frac{\kappa E^2}{\rho c_p} \tag{6}$$

2. Heat transfer into the rest of the material. When thermal conduction is the only heat transfer mechanism, the following applies:

$$\frac{dT}{dt} = G + \frac{\lambda \nabla^2 T}{\rho c_p} \tag{7}$$

where λ is the thermal conductivity of the material. The conditions calculated using G alone are thus the extrema of the problem. When convection acts; the problem is complicated by the presence of a velocity field for both particles and liquids. This is discussed further below.

The APV Baker heating unit involves a flow of a solid–liquid mixture past a series of electrodes. Even for plug flow of the mixture, solution of Eqs. (1), (2), and (7) to find the temperature field is not simple because the electric and temperature field solutions are coupled by the strong thermal dependence of electrical conductivity. If there is relative motion between the liquid and solids, or a liquid velocity profile in the heater, then the Navier–Stokes equation must be solved for the velocity field along with the other equations for the voltage and temperature fields.

The flow behavior of the material will also be a function of temperature, as the fluid viscosity will change. Within the liquid, the effects of forces and natural convection must also be modeled; it has been shown [19] that in a viscous liquid, significant differences in temperature may arise due to the effect of heat generation. In practice, the effect of flow may even out temperature differences in the liquid.

Full solution of the three fields is not feasible given the complexity of the problems and the coupling of the equations via the viscosity and electrical conductivity variation on temperature: Approximations must be sought. The next section considers ways of predicting and modeling heating rates in real systems.

3 EFFECTIVE ELECTRICAL CONDUCTIVITY OF MIXTURES

3.1 Percolation Approach

The conductivity of mixtures of different physical properties is a difficult mathematical problem (going back to Maxwell) that is important in a number of areas. In ohmic heating, it is obviously necessary to know the overall electrical conductivity of the mixture. This is key in determining the power consumption and mean heating rate of the process. Particles will be dispersed randomly in a fluid and will respond differently to an applied electric field.

This type of problem, which depends on the random distribution of electrical conductors, is currently typically approached via the mathematics of percolation. Percolation concepts can be used to calculate an effective conductivity for electrical transport, which facilitates the prediction of an expected temperature evolution for the mixture.

Percolation theory has been used to model many physical phenomena. There is an extensive literature on the subject, summarized in reviews such as that by Sahimi [20]. Percolation models are conceptually simple: Consider a medium consisting of two types of material, randomly distributed. The mixture can be thought of as a network, at each site of which there is a different material. Square (two dimensional) and simple cubic (three dimensional) networks lend themselves well to theoretical and numerical modeling and have been used here. The number of nearest-neighbor sites to any site in the network is called the *coordination number*, Z: In a square network, $Z = 4$, and in a simple cubic network, $Z = 6$. In the classic-site percolation problem, the sites of this network are either conducting elements, randomly and independently of each other, with probability p, or insulating elements with probability $1 - p$.

Two nearest-neighbor sites are said to be *connected* if they are both conducting, and a set of conducting sites is known as a *cluster*. Electrical transport (conduction) will only occur if there is a cluster of conducting sites spanning the region between the edges of the system. The critical value of p below which there is no conduction across the network is called the *percolation threshold*, p_c. Percolation quantities are characterized by universal scaling laws close to the percolation threshold; they are largely insensitive to network structure and microscopic detail. The accepted values of p_c for an infinite two-dimensional (2D) square network and for an infinite simple cubic network [20, pp. 9–22] are 0.5927 and 0.3116, respectively.

A random distribution for a 2D lattice is shown schematically in Figure 4a. Here, there is a percolating path between the two sides of the lattice for "conducting" squares, shown as white in the figure. The electrical conductivity will depend on the particle distribution. The particles could arrange themselves in two limiting cases, as shown in Figs. 4b and 4c:

1. In Figure 4b, all the material lies in parallel (i.e., the voltage gradient through each is the same).
2. In Figure 4c, all the material lies in series (i.e. the current flow through each is the same).

These represent limiting cases of a distributed mixture: They are statistically unlikely in practice but, as will be shown, represent the limits of electrical conductivity. Note that for Figure 4c, there is no percolating path between the electrodes, so if either material is an insulator, the mixture will not conduct, whereas in Figure 4b, percolating paths exist for both materials.

3.2 Effective Conductivity of a Two-Component Food Mixture

We now consider a range of possible particle arrangements.

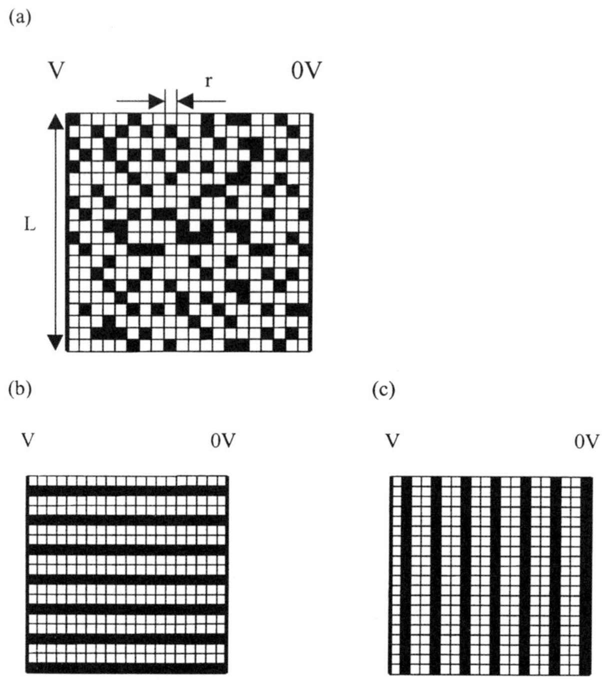

FIGURE 4 Configurations of the $L \times L$ grid for the electrical conductivity of a two-phase mixture: (a) random distribution, (b) parallel circuit, and (c) series circuit configuration.

3.2.1 Classic Percolation Threshold Problem

In the classic percolation threshold problem, one material has finite constant electrical conductivity and the other is zero [i.e., $\kappa_1 = \kappa_a$, $\kappa_2 = 0$, with $P(\kappa = \kappa_a) = p$, $P(\kappa = 0) = 1 - p$. Near the percolation threshold.

$$\kappa_e(p) \propto (p - p_c)^{\mu} \tag{8}$$

where μ is a largely universal transport exponent (i.e. independent of the microscopic details) and only depends on the dimensionality of the system. Its value is 1.3 for 2D systems and 2.0 for 3D systems [20, pp. 9–22]. Little electrical conduction—and therefore little heating—would be expected in a well-mixed mixture where more than the fraction $1 - p_c$ of the contents have very low electrical conductivity. Long, but finite, networks—and a system such as a food mixture, modeled by a network, will necessarily be finite in practice–change the value of p_c, but do not change this universality.

3.2.2 Effective Medium Approximation

To estimate a priori the macroscopic behavior of a random medium, such as a food mixture under electrical heating, we need to have a method of "averaging" its properties from its microscopic structure. For percolating systems with random conductances, the *effective medium approximation* (EMA), discussed by Kirkpatrick [21], is considered appropriate.

Effective medium approximation can determine the effective properties of a disordered medium by replacing it with a hypothetical uniform medium, with unknown properties, which is supposed to mimic the disordered system. The distribution of the voltage field in this uniform system is also uniform and can be calculated analytically.

For a regular network of coordination number Z and conductance distribution $f(\kappa_p)$, then EMA predicts that κ_e, the effective conductance of the network, is the solution of the equation

$$\int_0^\infty f(\kappa_p)\,\frac{\kappa_p - \kappa_e}{\kappa_p + (\gamma^{-1} - 1)\kappa_e}\,d\kappa_p = 0 \tag{9}$$

where $\gamma = 2/Z$, so that γ^{-1} is 2 for 2D systems and 3 for 3D systems.

Despite its simplicity, EMA can predict a nontrivial percolation threshold for various networks. For instance, for a network in which particles have conductivity κ_a with probability p and are nonconducting with probability $(1 - p)$, Eq. (9) predicts

$$\frac{\kappa_e}{\kappa_a} = \frac{p - 2/Z}{1 - 2/Z} = \frac{p - p_c}{1 - p_c} \tag{10}$$

(i.e., a straight line between $p = 1$ and $p = 2/Z = p_c$, where κ_e vanishes). This prediction is accurate for 2D bond networks, but not for 2D site networks and 3D networks. In general [20, pp. 59–61], regardless of the structure of $f(\kappa_p)$, EMA is very accurate if the system is not close to p_c, although predictions are more accurate for 2D networks than for 3D ones.

Figure 5 compares the theory for an infinite 2D system [Eq. (10) with $Z = 4$] with simulations from a 20×20 grid [22], showing the same trends but with statistical variation. It must be remembered that this type of model is statistically based: As in Figure 5, there is a small chance that percolation will arise at lower fractions than those predicted. Here, one material has been taken to be an insulator: This is not a feasible ohmic heating system for foods but shows how the approach can be used. Because food mixtures should contain components whose electrical conductivities are closely matched, to ensure thermal homogeneity, the value of p_c should not be crucial.

Consider the more general case where the two materials have nonzero constant conductivities κ_1 and κ_2, and the probabilities are such that at each

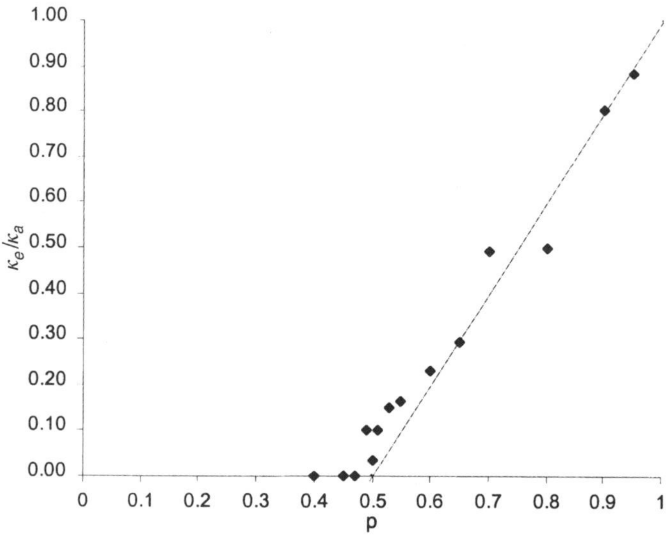

FIGURE 5 Graph of κ_e / κ_a against p, as predicted by Eq. (10), with some values from finite-element simulations overlaid.

site (square in grid) $P(\kappa = \kappa_1) = p$ and $P(\kappa = \kappa_2) = 1 - p$. Here, EMA can determine the macroscopic conductivity: The conductivity distribution can be substituted into Eq. (9) as

$$p \frac{\kappa_1 - \kappa_e}{\kappa_1 + (\gamma^{-1} - 1)\kappa_e} + (1 - p) \frac{\kappa_2 - \kappa_e}{\kappa_2 + (\gamma^{-1} - 1)\kappa_e} = 0 \tag{11}$$

This expression reduces to a quadratic in κ_e. The positive root of this equation is the desired effective medium conductivity:

$$\kappa_e = \frac{1}{2(\gamma^{-1} - 1)} \Big\{ (\gamma^{-1} - 2)\kappa_2 - (\gamma^{-1} p - 1)(\kappa_2 - \kappa_1) + \sqrt{[(\gamma^{-1} - 2)\kappa_2 - (\gamma^{-1} p - 1)(\kappa_2 - \kappa_1)]^2 + 4(\gamma^{-1} - 1)\kappa_1 \kappa_2} \Big\} \tag{12}$$

By writing $\kappa_2 = K\kappa_1$, the equation can be expressed as a conductivity ratio:

$$\frac{\kappa_e}{\kappa_1} = \frac{1}{2(\gamma^{-1} - 1)} \Big\{ (\gamma^{-1} p - 2)K - (\gamma^{-1} p - 1)(K - 1) \sqrt{[(\gamma^{-1} - 2)K - (\gamma^{-1} p - 1)(K - 1)]^2 + 4(\gamma^{-1} - 1)K} \Big\} \tag{13}$$

Then as $K \to 0$, Eq. (13) tends to Eq. (10). For large K, the ratio $\kappa_e/K\kappa_1$ also tends to eq. (10), where $P(\kappa = K\kappa_1 = \kappa_2) = 1 - p$, but in that case, the equation is only valid for $p < p_c$, with the condition $\kappa_e = \kappa_1$ for $p > p_c$.

This analysis implies that for food mixtures composed of a substrate plus an added component of different electrical conductivity, below certain (percolation) thresholds the mean electrical conductivity is determined by that of the lowest conducting fraction, whereas the highest conducting fraction has a negligible effect on the overall conductivity.

3.2.3 Demonstration of Mean Conductivity

To get an idea of how the conductivity of this effective medium fits into the range of possible overall conductivities, it is instructive to compare the EMA approach with two limiting configurations for a mixed two-component medium: parallel and series distributions.

Parallel and series distributions are illustrated schematically in Figures 4b and 4c, respectively. Two materials are distributed in an $L \times L$ box, with electrodes along the left and right sides at V volts and 0 volts respectively. The materials are distributed in a waferlike arrangement, with alternating layers of materials 1 and 2. Here, p represents the fraction of material 1 in the mixture.

An averaging argument [22] leads to the following expressions for mean electrical conductivity:

For parallel

$$\kappa_e = p\kappa_1 + (1-p)\kappa_2 \tag{14}$$

For series

$$\kappa_e = p\kappa_1\left(p + (1-p)\frac{\kappa_1}{\kappa_2}\right)^{-2} = (1-p)\kappa_2\left(p\frac{\kappa_2}{\kappa_1} + (1-p)\right)^{-2} \tag{15}$$

The difference between the limiting cases (i.e., the span of conductivities possible) can be illustrated for different K. Figure 6 shows different measures (series, parallel, EMA) of the mean electrical conductivity of a two-component system as a function of the content fraction, p, of material 1. In Figure 6a, the conductivity ratio $K = \kappa_2/\kappa_1 = 2/3$, relevant to a food material. The different assumptions predict a range of conductivities varying between about $\pm 10\%$ of the EMA. Note that for a much smaller K, a very wide range of conductivities are possible, as shown in Figure 6a, where $K = \kappa_2/\kappa_1 = 10^{-4}$. EMA values in Figure 6 are plotted for 2D and 3D random distributions. In each case, the parallel conductivity is highest, and the 3D EMA values are higher than the corresponding 2D ones; 3D EMA values are "closer" to the parallel values.* The analysis suggests that EMA is a

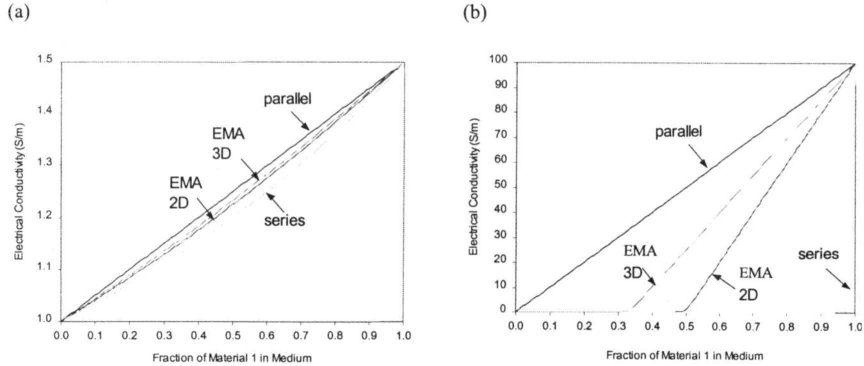

FIGURE 6 Comparison of measures of electrical conductivity of a distributed two-component mixture as a function of the content fraction of material 1, when the conductivity ratio $K = \kappa_2/\kappa_1$ is (a) 2/3 and (b) 10^{-4}. EMA values are plotted for both 2D and 3D random distributions.

good approximation to predict overall system behavior. However, the discussion has not addressed thermal effects: In practice, the different heat-generation rates within sections of the fluid will change the local conductivity distribution. It is thus necessary to consider temperature distributions at a particle length scale, as described below.

4 FINITE-ELEMENT MODELING OF OHMIC HEATING

4.1 Introduction

The above discussion shows how the conductivity of even simple systems can be difficult to predict. In practice, the commercial ohmic heater involves a flowing mixture where the electrical conductivities vary with temperature. Various approximations have been tested [23–26] to predict how flowing particle–liquid mixtures behave. Finite-element models have been used to develop solutions for ohmic heating of solid–liquid mixtures: this type of model can simulate both complex shapes and temperature-dependent physical properties. Early work [27] used in-house codes; more recently, commercial codes such as ANSYS [25,26] and FIDAP [28] have been used. Such commercial computational fluid dynamics (CFD) packages solve the Navier–Stokes equations but can also solve energy and potential field

* Note that for the n-dimensional food (the Hilbert carrot), where $\gamma^{-1} = n$, Eqs. (12) and (14) show that as $n \to \infty$, $\kappa_{\text{ema}} \to \kappa_{\text{parallel}}$: in the limit, the parallel approximation is best.

equations. In each simulation, the vessel geometry can be discretized by a paved mesh, and the material properties and boundary conditions given as input data.

In flow situations, the full Navier–Stokes equation for fluid flow must be solved together with the temperature and voltage field. The assumption of no fluid motion can simplify modeling considerably, shortening the computational run, and gives insight into the "worst-case scenario": When there is little convective heating, then the temperature differences between different regions of a food mixture will be more pronounced. Convection effects inside a fluid can be modeled using an enhanced thermal conductivity, λ^*, in Eq. (7) [17,19] to represent fluid mixing and reproduce experimental results. This approach negates the limitation suggested by Sastry and Salengke [29].

Simulations have studied the following:

- Heating patterns about simple solid–liquid combinations [27,30] to show how current channelling gives temperature variations between solid and liquid.
- Ways of representing flowing mixtures as a set of "unit cells" of solid spheres in a fully mixed fluid [25,26]; this model was then used as a basis for a model of the process, as described in the next section.
- The effect of inclusions of different sizes and electrical conductivities [19,28,31].

The aim of simulations is to demonstrate the nature and extent to which effects might arise in real situations. The finite-element (FE) method is expensive in computer time (although advances in hardware have made it possible to use high-end PCs rather than workstations), but if there is symmetry in the system, only parts of the whole need to be modeled. It is best as a design and research tool rather than as the basis for a control system. Here, work on the validation of the technique is described; more detail is given in Ref. 28.

4.2 Validation

It is important to demonstrate the accuracy of any model. As already noted, there are few situations where an analytical solution is possible. Two limiting cases for the electrical heating of a single rectangular particle in a rectangular static ohmic heating vessel can be modeled:

- When the particle lies along the vessel from electrode to electrode, the system can be treated as a simple electrical circuit with resistors in parallel.

Table 1 Material Properties Used in Simulations

Material property	Perspex	Water	Albumen
Density ρ (kg/m^{-3})	1170	1000	1000
Specific heat c_p (J/kg/K)	1500	4180	4180
Thermal conductivity λ (W/m/K)	0.17	4[a]	0.6
Electrical conductivity ρ (S/m)	10^{-12}	0.0412 $[T(^\circ\text{C}) - 20]$ $+1.7719$	0.0952 $[T(^\circ\text{C}) - 20]$ $+4.0635$

[a] Enhanced thermal conductivity, λ^*, used for water, to take account of convection effects.
Source: Ref. 28.

- When the particle lies across the whole vessel, the system can be treated as a simple electrical circuit with resistors in series.

If the system is thermally and electrically insulated, the following first-order ordinary differential equation results (neglecting thermal conduction):

$$\rho c_p \frac{dT}{dt} = \kappa \left(\frac{dV}{dx} \right)^2 \tag{16}$$

Davies et al. [28] simulate systems whose physical properties were taken as those of (1) an egg albumen particle in salt water and (2) a perspex particle in agar gel. Physical properties are given in Table 1, and the geometrical data for the parallel case is detailed in the diagram in Figure 7a. In this case, if the vessel and particle are of length L, then the voltage gradient is $-V/L$ everywhere. Where the electrical conductivity is of the form $\kappa = \kappa_0 + m(T - T_0)$, where κ_0, m, and T_0 are constants, Eq. (16) can be separated and integrated to give

$$T = \frac{\kappa_0}{m} \left\{ \exp \left[\left(\frac{m}{\rho c_p} \right) \left(\frac{V}{L} \right)^2 t \right] - 1 \right\} + T_0 \tag{17}$$

where T_0 is the starting temperature (°C). In practice, the electrical conductivity of perspex is so low that T is about T_0 throughout. Two sets of simulations were conducted using FIDAP: one with realistic thermal conductivities and the other using $\lambda = 0.01$ W/m/K to approximate the theoretical case of no thermal conduction. The voltage gradient is constant

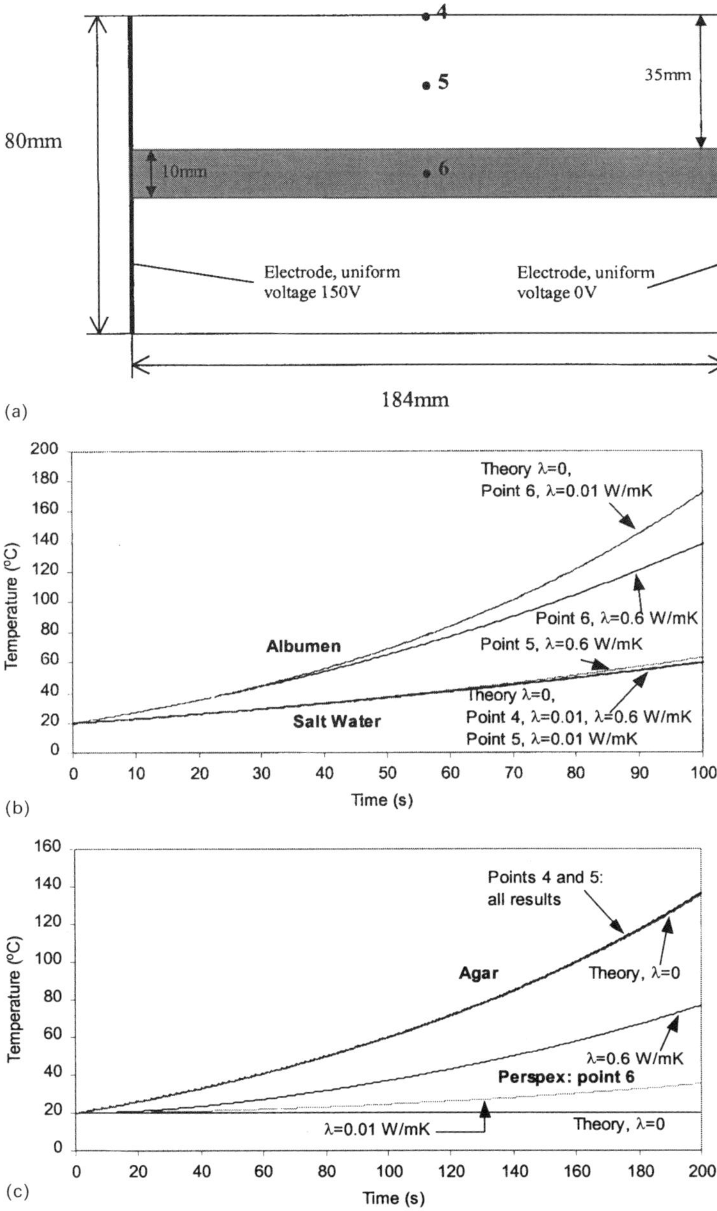

FIGURE 7 Computational model validation. (a) Parallel configuration, together with analytical and computational calculations for the heating of (b) a block of egg albumen in salt water and (c) a block of perspex in agar gel. Temperature readings are recorded at points 4, 5, and 6, shown in (a). (From Ref. 28.)

in each region. Thermal conduction will result in the following:

- For case (1), temperatures in the egg albumen lower than those predicted by Eq. (17)
- For case (2), temperatures in the perspex higher than T_0

Analytical and computational results are compared in Figures 7b and 7c. Simulated temperatures at points 4, 5, and 6 were recorded; these are 0 mm, 20.4 mm, and 40 mm from the upper edge of the vessel respectively (see Figure 7a. There is very good agreement between theory and FIDAP for the temperature of both the "salt water" and the 'agar gel' at point 4 (Figures 7b and 7c, respectively). This is expected, as simulated temperatures were taken well away from the interface, minimizing the effects of thermal conduction. At point 5, in Figure 7b, the salt water in the $\lambda = 0.01$ W/m/K model compares well with the theoretical curve; however, higher thermal conductivity gives a different heating rate.

The effect of thermal conduction between the two phases is most significant around the particle. Temperatures were recorded at point 6; here, Figure 7b shows an albumen temperature 34°C above the no-conduction theory after 100 sec. However, if a thermal conductivity of 0.01 W/m/K is assumed, very good agreement with theory is found. The effects of conduction are also pronounced for perspex (Figure 7c) with natural thermal conductivity, the temperature is 16.8°C higher than theory after 100 sec. The 0.01 W/m/K model shows a temperature gain of 4.38°C after 100 sec, not as good as that for the albumen case.

The temperature profile given by theory and FIDAP across the two phases is shown in Figure 8. Overall, the simulations demonstrate both the accuracy and the limitations of any computational model of this type. It is possible to validate the software by comparing with systems which approximate simple circuits; theoretical predictions may be approached for some materials by using low thermal conductivities. The numerical routines will be least accurate at the interfaces between materials and in regions of high-temperature gradients.

4.3 Use of Model to Study "Shadow Effects" Around Particles

When electrical conductivities are significantly different from one another Fryer *et al.* [19] show that local "shadow regions" of low electric field strength can arise, resulting in large differences in temperature between the liquid and the solid. Within the liquid, these changes in temperature can be minimized by convective mixing, an effect largest for low viscosity fluids such as water. However, commercial food processes commonly use

(a)

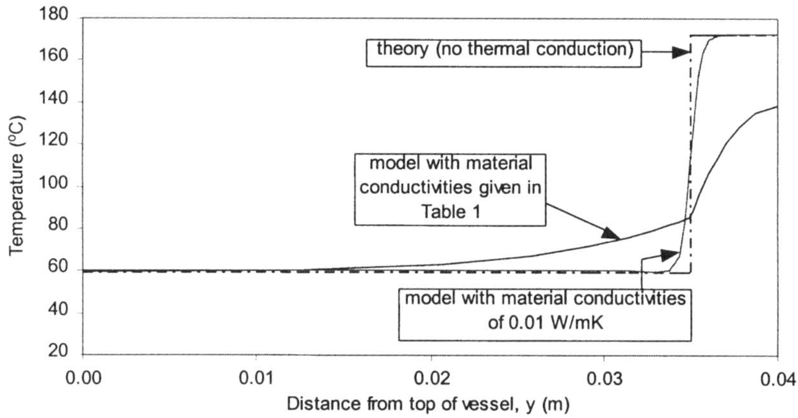

(b)

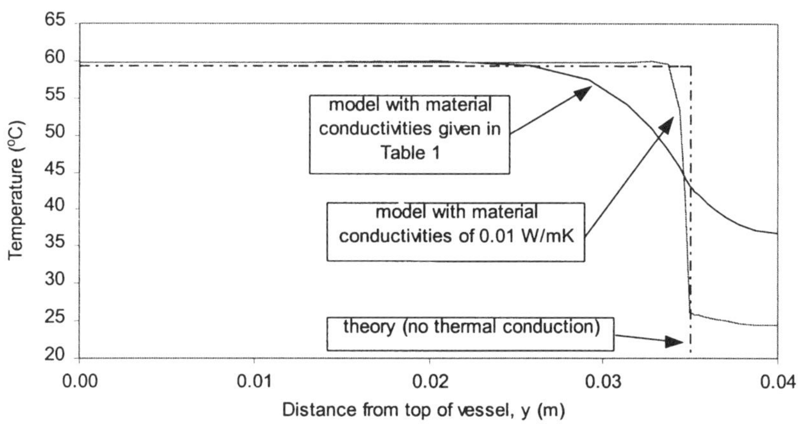

FIGURE 8 Temperature distribution along vertical centreline of vessel, for the computational model validation detailed in Figure 7a. Distribution shown from top of vessel to center of particle, after 100 sec for the heating of (a) a block of egg albumen in salt water and (b) a block of perspex in agar gel. (From Ref. 28.)

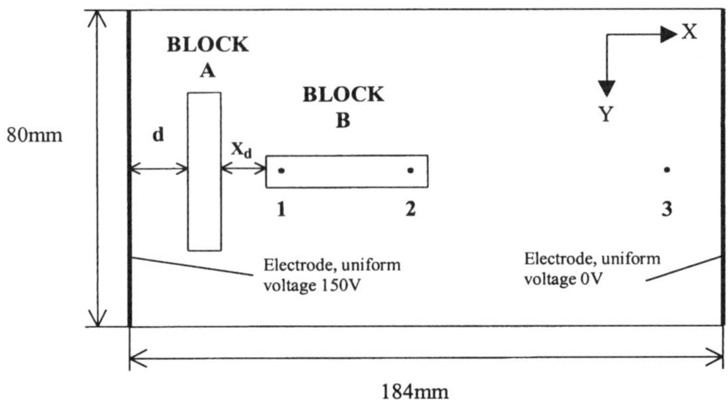

FIGURE 9 Plan view of the experimental system in the static ohmic heater box. The walls of the box are thermally and electrically insulated. Block A is perspex, and block B is egg albumen. Both have dimensions 10 × 35 × 50 mm. Distanced d is fixed at 10 mm, and x_d is varied between 0 mm and 20 mm. Temperature readings are recorded at points 1, 2, and 3 along the centerline. (From Ref. 28.)

high-viscosity carrier fluids (such as starch) in which convective processes will be significantly slower. Experiments and models have been compared to study to what extent these shadows might cause processing problems in a real industrial situation (full details are given in Ref. 28). The situation simulated is shown in Figure 9. In all experiments, the electrically insulating block A will cast a cold "shadow" in the fluid or gel adjacent to it in the vessel. The effect of the electric field shadow is minimal where convection or thermal mixing are significant, but it is clear in carrier fluids of higher viscosity. Experiments were performed to determine the effects of viscosity (and thus the fluid mixing) on the heating rates in the heater. Experiments used two test fluids:

1. Carboxymethylcellulose (CMC) solution of viscosity 1200 cP at 20°C, which was modeled using an enhanced thermal conductivity to represent fluid mixing
2. A 1.5% agar gel, to approximate to the case of conduction alone

Experiments were performed on a system where block B (10 × 35 × 50 mm) was constructed from egg albumen dosed with salt water to an initial electrical conductivity twice that of the contact solution and positioned at $x_d = 0$, 5, 10, and 20 mm. Under normal circumstances, the block will overheat the liquid; however, when it is in the shadow, it will heat solely by conduction. The shadow effect seen here is due to differences in the

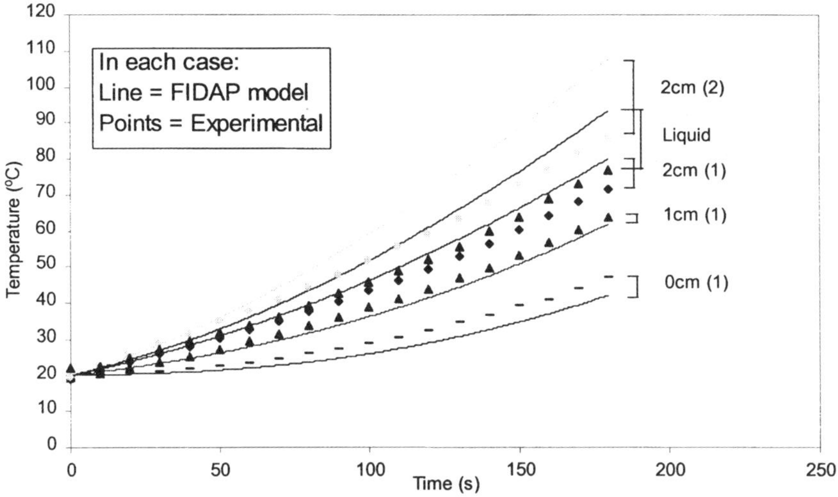

FIGURE 10 Change in time–temperature history at point (1) within block B with increasing distance x_d. A typical liquid heating rate is also shown, as is the heating rate at point 2 within block B when $x_d = 20$ mm (i.e., the heating rate within the solid affected minimally by the thermal shadow region.) (From Ref. 28.)

current inside the particle. Fluid mixing may equalize the fluid temperatures, but it will not affect the temperature within block B over the time-scale of an ohmic process.

One key characteristic of the ohmic process is the ability of particles to heat faster than the surrounding fluid. Here, as x_d is increased, the heating rate within the albumen particle surpasses that of the liquid (point 3); one-half of the block *overheats* the liquid and the other half *underheats*. This can be seen in Figure 10, which summarizes some of the computational and experimental data. It shows the change in heating rate at point 1 within the albumen for increasing x_d, together with data for point 3 when $x_d = 0$ mm (a typical temperature evolution for the liquid bulk, labeled "Liquid" in Figure 10) and the heating rate at point 2 when $x_d = 20$ mm. The model predicts the magnitude of the temperatures well.

The 2D geometry has been simulated in FIDAP. Data for the thermal field for different block positions and after 170 sec are given in Figure 11. The simulations have two major differences compared with experiment. As the cell is open to atmosphere, the maximum temperature in the solution or agar gel is approximately 100°C: however, no such limitation applies within the simulation. Latent heat has been approximated by increasing the specific heat capacity by a factor of 10^3 above 110°C; this approximates the effects

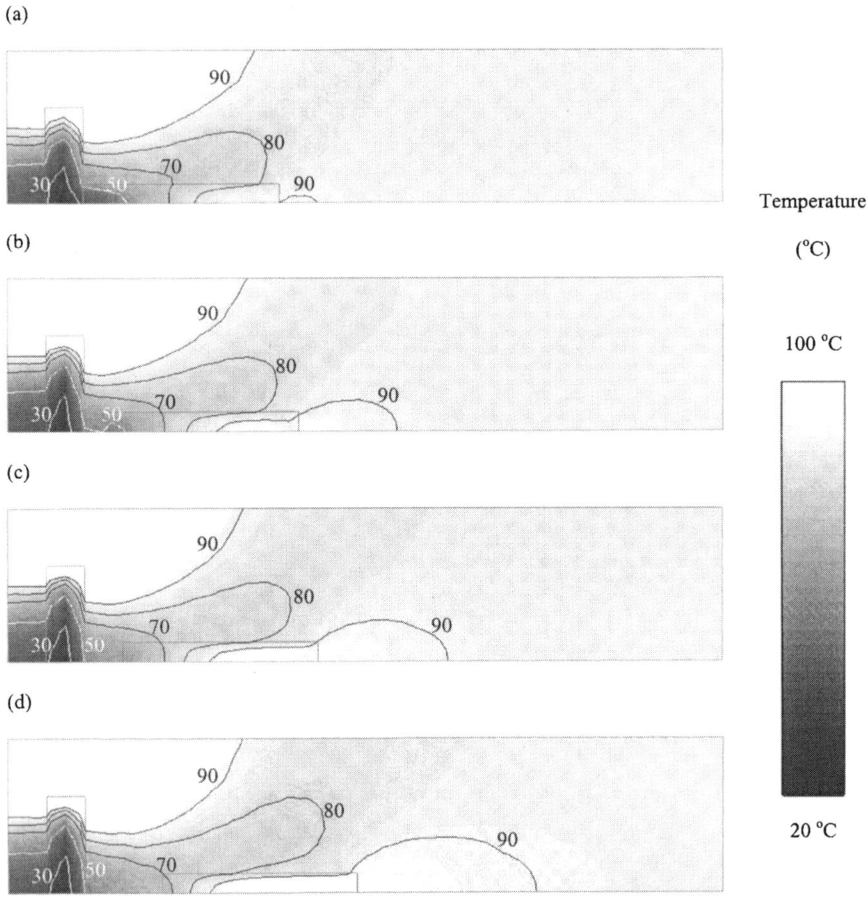

Figure 11 Modeled temperature distribution for the system illustrated in Figure 9 after 170 sec of electrical heating (upper half of vessel shown, only). The distances between the blocks in each diagram are (a) 0 mm, (b) 5 mm, (c) 10 mm, and (d) 20 mm. (From Ref. 28.)

of latent heat to ensure the system has an effective maximum temperature. Second, the simulations use an enhanced thermal conductivity term $\lambda*$ in the heat equation (7) for the solution/gel regions, as explained in Section 4.1.

The simulations show both the overheating of the particle when it is as far from the insulator and the situation where the particle both underheats and overheats, seen most clearly in Figures 11b and 11c. The "hot shadow" cast by the albumen particle is also seen: the high current flow through the particle leads to a region of high temperature immediately behind it. Overall, the simulations show that it is key to match the conductivities of the two phases, as shadow regions occur which have the same order of magnitude as that of the inclusion. Internal mixing within the liquid is useful; if the fluid temperature is uniform, the effect of the shadow on the fluid is minimized. However, this may have little effect on the particle behavior. The effect of the particle on the surrounding thermal field depends on the size of the particle, the thermal conductivities of both phases, and the timescale of the experiment. These results show that centimeter-scale inclusions can create inhomogeneities which cannot be removed by thermal conduction over the timescales typical of ohmic processes; care must be taken with the preparation of formulations.

The FE approach is useful in a number of ways; it can be used to study local behavior and predict how formulations will behave. It also shows the types of effect that can be found in practice. This type of model is basically a resource tool: Modeling gives a way of studying local effects and identifying problems. To study how formulations might behave in a commercial system, a simple model of the heating rates is needed. This is discussed next.

5 MODELS FOR FLOW AND HEAT GENERATION

To model the effects of flow and heat generation together in a real system, it is necessary to solve the Navier–Stokes equation together with Eqs. (1), (2), and (7). This is not possible for particle–liquid mixtures at the moment, because of both the lack of information on the basic physics and the complexity of the calculations. Extensive experimental work has been done to study two-phase food flows in conductive/convective heating processes; for reviews, see Refs. 32 and 33.

Some work has been done on the flow of single-phase foods undergoing electrical heating. Although of limited practical interest, such models show the effects found in real systems. Zhang and Fryer [34] modeled simple laminar flows to demonstrate the type of effect which could be found during the simultaneous flow and heat transfer of real food fluids. Other workers [35] have showed that complex flow and temperature profiles can result.

Single-phase flows may be very complex; two-phase mixtures are less well understood, although high solids fractions do, however, approximate to plug flow [36,37]. The rate of sterilization of a material is highly temperature dependent, and so it is important that any model be as accurate as possible [38]. Sastry [24] developed an approach in which the fluid and solid within an element of the mixture were represented as a set of series and parallel resistances. The model is simple and easy to compute, but it is not suited to studying local heating patterns. At this length scale, however, no such approach is needed.

5.1 Modeling the Ohmic Heater

The two-phase system has been modeled using a series of approximations to simplify the problem to the point where it can be solved computationally [25,26]:

- That heat transfer and generation equations for the solid and liquid phases can be modeled by a one-dimensional heat balance; differences between the velocities of the phases can be accommodated.
- That at each point heat generation, temperatures and velocities of the phases are uniform. Reference 19 shows that uniform heating in the liquid is possible and Ref. 25 shows that a rotating particle can heat uniformly.
- Heating in the electrode housing is neglected. Boundary conditions of uniform voltage across the entrance and exit of the tube are defined for solution of the electrical field.

As the mixture flows up the tube, the temperature—and thus the electrical conductivity—of the two phases changes. The voltage gradient will thus vary up the tube, and an iterative calculation is necessary.

Zhang and Fryer [25,26] modeled the solid–liquid mixture as a uniform lattice of spherical particles on a regular grid. A "unit cell" of the material was modelled (using an FE approach) as representative of the whole, an approach common in symmetrical systems, such as crystallography. The model assumes uniform temperature in each phase (i.e., full mixing of the fluid). The circuit approach of Sastry [24] is an alternative to this.

Using the above assumptions, the response of the system can be modeled as a set of thermal balances. An enthalpy balance for a particle traveling at velocity v_S can be written:

$$-ha(T_S - T_L) + Q_S = v_S(\rho c_p)_S \frac{dT_S}{dx} \tag{18}$$

where h is the convective heat transfer coefficient between particle and liquid, a is the area of a particle per unit volume ($6/d_p$ for a sphere of diameter d_p), and W_S is the heat generation rate per unit volume of solid. This can be rewritten as

$$-H_S(T_S - T_L) + G_S = v_S \frac{dT_S}{dx} \tag{19}$$

where $H_S = ha/(\rho c_p)_S$ is a modified heat transfer term and G_S has already been defined. Assuming no heat loss, the liquid-phase heat balance can be written

$$\frac{\phi}{1-\phi} ha(T_S - T_L) + Q_L = v_L(\rho c_p)_L \frac{dT_L}{dx} \tag{20}$$

where ϕ is the fraction of the volume of the system occupied by the solid, v_L is the liquid velocity, and Q_L is the heat-generation rate per unit volume of liquid. This can be written

$$H_L(T_S - T_L) + G_L = v_L \frac{dT_L}{dx} \tag{21}$$

where

$$H_L = \frac{\phi}{1-\phi} \frac{ha}{(\rho c_p)_L} \quad \text{and} \quad G_L = \frac{Q_L}{(\rho c_p)_L}$$

To find the temperatures of the two phases, Eqs. (19) and (21) must be solved together. Either of the approaches for determining heating rates discussed above can be used. Q has been calculated using the unit cell model for a range of R_κ and solids fraction ϕ, these data were then fitted to polynomials which were incorporated into the computational scheme [25], which then ran rapidly on a PC. The equations can then predict temperatures, including any thermal crossover between the two phases. Figure 12 plots the variation in the temperature of the solid and liquid for a case where the solid is originally less conductive than the liquid but becomes more conductive with temperature.

Such a model shows how the temperature of the two phases may change during electrical processing. It does not indicate how local changes in temperature may affect the sterility pattern: the FE model should be used to check that. Neither the unit cell model nor the circuit approach considers how foods of complex shapes heat; local current concentrations can arise, for example, at sharp edges or corners of particles. At this scale, this sort of process model is a useful basis for process simulation and control, but experimental verification will be necessary.

5.2 Holding and Cooling Sections

From the heater, material will pass to the holding section and then to cooling. In a conventional process the holding section allows thermal equilibration between particle and liquid and holds the material at the temperature long enough for the required level of sterility [12–14]. In electrical heating, equilibration may not be needed. For process control, it is useful if the

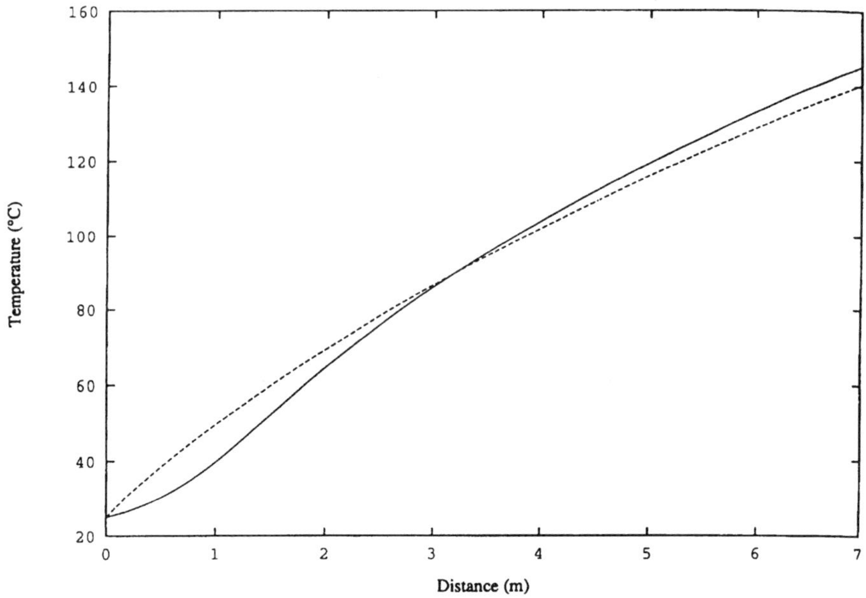

FIGURE 12 Variation in temperature of a solid (——) and a liquid (- - - -). (From Ref. 26.)

particle temperature at the end of the heating section exceeds the liquid, as the liquid temperature is simplest to measure. However, any overheating of particles may impair quality, because it is necessary to remove heat by conduction during cooling. Here, conventional models for heat transfer must be used. Heat balances for the two phases can be written as before. If it is assumed that the two phases move at uniform (but not necessarily the same) velocity in a cooling tube, then, for the liquid,

$$(wc_p)_L \frac{dT_L}{dx} = \pi d_t h_W (T_W - T_L) = \frac{a\pi d_t^2}{4} h_p (T_L - T_S) \tag{22}$$

where h_W is the wall heat transfer coefficient, T_W is the wall temperature, d_t is the tube diameter, and w is the mass flow. This must be solved together with non-steady-state balances for conduction within the solids, such as for spheres,

$$\frac{dT_S}{dt} = \frac{1}{R_p^2} \frac{d}{dr} \left(\frac{\lambda_S}{\rho c_p} r^2 \frac{dT_S}{dr} \right) \tag{23}$$

In practice, there can be significant variation in the velocity of both phases' Ref. 37, amongst many others, shows this effect. Any variation in velocity by changing the residence time in the system will change the sterility and quality of the food.

More data will be needed before a fully accurate model is available. The question is how accurate a model is required for a study of the whole process. One use of this type of model is to optimize the whole process (i.e., consider the effect of temperatures in the heater on the subsequent process). Overheating of particles is easiest for control purposes. In practice, it might be best for the solids to underheat the liquid slightly during heating. As it is easier to cool the liquid phase; solids, even if they are initially cooler than the liquid, will spend more time at high temperature than the liquid and thus will be sterilized during the hold and cool sections.

6 THERMAL HOMOGENEITY FOR DIFFERENT PROCESSES: CONCEPTUAL MODELS

The above has shown how modeling can study the local effects of ohmic heating and the behavior of commercial processes. The advantages of ohmic heating depend critically on the design of the process and the thermal homogeneity that results. In any thermal process, the process is set by the need to process the coldest point; if the rest of the material is overprocessed, then the advantages of rapid heating are lost. It is important to determine the factors which control homogeneity, to devise processes which exploit the advantages of ohmic heating fully. Here, we elaborate on some current research activity [22] and some conceptual models are developed, which demonstrate the problem and how it can be approached.

Ideally, process conditions should be such that the system can be considered macroscopically homogeneous (i.e., that heating will be uniform throughout). With ohmic heating, there are two competing mechanisms governing the temperature evolution: internal heat generation and thermal conduction. Because the medium is not uniform, heat generation due to electrical resistance will increase the temperature differences between components of the mixture, whereas heat conduction is a smoothing effect, acting to decrease temperature differences.

Different heating behaviors will dominate on the microscopic and macroscopic levels. Thermal conduction will dominate at particle length (microscopic) scales; at some length scale, local temperature differences arising from heat generation across the components of a mixture will be eliminated by conduction over the timescale of a process. For larger length scales, the timescale of a process may, however, be such that on a macroscopic level, the temperature of the system is not uniform.

Therefore, we need to determine the process conditions under which conduction dominates locally while heat generation dominates globally. The scale of the system can be studied by rewriting the heat balance:

$$\frac{\partial T}{\partial t} = \frac{\lambda}{\rho c_p} \nabla^2 T + \frac{\kappa}{\rho c_p} (\nabla V)^2 \tag{24}$$

In this equation, thermal conduction is given by the $(\lambda/\rho c_p)\nabla^2 T$ term, and internal heat generation comes from the $(\kappa/\rho c_p)(\nabla V)^2$ term. We can consider the orders of magnitudes of physical properties associated with single particle of the mixture (i.e., consider the local behavior at the individual particle level) and examine the associated orders of magnitude of physical quantities [22]. Consider a typical length scale (particle size), r, for a component of the mixture. This is the length scale over which heat conduction must occur between two particles in the mixture for the system to be considered macroscopically homogeneous. Under these conditions, the potential gradient ∇V will be of order V/L throughout. If the electrical conductivity of the particle is of the order of size S over the temperature range considered, then we can now use the following scalings.

Let $T = \tau \underline{T}$, $t = t_0 \underline{t}$, $x_i = r\underline{x_i}$, $\nabla V = (V/L)\underline{\underline{\nabla}}\,\underline{\Phi}$, and $\kappa = S\underline{\kappa}$, so that

$$\frac{\partial \underline{T}}{\partial \underline{t}} = \frac{t_0}{\rho c_p} \frac{\lambda}{r^2} \underline{\nabla}^2 \underline{T} + \frac{t_0}{\rho c_p} \frac{S}{\tau} \left(\frac{V}{L}\right)^2 \kappa(\underline{\underline{\nabla}}\,\underline{\Phi})^2 \tag{25}$$

where τ, t_0, r, V/L, and S are constants representing typical magnitudes of the scaled quantities, and the nondimensionalized variables (denoted by underscores) are all of order 10^0.

For thermal conduction to dominate at one length scale (i.e. for approximate thermal uniformity at any time t), we require $|\lambda/r^2| \gg |(S/\tau)(V/L)^2|$. Typically, λ is $O(10^{-1})$ to $O(10^0)$ and S is $O(10^{-2})$ to $O(10^1)$ for mineral solutions and foodstuffs with a significant moisture content. Often, we have $\kappa = \kappa_0 + mT$, where κ_0 and m are constants, so that $S/\tau = m = (10^{-2})$. To simplify the requirement for the purposes of illustration, let us assume that the particle is a foodstuff for which $S = O(\lambda)$. Then, for thermal conduction to dominate, the requirement is

$$\left|\frac{1}{r^2}\right| \gg \left|\frac{1}{\tau}\left(\frac{V}{L}\right)^2\right| \quad \text{(a)}$$

or (26)

$$\left|\frac{L^2}{r^2}\right| \gg \left|\frac{1}{\tau} V^2\right| \quad \text{(b)}$$

that is, written as form (a), we require that the magnitude of the reciprocal of the individual particle size be significantly greater than that of $1/\tau$ times the voltage gradient, where $\tau = O(10^2)$ is a typical change in temperature in a mixture undergoing ohmic heating; as form (b), we require that the ratio of the macroscopic to the microscopic system length be significantly greater than the magnitude of $1/\tau$ times the applied voltage.

To check the conditions for heat generation dominating globally, the previous nondimensionalized variable scalings can be substituted into Eq. (24), together with a new length scaling $x_i = Lx_i$, where L is the macroscopic length scale over which heat generation takes place due to the potential difference across the mixture [22]:

$$\frac{\partial \underline{T}}{\partial \underline{t}} = \frac{t_0}{\rho c_p} \frac{\lambda}{L^2} \underline{\nabla}^2 \underline{T} + \frac{t_0}{\rho c_p} \frac{S}{\tau} \left(\frac{V}{L}\right)^2 \underline{\kappa} (\underline{\nabla} \underline{\Phi})^2 \tag{27}$$

So heat generation dominates globally (i.e., there is negligible thermal conduction) when $|\lambda| \ll |(S/T)V^2|$. Note that this condition is independent of length scales.

The above analysis puts bounds on the scales over which homogeneity will occur. The problem can be illustrated using FIDAP to study a conceptually simple situation. The case where different phases are considered to be in parallel has already been discussed. Consider an arrangement of slices of material, where each layer is of thickness x, as illustrated in Figure 13. Across the system, the voltage gradient is $V/L = E$. Inequality 26(a) showed us that the magnitude of x must be much smaller than that of E to achieve thermal homogeneity, and in that case the temperature evolution is given by Eq. (14) with $p = 0.5$.

If the order of magnitude of x were increased from a suitably small value, we would expect the following:

- At low x, the system will have the same temperature.

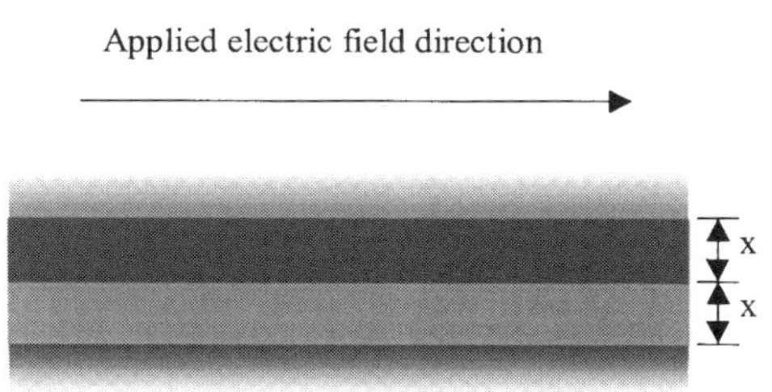

Figure 13 Two-component wafer, with each component of thickness x, heated using a constant, uniform potential gradient E.

- At high x, the temperature inhomogeneity will be such that the temperature in the centre of each layer is largely unaffected by thermal conduction over typical process times. The temperature in the center of each layer can then be calculated from Eq. (14) by substituting $p = 1$ for material 1, and $p = 0$ for material 2.

Simulations were performed using FIDAP to illustrate this transformation while keeping the potential gradient constant and uniform. Equations (1), (2), and (7) were solved for a section illustrated in Figure 13. The upper and lower boundaries of the simulated region were horizontal centerlines of the slices, giving symmetry (no-flux) boundary conditions. Zero thermal flux boundary conditions were also imposed at the electrodes. Material properties used were $\rho_1 = \rho_2 = 1000\,\mathrm{kg\,m^{-3}}$, $c_1 = c_2 = 4180\,\mathrm{J/kg/K}$, $\lambda_1 = 0.5\,\mathrm{W/m/K}$, $\lambda_2 = 0.7\,\mathrm{W/m/K}$, with constant electrical conductivities $\kappa_1 = 1\,\mathrm{S/m}$, $\kappa_2 = 2\,\mathrm{S/m}$. These values are in the typical range for fruit and vegetable foodstuffs.

Figures 14a and 14b show how the temperature in the center of two adjacent layers varies with the log of the thickness length scale, x. In Figure 14a, $E = 1000$ V/m (a voltage gradient comparable to commercial systems). The temperature is homogeneous throughout for $x < 10^{-3}$ m and is that which is given by solving Eqs. (6) and (14), substituting for material properties and with $p = 0.5$: $T - T_0 = 0.359t$. However, for $x > 5 \times 10^{-2}$ m, two distinct temperature curves are found, as given by Eqs (6) and (14) with $p = 1$ for material 1 and $p = 0$ for material 2: $T_1 - T_0 = 0.239t$, $T_2 - T_0 = 0.478t$. The transition occurs over the range $10^{-3} < x < 5 \times 10^{-2}$ m.

It would be expected that the slower the heating rate, the larger the length scale over which homogeneity would be achieved. Simulations were carried out (Figure 14b) for the same physical conditions, save that $E = 100$ V/m. The temperature curves are the same basic shape as for the higher-voltage gradient but are shifted along the length scale axis by one order of magnitude (note that the rate at which the heating takes place is much slower, so conduction has more time to act). This follows from the above dimensional analysis, which relates the system behavior to the particle length scale, the macroscopic length scale, and the voltage. Whereas inequalities (26) give conditions for thermal homogeneity and show how these quantities must be related, dimensional analysis yields a more general relationship between them, which can be used to express a "homogeneity factor". That is to say, because these quantities are considered the principal ones which determine the extent of homogeneity, Eq. (25) shows that they can be grouped together as $\chi = (V/L)r$. The greater the value of λ, then the greater the inhomogeneity. In this two-component wafer example, $r = x$, and if we reduce V/L by one order of magnitude, then we will get the

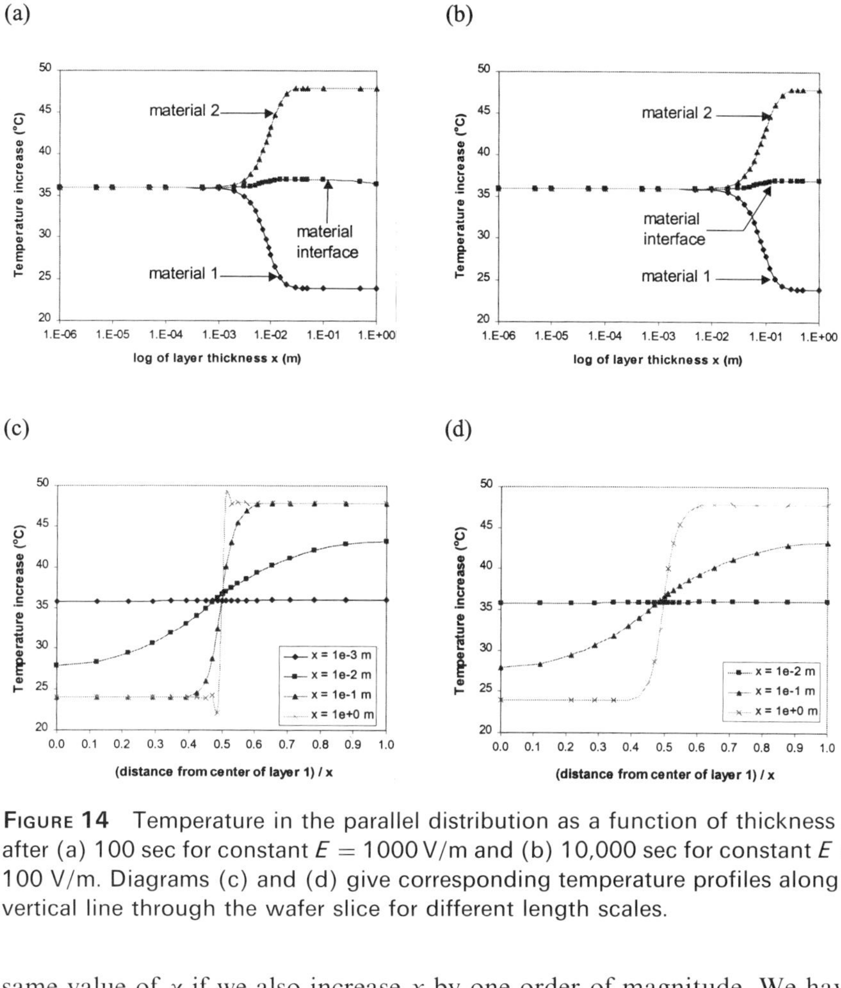

FIGURE 14 Temperature in the parallel distribution as a function of thickness x after (a) 100 sec for constant $E = 1000$ V/m and (b) 10,000 sec for constant $E = 100$ V/m. Diagrams (c) and (d) give corresponding temperature profiles along a vertical line through the wafer slice for different length scales.

same value of χ if we also increase x by one order of magnitude. We have also found in this example that for $\chi < 1$, there is homogeneity, and for $\chi > 50$, there is maximum temperature range in the mixture. Note that consideration of the series distribution, rather than the parallel distribution, would have lead to the same conclusions.

Where conductivities and temperatures are expected to have a greater variation, then we can rewrite χ as a general "dimensionless homogeneity factor," also derived from Eq. (25):

$$\chi = \frac{S}{\lambda T_0}\left(\frac{V}{L}\right)^2 r^2 \tag{28}$$

The distribution of temperature within the two-phase region can be calculated: Figures 14c and 14d show temperature profiles along the vertical centerline of the wafer, between the centers of two adjacent layers, for different length scales. With a potential gradient of 1000 V/m (Figure 14c), we find that for $x = 10^{-3}$ m, the temperature is homogeneous along the vertical centerline, but the variation is significant at larger scales. For the limiting case of $x = 10^{+0}$ m, the temperature gradient is so steep across the interface between the two components that some computational error can be seen. Again, Figure 14d, with a potential gradient of 100 V/m, shows the same behavior with an increase of one order of magnitude on the particle length scale, as predicted by Eq. (28), and two orders of magnitudes on the time scale.

This type of calculation is useful in determining the parameters for a successful ohmic process. It may be, for example, that the natural variation of foods is such that there is some physical maximum to the heating rate to give a range of temperatures which keeps the high quality of the product. The use of modeling in this case is conceptual: It is not possible to simulate a "real" situation in detail, but we must consider a case which gives an understanding of real situations.

7 DISCUSSION AND CONCLUSIONS

This chapter has considered different ways of modeling ohmic heating; different types of model are needed for different situations. The aim of all of the models and the groups producing them is to show how this process can best be used by the food industry. Ohmic heating can potentially overcome the limitations of thermal conduction in the heating of food particles. The process works by passing electrical current through a food system; as has been shown earlier, this can give rapid heating of a mixture. To get the full advantages out of the process requires an understanding of how it works and how to produce a high-quality product.

Models are required on several levels. There is a need to convince process regulators (and the people who buy new processes!) that ohmic heating is understood. There is also a need to produce a product which has the best achievable quality. One key area of modeling is to demonstrate the physical principles at work; much of the operation of ohmic heaters is not intuitive. The current flow is a function of the shape and physical properties of the media; current will be diverted into a conductor and round an insulator. As shown by experiment and model, particles can overheat or underheat the surrounding fluid, depending on the geometry and properties of the system.

Care must be taken in the design of formulations to eliminate temperature changes. The closer the match between the electrical conductivities, the more even the current flow will be; this still may not lead to uniform heating because the thermal capacity of the phases may be different, and the temperature dependence of the thermal conductivities may also differ. A useful first step in the analysis of any system is to identify the inherent heating rate, G [defined in eq. (6)], of each species across the required temperature range. This emphasizes the need to know the physical properties of the system over the whole range of processing. If there is significant variation between components of a mixture, then there might be problems in processing the formulation.

It is possible to model ohmic heating using finite-element programs to represent the geometry of real systems. Here, we have shown the use of the technique to model the temperature field around inclusions of low electrical conductivity. This is an extreme test of the model, representing a case where a foreign body is being processed by accident. This need not be a piece of plastic; it might be bone or fat instead of meat tissue. The model shows the types of effects that are found by experiment. The mixing in the fluid due to convection can be taken into account by using an enhanced thermal conductivity. This type of model can be used to model particular situations: It is not sensible to use it to model a full flowing system.

To show an understanding of how the process will work in practice and of how to control it, various models of the full heater have been proposed. All of these make approximations about the flow patterns of the mixture and the ways in which it will heat. It is not going to be possible ever to model a working process completely; the need is to be able to predict approximate temperatures to show how systems will behave. One major difficulty is predicting the mean conductivity of a mixture; although the EMA approximation may be appropriate, real systems will vary because of the nature of the distribution of particles in a medium.

The key challenge is to optimize all three stages of a process: heating, holding, and cooling. The length of the holding and cooling system required depends critically on the process temperature and the difference between solid and liquid temperatures during heating. It is not clear what the maximum feasible heating rate in a system will be. Any overheating of a section will lead to overprocessing, whereas underheating will give possible sterility problems. In conventional heating, the timescale is sufficiently slow that temperature differences are evened out by thermal conduction. In ohmic heating, it is possible that heating will be so rapid that this evening out will not occur over the timescale of the process. At a cellular level within the food, there will be local differences in conductivity which have to be evened out by conduction; at a process level, however, it will be necessary to ensure homogeneity over a centimeter scale, as there is not enough time to ensure evenness of temperature by conduction. Some models of how thermal homogeneity occurs over different timescales and length scales have been shown here. Again, the need is to look at heating, holding, and cooling stages. It may be that if the heating rate is too large, then the differences in temperature between the solid and liquid are so large that quality is lost during holding and cooling. A lower heating rate, with smaller temperature differences, may well be a better practical solution that gives a better product. This is a useful area for future work.

NOMENCLATURE

a	Area of the particle per unit volume
c_p	Specific heat
d	Dimension
d_t	Tube diameter
E	Field strength
E_0	Field strength a distance from the particle
f	Distribution function
G	Inherent heating rate of the material$=\frac{Q}{\rho c_p}$
h	Convective heat transfer coefficient between particle and liquid
H_L	Modified heat transfer term for liquid $=\frac{\phi}{1-\phi}\frac{ha}{(\rho c_p)_L}$
H_S	Modified heat transfer term for solid $=\frac{ha}{(\rho c_p)_S}$
h_W	Wall heat transfer coefficient
I	Electric current
K	Ratio of solid to liquid conductivity κ_S/κ_L
L	Length
m	Electric conductivity coefficient S/m/K
p	Probability, material fraction
Q	Heat-generation rate
R	Electrical resistance
r	Radius, particle length
S	Characteristic process electrical conductivity
T_0	Base/reference temperature
T	Temperature
t	Time
V	Voltage
v	Velocity
V_0	Voltage a distance from the particle
w	Mass flow rate
x	Length
$\mathbf{x_d}$	Length

Z	Coordination number
γ^{-1}	dimension
ϕ	Solid fraction
$\underline{\Phi}$	Dimensionless voltage
κ	Electrical conductivity
κ_0	Base temperature electrical conductivity
κ_e	Mean system electrical conductivity
λ	Thermal conductivity
λ^*	Enhanced thermal conductivity
μ	Exponent
θ	Angle
ρ	Density
τ	Characteristic process temperature

Subscript

0	Initial, characteristic value
1, 2	Index
a, b	Index
c	Critical value
e	Effective medium
L	Liquid phase
S	Solid phase

Note: Underscores indicate dimensionless variables.

ACKNOWLEDGMENTS

Work on ohmic heating described here has been funded by a number of companies. LJD wishes to acknowledge financial support from BBSRC and Unilever.

REFERENCES

1. SD Holdsworth. Aseptic Processing and Packaging of Food Products. London: Elsevier, 1993.
2. AC Metaxas. Foundations of Electroheat, A Unified Approach. New York: John Wiley & Sons, 1996.
3. AAP de Alwis, PJ Fryer. The use of direct resistance heating techniques in the food industry. J Food Eng 11:3–27, 1990.

4. J Yongsawatdigul, JW Park, E Kolbe, YA Dagga, MT Morissey. Ohmic heating maximises gel functionality of pacific whiting surimi. J Food Sci 60:10–14, 1984.
5. PJ Skudder, CH Biss. Aseptic processing of food products using ohmic heating. Chem Eng 433:26–28, 1987.
6. CH Biss, SA Coombes, PJ Skudder. In: RW Field, JA Howell, ed. Process Engineering for the Food Industry. London: Elsevier, 1987, pp 17–27.
7. R Stirling. Ohmic heating—A new process for the food industry. Power Eng J 1(6):365–371, 1987.
8. DL Parrott. Use of ohmic heating for aseptic processing of particles. Food Technol 46:68–72, 1992.
9. PJ Fryer. Electrical resistance heating of foods. In: G Gould, ed. New Methods of Food Preservation. Glasgow: Blackie, 1995, pp 205–235.
10. HJ Kim, YM Choi, TCS Yang, IA Taub, P Tempest, P Skudder, G Tucker, DL Parrott. Validation of ohmic heating for quality enhancement of food products. Food Technol 50:253–261, 1996.
11. HJ Kim, YM Choi, A Yang, IA Taub, J Giles, C Ditusa, S Chall, P Zoltai. Microbiological and chemical investigation of ohmic heating of particulate foods using a 5 kW ohmic system. J Food Proc Preserv 20:41–58, 1996.
12. SK Sastry, CA Zuritz. Review of particle behaviour in tube flow: applications to aseptic processing. J Food Proc Eng 10:27–52, 1987.
13. C Skjoldebrand, T Ohlsson. Computer simulation program for evaluation of the continuous heat treatment of particulate food products. Part 1: Design. J Food Eng 20:149/166, 1993.
14. C Skjoldebrand, T Ohlsson. Computer simulation program for evaluation of the continuous heat treatment of particulate food products. Part 2: Utilization. J Food Eng 20:167–182, 1993.
15. S Mankad, CA Branch, PJ Fryer. The effect of particle–liquid slip on the sterilisation of solid–liquid food mixtures. Chem Eng Sci 50:1311–1321, 1995.
16. AAP de Alwis, K Halden, PJ Fryer. Shape and conductivity effects in the ohmic heating of foods. Chem Eng Res Des 67:159–168, 1989.
17. AAP de Alwis, PJ Fryer. Operability of the ohmic heating process: Electrical conductivity effects. J Food Eng 15:21–48, 1992.
18. SK Sastry, S Palaniappan. Influence of particle orientation on the effective electrical resistance and ohmic heating rate of a liquid–particle mixture. J Food Proc Eng 15:213–227, 1992.
19. PJ Fryer, AAP de Alwis, B Koury, AGF Stapley, L Zhang. Ohmic processing of solid–liquid mixtures: Heat generation and convection effects. J Food Eng 18:101–125, 1993.
20. M Sahimi. Applications of Percolation Theory. London: Taylor & Francis, 1994.
21. S Kirkpatrick. Percolation and conduction. Rev Mod Phys 45:574–588, 1973.
22. LJ Davies. The effect of ohmic heating on particles in food mixtures. PhD dissertation, The University of Birmingham, 1999.
23. SK Sastry, S Palaniappan. Mathematical modelling and experimental studies on ohmic heating of liquid–particle mixtures in a static heater. J Food Proc Eng 15:241–261, 1992.

24. SK Sastry. Model for heating of liquid–particle mixtures in a continuous flow ohmic heater. J Food Proc Eng 15:263–278, 1992.
25. L Zhang, PJ Fryer. Models for the electrical heating of solid–liquid food mixtures. Chem Eng Sci 48:633–643, 1993.
26. L Zhang, PJ Fryer. Food sterilization by electrical heating: Sensitivity to process parameters. AIChE J 40:888–898, 1994.
27. AAP de Alwis, PJ Fryer. A finite element analysis of heat generation and transfer during ohmic heating of foods. Chem Eng Sci 45(6):1547–1560, 1990.
28. LJ Davies, MR Kemp, PJ Fryer. The geometry of shadows: Effects of inhomogeneities in electrical field processing. J Food Eng 40:245–258, 1999.
29. SK Sastry, S Salengke. Ohmic heating of solid-liquid mixtures: A comparison of mathematical models under worst-case heating conditions. J Food Proc Eng 21:441–458, 1998.
30. AAP de Alwis, L Zhang, PJ Fryer. Modelling sterilization and quality in the ohmic heating process. In: RK Singh, PB Nelson, eds. Advances in Aseptic Processing Technologies. London: Elsevier, 1992, pp 103–142.
31. L Zhang, PJ Fryer. Electrical resistance heating of foods. Trends Food Sci Technol 4(11):364–369, 1993.
32. M Barigou, S Mankad, PJ Fryer. Heat transfer in two-phase solid–liquid food flows: A review. Trans IChemE Part C 76:3–29, 1998.
33. C Lareo, PJ Fryer, M Barigou. The fluid mechanics of two-phase solid–liquid food flows: A review. Trans IChemE Part C 75:73–105, 1997.
34. L Zhang, PJ Fryer. Heat transfer and generation in electric heating of a laminar flow of food. In: T Yano, R Matsuno, K Nakamura, eds. Developments in Food Engineering. London: Blackie, 1994, pp 760–762.
35. AL Quarini. Thermalhydraulic aspects of the ohmic heating process. J Food Eng 23:561–574, 1994.
36. C Lareo, CA Branch, PJ Fryer. Particle velocity profiles for solid–liquid food flows. I: Single particles. Powder Technol 93:23–34, 1997.
37. C Lareo, RM Nedderman, PJ Fryer. Particle velocity profiles for solid–liquid food flows. II: Multiple particles. Powder Technol 93:35–46, 1997.
38. L Zhang, PJ Fryer. Alternative formulations for the prediction of electrical heating rates of solid–liquid food materials. J Food Proc Eng 18:85–98, 1995.

9

Stochastic Finite-Element Analysis of Thermal Food Processes

Bart M. Nicolaï, Nico Scheerlinck, Pieter Verboven, and Josse De Baerdemaeker
Katholieke Universiteit Leuven, Leuven, Belgium

1 INTRODUCTION

For the design of thermal food process operations, the temperature in the thermal center of the food during the process must be known. Whereas traditionally this temperature course is measured using thermocouples, there is a growing interest toward the use of mathematical models to predict the food temperature during the thermal treatment [1–4]. Advantages of such an approach include the computation of heat-penetration curves corresponding to arbitrary process conditions and container shapes (e.g., glass jars [2]), the ability to predict overshoot [5], rapid on-line evaluation of unscheduled process deviations [6], and optimization of thermal processes [7,8].

In the case of conduction-heated foods, the heat transfer process is described through the Fourier equation. For complicated geometries and time-dependent boundary conditions, usually no analytical solutions are available for the Fourier equation, and a numerical solution becomes mandatory. Several methods, including finite differences [1,3,4] and finite elements [2] have been applied successfully by numerous investigators to the numerical solution of the Fourier equation. Although commercial codes (mostly based on the finite-element or finite-volume method) are now widely available, they require that the product and process parameters are accurately known. However, in reality, these parameters may vary quite extensively, due to biological variability or unpredictably changing conditions such as the ambient temperature. Consequently, the temperatures inside the product are stochastic quantities, which must be characterized by statistical means.

Nicolaï et al. [9] reviewed the sources of uncertainty in thermal sterilization processes. The CV (coefficient of variation) of the f value, and, hence, the thermal diffusivity, is typically 3–15%, although values as high as 26% have been observed [10]. Meffert [11] concluded that the possible maximum error in the experimental determination of the thermal conductivity can be as high as 30–50% at the 95% confidence level. Sheard and Rodger [12] compared the time required for vacuum-packed and non-vacuum-packed potato slabs to establish a temperature increase from 20°C to 75°C with an oven setpoint of 80°C at different positions in commercial steamers and for different oven types. They found substantial differences in heating time between packs of the same shelf. According to these authors, the observed heating time variations were due to the intermittent inputs of the steam used to maintain temperatures below 100°C. The standard deviation of the temperature of a well-controlled retort is typically 1°C [13]. In a more recent publication [14], it was reported that the retort temperature variability is normally less than ±0.5°C. Ramaswamy et al. observed that the maximum difference between different positions during the holding phase was between 2.6°C and 3.5°C [15]. The average of the standard deviations of the retort temperatures at each time was 1.3°C. From experiments inside a pilot-scale water cascading retort, it was found [16] that the average of the standard deviations of the temperatures at different positions was equal to 0.7°C during the entire cook period. The overall standard deviation during the cook period was 0.9°C, and the maximum temperature difference between positions 1 min after the coming-up period was equal to 3.2°C. Little information is available on the variability of the surface heat transfer coefficient in thermal food processes. Martens [13] used a CV of 10% and 25% for his Monte Carlo analyses.

As the thermal inactivation of microorganisms is highly dependent on the temperature, it is very possible that this uncertainty will result in a situation where some foods of the same batch are microbiologically safe while others are not. The uncertainty involved in thermal food process design has therefore been addressed by several authors [13,17–21] by means of Monte Carlo analyses. In this method a large number of samples of the random parameters are generated by the computer, and for every set, the thermal problem is solved. In the end, statistical parameters such as the mean value and the variance of the temperature at the thermal center can be calculated using statistical inference. The drawback of the Monte Carlo method is the large amount of computer time, particularly when the thermal problem is to be solved numerically. Alternative algorithms have therefore been suggested to calculate the propagation of parameter fluctuations in space and/or time [22–25].

In this chapter, the use of stochastic finite-element methods to calculate statistical characteristics of the temperature field inside conduction heated foods will be described. The main features of the Monte Carlo and variance propagation algorithms will be illustrated by a numerical example.

2 NUMERICAL COMPUTATION OF CONDUCTION HEAT TRANSFER

Transient linear heat transfer in solid foods subjected to convection boundary conditions is governed by the Fourier equation

$$k\nabla^2 T + Q = \rho c \frac{\partial T}{\partial t} \tag{1}$$

$$k \frac{\partial}{\partial n} T = h(T_\infty - T) \quad \text{on } \Gamma \tag{2}$$

$$T = T_0 \quad \text{at } t = t_0 \tag{3}$$

where T is the temperature (°C), k is the thermal conductivity (W/m °C), ρc is the volumetric heat capacity (J/m^3 °C), T_∞ is the (known) process temperature (°C), n is the outward normal to the surface, h is the convection coefficient (W/m^2 °C), Γ is the boundary surface, Q is the heat generation (W/m^3), and t is the time (sec).

For many realistic heat-conduction problems, no analytical solutions of Eq. (1) subjected to Eqs. (2) and (3) are known. In this case, numerical discretization techniques such as the finite-difference or finite-element method can be used to obtain an approximate solution. The finite-element method in particular is a very flexible and accurate method for solving partial differential equations such as the Fourier equation. In the framework of the finite-element method, the continuum is subdivided in elements of variable size and shape which are interconnected in a finite number of nodal points, n_{nod}. In every element j, the unknown temperature is approximated by a low-order interpolating polynomial

$$u^j(T) = \boldsymbol{\phi}^{j^T} \mathbf{u}^j(t) \tag{4}$$

where $u^j(t)$ is the approximate temperature in element j, $\mathbf{u}^j(t)$ is the vector containing the nodal temperatures in element j, and $\boldsymbol{\phi}^j$ is the vector of shape functions corresponding to element j. The application of a suitable spatial discretization technique such as the Galerkin weighted residual method to Eq. (1) subjected to Eqs. (2) and (3) results in the following differential system [26]:

$$\mathbf{C}\frac{d}{dt}\mathbf{u} + \mathbf{K}\mathbf{u} = \mathbf{f} \tag{5}$$

$$\mathbf{u}(t=0) = \mathbf{u}_0 \tag{6}$$

with $\mathbf{u} = [u_1 \ u_2 \ \cdots \ u_{n_{\text{nod}}}]^T$ the overall nodal temperature vector, $\mathbf{C}$ is the capacitance matrix and $\mathbf{K}$ is the conductance matrix, both $n_{\text{nod}} \times n_{\text{nod}}$ matrices, and $\mathbf{f}$ is a $n_{\text{nod}} \times 1$ vector. The system (5) can be solved by finite differences in the time domain. For the construction of the global finite-element matrices $\mathbf{C}$, $\mathbf{K}$ and $\mathbf{f}$, it is most convenient from the programming point of view to first assemble the contributions of each element (the "element matrices") $\mathbf{C}^j$, $\mathbf{K}^j$, and $\mathbf{f}^j$:

$$\mathbf{C}^j = \int_{V^j} \rho c \boldsymbol{\phi}^j \boldsymbol{\phi}^{j^T} \, dV \tag{7}$$

$$\mathbf{K}^j = \int_{V^j} k \mathbf{B}^j \mathbf{B}^{j^T} \, dV + \int_{S^j} h \boldsymbol{\phi}^j \boldsymbol{\phi}^{j^T} \, dS \tag{8}$$

$$\mathbf{f}^j = \int_{S^j} h T_\infty \boldsymbol{\phi}^j \, dS + \int_{V^j} Q \boldsymbol{\phi}^j \, dV \tag{9}$$

with

$$\mathbf{B}^j = \frac{\partial \boldsymbol{\phi}^j}{\partial \mathbf{z}},$$

and S_j and V_j the boundary surface and volume of element j, respectively. The element matrices are then incorporated in the global matrices. The matrices $\mathbf{K}$ and $\mathbf{C}$ are sparse and this property can be exploited advantageously for reducing the CPU time required for the solution of Eq. (5).

The finite-element method has been successfully used in a number of thermal food processing applications such as sterilization of baby food jars [2] and cooling of broccoli stalks [27] and tomatoes [28].

3 DESCRIPTION OF UNCERTAINTY

In the finite-element method, it is assumed that all parameters are deterministic and known. However, in reality, this is certainly not always the case. In the stochastic finite-element method, knowledge about the uncertainty of the material and process parameters is explicitly incorporated in the calculations. It is therefore required that an appropriate mathematical description of the random parameters is available. In this section, the random variable model, along with its multidimensional extensions such as random process, field, and wave, will be introduced. For a more precise description of these concepts, the reader is referred to the literature [29].

3.1 Random Variables

The most simple uncertainty model is that of a random variable. A random variable X is a real-numbered variable whose value is associated with a random experiment. For example, the heat capacity of a potato is a random variable which can vary between different potatoes. A random variable X can be characterized by its probability density function $f(x)$ and statistical moments such as the mean value $\bar{X}$ and the variance σ^2, if existent and known:

$$\bar{X} \triangleq \mathrm{E}(X) \tag{10}$$

$$\triangleq \int_{-\infty}^{\infty} xf(x)\,dx \tag{11}$$

$$\sigma^2 \triangleq \mathrm{E}(X - \bar{X})^2 \tag{12}$$

$$\triangleq \int_{-\infty}^{\infty} (x - \bar{X})^2 f(x)\,dx \tag{13}$$

with E the expectation operator.

Sometimes, an experiment will yield values for two or more physical parameters. Assume, for example, that both the thermal conductivity and the volumetric heat capacity of a material are measured simultaneously. In this case, the outcome of the experiment is called a bivariate (2) or multivariate (more than 2) random variable. The random variables X_1 and X_2 can then be stacked conveniently in a random vector $\mathbf{X}$. Similar to the univariate case, we can then define the mean value $\bar{X}$ and covariance matrix $\mathbf{V}$ of the random vector as

$$\bar{X} \triangleq \mathrm{E}(\mathbf{X}) \tag{14}$$

$$\mathbf{V} \triangleq \mathrm{E}[(\mathbf{X} - \mathbf{X})(\mathbf{X} - \mathbf{X})^T] \tag{15}$$

The ith diagonal entry of $\mathbf{V}$ is the variance $\sigma^2_{X_i}$ of random variable X_i; the (i, j)th entry of $\mathbf{V}$ is the covariance σ_{X_i,X_j} of random variables X_i and X_j.

As expected, the probability density function $f(X_1, X_2)$ of a bivariate random variable is a function of two variables. The bivariate Gaussian density function is often used to describe bivariate random variables. It is defined as

$$f(X_1, X_2) = \frac{(1 - R^2_{X_1,X_2})^{-1/2}}{2\pi\sigma_{X_1}\sigma_{X_2}} \exp\left\{ -\frac{1}{2(1 - R^2_{X_1,X_2})}\left[\left(\frac{X_1 - \bar{X}_1}{\sigma_{X_1}}\right)^2 - 2R_{X_1,X_2}\left(\frac{X - \bar{X}_1}{\sigma_{X_1}}\right)\left(\frac{X_2 - \bar{X}_2}{\sigma_{X_2}}\right) + \left(\frac{X_2 - \bar{X}_2}{\sigma_{X_2}}\right)^2\right]\right\} \tag{16}$$

R is called the correlation coefficient and $-1 \leq R \leq 1$.

3.2 Random Processes

If a parameter changes in an unpredictable way as a function of the time coordinate, it can be described conveniently by means of a random process. The mean $\bar{X}$ and covariance V of a stationary process X with probability density function $f(x, t)$ are defined by

$$\bar{X} = \mathrm{E}(X) \tag{17}$$

$$\triangleq \int_{-\infty}^{\infty} xf(x, t)\, dx \tag{18}$$

$$V(\tau) = \mathrm{E}\{[X(t) - \bar{X}][X(t + \tau) - \bar{X})]\} \tag{19}$$

The covariance function describes how much the current value of the random function will affect its future values. By definition of a *stationary* random process, the mean of the process does not change in time and its covariance function is only a function of the separation time τ. The correlation function R is found by normalization of the covariance function:

$$R(\tau) = \frac{V(\tau)}{\sigma^2} \tag{20}$$

A Gaussian stationary white-noise process W with covariance

$$V_{W,W}(\tau) = \sigma^2_{W,W}\delta(\tau), \tag{21}$$

where δ is the Dirac delta, can be used to describe very rapid unpredictable fluctuations. Sample values of W are uncorrelated no matter how close together in time they are. However, white noise does not exist in reality, as it has an infinite energy content and variance. Autoregressive processes provide a tool to incorporate fluctuations which change more smoothly as a function of time. An autoregressive random process of order m [i.e., AR(m)] is defined by the following stochastic differential equation:

$$\frac{d^m}{dt^m} X(t) + a_1 \frac{d^{m-1}}{dt^{m-1}} X(t) + \cdots + a_m X(t) = W(t) \tag{22}$$

where $a_1, a_2, \ldots, a_m$ are constants, $m \geq 1$ and $W(t)$ is a stationary Gaussian white-noise process with mean $\bar{W} = a_m \bar{X}$. The timescale of the fluctuations is dependent on the coefficients $a_1, \ldots, a_m$, and their high-frequency content decreases with increasing order m. The (Gaussian) random variable initial condition corresponding to the stochastic differential equation (22) is defined as

$$\mathrm{E}[X(t_0)] = \bar{X} \tag{23}$$

$$\mathrm{E}[X(t_0) - \bar{X}]^2 = \sigma^2 \tag{24}$$

Note that a *random variable* parameter X can be modeled as a trivial case of an AR(1) process:

$$\frac{d}{dt} X = 0 \tag{25}$$

AR(m) processes are a special case of the class of *physically realizable* stochastic processes which comprise most of the random processes seen in practice [30].

In order to describe the smoothness of a random process by means of a single measure, Vanmarcke [29] introduced the concept of *scale of fluctuation*, which is defined as

$$\theta = \int_{-\infty}^{+\infty} R(\tau)\, d\tau$$

It gives an indication of the time beyond which a future value of a random process will not be affected anymore by its current value. In Table 1, the variance, the autocovariance function, and the scale of fluctuation are given for AR(1) and AR(2) processes. For the latter, the characteristic polynomial

$$\xi^m + a_1 \xi^{m-1} + \cdots + a_m = 0 \tag{26}$$

has two real or two complex conjugate roots, resulting in nonoscillating or oscillating correlation functions, respectively. In Figure 1 some correlation functions and corresponding realizations of an AR(2) process with different scales of fluctuation are compared. If $\theta \to 0$, then the process approximates a white-noise process. On the other hand, if $\theta \to +\infty$, then the values of the realization at arbitrary points are completely correlated. In this case, the random process concept is far too sophisticated to describe the physical quantity because all the meaningful probabilistic features of the quantity can be captured by a simple random variable model.

TABLE 1 Autocovariance Function and Scale of Fluctuation of AR(1) and AR(2) Processes

AR(1)	$V_{x,x}(\tau) = \sigma_x^2 e^{-a_1 \lvert\tau\rvert}$ $\theta = 2/a_1$
AR(2) real roots	$V_{x\,x}(\tau) = \sigma_x^2(\xi_2 e^{\xi_1\lvert\tau\rvert} - \xi_1 e^{\xi_2\lvert\tau\rvert})[(\xi_2 - \xi_1)]^{-1}$ with ξ_1 and ξ_2 the roots of $\xi^2 + a_1\xi + a_2 = 0$ $\theta = 2a_1/a_2$
AR(2) complex roots	$V_{x\,x}(\tau) = \sigma_x^2 e^{-a_1\lvert\tau\rvert/2}[\cos(p\lvert\tau\rvert) + (a_1/2p)\sin(p\lvert\tau\rvert)]$ with $p = (a_2 - a_1^2/4)^{1/2}$ $\theta = 2a_1/a_2$

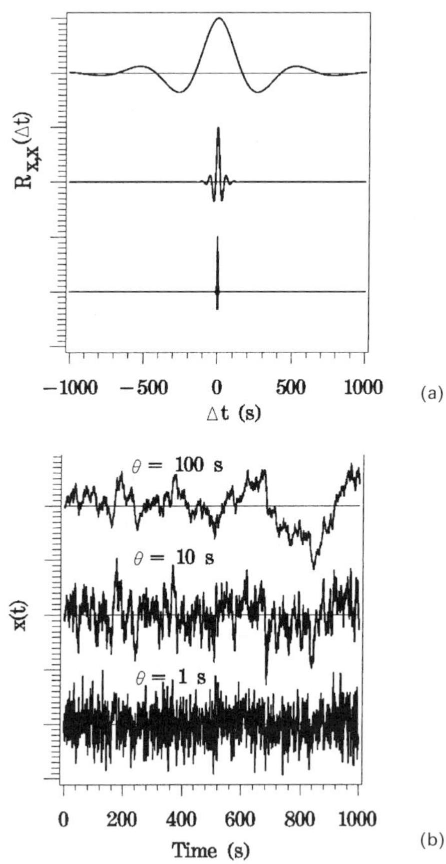

FIGURE 1 Correlation functions (a) and corresponding realizations (b) of a AR(2) process with different scales of fluctuation.

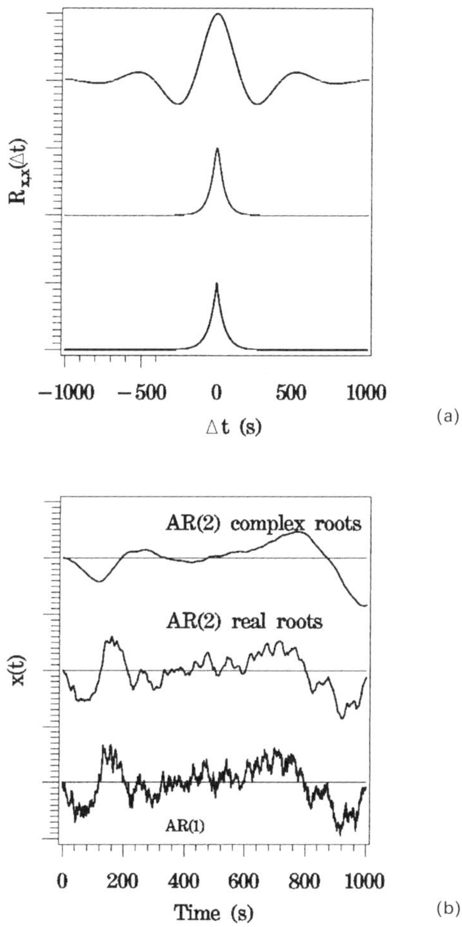

FIGURE 2 Correlation functions (a) and corresponding realizations (b) of several types of autoregressive processes with the same scale of fluctuation.

The correlation functions and corresponding realizations of different types of random processes are shown in Figure 2. Clearly, the scale of fluctuation is a measure of how frequent the process wiggles around the mean axis, irrespective of the order of the process.

It is convenient to write the autoregressive process (22) in the following *state-space form* [31]:

$$\frac{d}{dt}\mathbf{X}(t) = \mathbf{A}\mathbf{X}(t) + \mathbf{B}W(t) \tag{27}$$

where

$$\mathbf{X} = \begin{bmatrix} X \\ \dfrac{dX}{dt} \\ \ddots \\ \dfrac{d^{m-2}X}{dt^{m-2}} \\ \dfrac{d^{m-1}X}{dt^{m-1}} \end{bmatrix}, \qquad \mathbf{B} = \begin{bmatrix} 0 \\ 0 \\ \vdots \\ 0 \\ 1 \end{bmatrix}$$

$$\mathbf{A} = \begin{bmatrix} 0 & 1 & \dots & 0 \\ \ddots & & \ddots & \vdots \\ 0 & 0 & \dots & 1 \\ -a_m & -a_{m-1} & \dots & -a_1 \end{bmatrix}$$

The vector **X** is called the state vector, the matrix **A** is the companion matrix, and the vector **B** is an auxiliary vector.

3.3 Random Fields and Random Waves

Often a physical quantity varies randomly as a function of the time and/or space coordinates. Examples include the temperature in an oven, the thermophysical properties of heterogeneous materials such as foods, hydraulic properties of soils, elastic properties of construction materials, and so forth. The random field concept provides a convenient mathematical framework to describe such phenomena [29].

A parameter which fluctuates in both space and time can be described by means of random waves. The random-wave model is a straightforward extension of the random field model combined with the random process model. A full account of random fields and random waves is beyond the scope of this chapter, and the reader is referred to the literature [29].

4 THE MONTE CARLO METHOD

4.1 Description

The Monte Carlo method was introduced by John von Neumann and Ulam during World War II for studying random neutron diffusion problems in fissionable material which arose in the development of the atomic bomb [32]. The code name of the project was *Monte Carlo*, from where the method inherits its name.

In the Monte Carlo finite-element method, samples of the random parameters are generated by means of a random generator. For every parameter set, the heat-conduction problem is solved by analytical or numerical means, and the solution is stored for future use. This process is repeated a large number of times n, and in the end, the statistical characteristics are estimated. For the mean and the variance of the solution T at arbitrary space–time coordinates, the following nonbiased estimation formulas can be applied:

$$\hat{\bar{T}} = \frac{1}{n} \sum_{j=1}^{n} T^j \tag{28}$$

$$\hat{\sigma}^2 = \frac{1}{n-1} \sum_{j=1}^{n} (T^j - \bar{T})^2 \tag{29}$$

where T^j is the solution in the jth Monte Carlo run and the "hat" symbol (ˆ) means "estimate of."

If T is a linear function of the random parameters (e.g., ambient temperature) and if the latter are normally distributed, then T is also normally distributed. In this case, the confidence intervals for $\hat{\bar{T}}$ and $\hat{\sigma}^2$ are given by

$$\hat{\bar{T}} - t_{0.975} \frac{\hat{\sigma}}{\sqrt{n}} \leq \bar{T}_i \leq \hat{\bar{T}} + t_{0.975} \frac{\hat{\sigma}}{\sqrt{n}} \tag{30}$$

$$\frac{\chi^2_{0.025}}{n-1} \leq \frac{\hat{\sigma}^2}{\sigma^2} \leq \frac{\chi^2_{0.975}}{n-1} \tag{31}$$

where the t and χ^2 are Student t and χ^2 distributed and are to be evaluated with $n - 1$ degrees of freedom [33]. For $n \leq 30$, they are tabulated in all textbooks on introductory statistics; for $n \geq 30$, it can be shown [33] that $\sqrt{2\chi^2} - \sqrt{2n-1}$ is normally distributed with zero mean and unit variance. The Student's t distribution then approximates a Normal distribution.

The formulas in Eqs. (30) and (31) are only valid if the temperature T is linearly dependent on the random parameter(s). Even if this is not so (e.g., in the case of random thermal conductivity), these formulas can be used as a first-order approximation of the real confidence intervals if the variability of the random parameters is not too large.

For n equal to 100 and 1000, Eq. (31) becomes

$$0.74 < \frac{\hat{\sigma}^2}{\sigma^2} < 1.28 \quad \text{for } n = 100 \tag{32}$$

$$0.91 < \frac{\hat{\sigma}^2}{\sigma^2} < 1.09 \quad \text{for } n = 1000 \tag{33}$$

This means that, even for $n = 1000$, the relative confidence band is almost 20% wide. It can therefore be concluded that the large number of repetitive simulations necessary to obtain an acceptable level of accuracy is a major drawback of the Monte Carlo method, particularly when, in each run, a finite-element problem must be solved. If the random parameters are of the random field type, the time required to generate the parameter *samples* can outweigh by far the actual CPU time required to solve the finite-element problem. A careful choice of the algorithms to generate the random samples is therefore imperative, as it may considerably reduce the total CPU time.

A further drawback of the Monte Carlo method is the fact that the stochastic parameter set must be completely specified in the probabilistic sense, including (joint) probability density functions.

4.2 Generation of Random Variables and Processes

Uniformly distributed random numbers are now most commonly generated by means of a *congruential generator*. In this method, a discrete random number x_{i+1} is derived from a previous one, x_i, based on a fundamental congruence relationship:

$$x_{i+1} = (ax_i + c) \mod m \qquad i = 0, 1, \ldots \tag{34}$$

where the *multiplier* a, the *increment* c, and the *modulus* m are non-negative integers. The modulo (mod) is defined as the remainder of the integer division (e.g., 5 mod 3 is equal to 2). The recursion is started with a starting value x_0, the *seed*. It has been shown statistically that the x_i are uniformly distributed on the interval $[0, m)$. Uniformly distributed random numbers on the unit interval $[0, 1)$ can be obtained by dividing the x_i by m. Obviously, after at most m recursions, the random sequence will repeat itself. Conditions on a, c, and m can be found such that the period after which the sequence will repeat itself is maximal [32]. The following values give random numbers of reasonable quality [34] and do not cause integer overflow on most systems: $m = 233{,}280$, $c = 49{,}297$, and $a = 9{,}301$.

It is emphasized here that the implementation of a good random number generator is by no means a trivial task. Inappropriate constants a, m, and c can lead to numbers which are highly correlated. It is therefore suggested to use random number generators which are provided with standard mathematical packages such as Nag (The Numerical Algorithms Group, Oxford, U.K.) or IMSL (Visual Numerics, Inc., Houston, TX). The generators which are included in compilers must be used with special care; for example, those implemented in some commercial C compilers generate random numbers of very poor quality. For a discussion, see Ref. 34.

Random variates with nonuniform probability density function $f(x)$ can be obtained from uniformly distributed random numbers on [0, 1) by several methods, including the transformation method and the acceptance–rejectance method. For example, consider the following transformation due to [35]

$$Z_1 = (-2 \ln U_1)^{1/2} \cos 2\pi U_2 \tag{35}$$

$$Z_2 = (-2 \ln U_1)^{1/2} \sin 2\pi U_2 \tag{36}$$

It can be shown that if U_1 and U_2 are two uniformly random numbers in $[0, 1)$, then Z_1 and Z_2 are two uncorrelated standard ($\mu = 0, \sigma^2 = 1$) Gaussian random numbers. The histogram in Figure 3 was produced from 1000 Gaussian numbers according to Eqs. (35), (36), and (34) with the above given numerical values of a, c, and m.

Algorithms for other probability density functions are described in Ref. 32.

Samples (or *realizations*) of an AR(m) random process can be generated recursively by time discretization of the corresponding differential equation. For example, for an AR(1) process, we have that

$$\frac{d}{dt} X + a_1 X = W \tag{37}$$

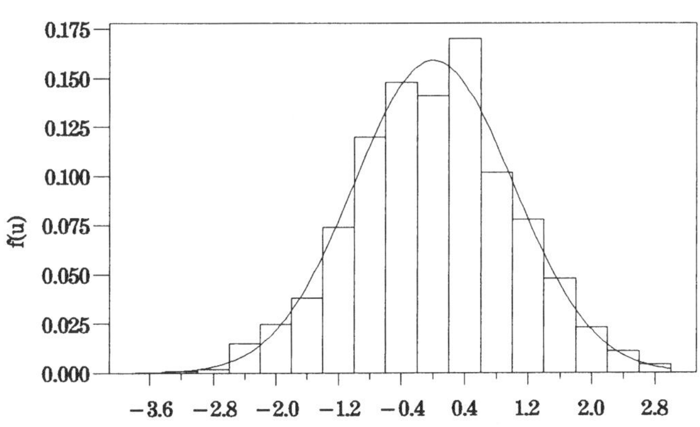

FIGURE 3 Histogram of a measurement experiment of a random variable U. The solid curve is the limiting probability density function.

By applying an implicit Euler discretization, we obtain the following time series:

$$(1 + a_1 \Delta t)X(t + \Delta t) = X(t) + \Delta t Z(t + \Delta t) \tag{38}$$

or

$$X(t + \Delta t) = (1 + a_1 \Delta t)^{-1}[X(t) + \Delta t Z(t + \Delta t)] \tag{39}$$

with Z discrete-time white-noise (a sequence of Gaussian random numbers). It can be shown [36] that the variance σ_Z^2 of Z is equal$\sigma_W^2/\Delta t$. The algorithm is bootstrapped with a random value x of the process as specified in Eqs. (23) and (24). A sufficiently small time step should be selected because otherwise the variance of the generated sample will be smaller than the target variance. Methods to generate samples of AR(m) processes of arbitrary order are compared in [37].

5 THE VARIANCE PROPAGATION ALGORITHM

The major drawback of the Monte Carlo method is the considerable amount of CPU time required to obtain accurate estimates of the stochastic characteristics of the temperature field. The variance propagation algorithm is an alternative to the Monte Carlo method. For a full account of the algorithm, the reader is referred to [36].

5.1 Lumped-Capacitance Heat-Conduction Problems

In order to explain the variance-propagation algorithm, we will consider the following simple *lumped-capacitance* heat transfer problem [38]. Consider a sphere of radius r_0 with thermal capacity c and density ρ. The sphere is initially at a uniform temperature T_0. At time $t = 0$, the sphere is immersed in a water bath at temperature T_∞. The temperature of the sphere will approach T_∞ with a rate which depends on the surface heat transfer coefficient h at the solid–liquid interface. In the lumped-capacitance method, it is assumed that, because of the high thermal conductivity of the solid medium, the temperature inside the solid is uniform at any instant during the transient heat transfer process. This hypothesis holds if the Biot number, Bi, satisfies the following constraint:

$$\mathrm{Bi} = \frac{hL}{k} < 0.1 \tag{40}$$

where L is the characteristic length of the solid which, in the case of a sphere, is usually defined as $L = 3r_0$ [38]. It is easy to show that applying an overall energy balance leads to the following differential equation:

$$\rho c \frac{d}{dt} T = \frac{3h}{r_0} (T_\infty - T) \tag{41}$$

After integration, the following formula for the temperature course is found:

$$T = T_\infty + (T_0 - T_\infty) \exp\left(-\frac{3h}{\rho c r_0} t\right) \tag{42}$$

For simplicity, we will assume that T_∞ is an AR(1) random process described by means of the following differential equation:

$$\frac{d}{dt} T_\infty + a_1 T_\infty = W \tag{43}$$

with W a white-noise process with mean $\bar{W} = a_1 \bar{T}_\infty$. We can combine Eqs. (43) and (41) into the following global system:

$$\frac{d}{dt} \mathbf{x} = \mathbf{g}(\mathbf{x}) + \mathbf{h}(W - \bar{W})(t) \tag{44}$$

with

$$\mathbf{x} = \begin{bmatrix} T \\ T_\infty \end{bmatrix} \tag{45}$$

$$\mathbf{g} = \begin{bmatrix} \frac{3h}{\rho c r_0} T_\infty - \frac{3h}{\rho c r_0} T \\ \bar{W} - a_1 T_\infty \end{bmatrix} \tag{46}$$

$$\mathbf{h} = \begin{bmatrix} 0 \\ 1 \end{bmatrix} \tag{47}$$

It can be shown [36] that first-order approximate expressions for the mean vector and the covariance matrix of the solution of Eq. (44) are given by

$$\frac{d}{dt} \bar{\mathbf{x}} = \mathbf{g}[\bar{\mathbf{x}}] \tag{48}$$

$$\frac{d}{dt} \mathbf{V}_{\mathbf{x},\mathbf{x}} = \frac{\partial \mathbf{g}}{\partial \mathbf{x}} \mathbf{V}_{\mathbf{x},\mathbf{x}} + \mathbf{V}_{\mathbf{x},\mathbf{x}} \left(\frac{\partial \mathbf{g}}{\partial \mathbf{x}}\right)^T + \mathbf{h} \sigma_W^2 \mathbf{h}^T \tag{49}$$

with

$$\frac{\partial \mathbf{g}}{\partial \mathbf{x}} \triangleq \left. \frac{\partial \mathbf{g}[\mathbf{x}(t),\ t]}{\partial \mathbf{x}(t)} \right|_{\bar{\mathbf{x}}(t)}$$

Equations (48) and (49) are called the *variance-propagation algorithm*. Equation (49) is a matrix differential equation of the *Lyapunov* type.

If we combine Eqs. (44)–(49), we obtain the following system:

$$\frac{d}{dt}\bar{T} = \frac{3h}{\rho c r_0}(\bar{T}_\infty - \bar{T}) \tag{50}$$

$$\frac{d}{dt}\bar{T}_\infty = 0 \tag{51}$$

$$\frac{d}{dt}\mathbf{V}_{\mathbf{x},\mathbf{x}} = \begin{bmatrix} -\dfrac{3h}{\rho c r_0} & \dfrac{3h}{\rho c r_0} \\ 0 & -a_1 \end{bmatrix} \mathbf{V}_{\mathbf{x},\mathbf{x}} + \mathbf{V}_{\mathbf{x},\mathbf{x}} \begin{bmatrix} -\dfrac{3h}{\rho c r_0} & 0 \\ \dfrac{3h}{\rho c r_0} & -a_1 \end{bmatrix} + \begin{bmatrix} 0 & 0 \\ 0 & \sigma_w^2 \end{bmatrix} \tag{52}$$

with

$$\mathbf{V}_{\mathbf{x},\mathbf{x}} = \begin{bmatrix} \sigma_T^2 & \sigma_{T,T_\infty} \\ \sigma_{T,T_\infty} & \sigma_{T_\infty}^2 \end{bmatrix} \tag{53}$$

and σ_{T,T_∞} the covariance of T and T_∞. The initial conditions are given by

$$\bar{T}(t=0) = T_0 \tag{54}$$

$$\bar{T}_\infty(t=0) = \bar{T}_\infty \tag{55}$$

$$\mathbf{V}_{\mathbf{x},\mathbf{x}}(t=0) = \begin{bmatrix} \sigma_{T_0}^2 & 0 \\ 0 & \sigma_{T_\infty}^2 \end{bmatrix} \tag{56}$$

Equation (50) expresses that the mean solution can be found by solving the original differential equation for the mean value of the random parameter. Equation (51) confirms that the mean value of the random parameter is constant (which we expected because an autoregressive process is stationary).

Equation (52) can be elaborated further to yield

$$\frac{d}{dt}\sigma_T^2 = -\frac{6h}{\rho c r_0}\sigma_T^2 + \frac{6h}{\rho c r_0}\sigma_{T,T_\infty} \tag{57}$$

$$\frac{d}{dt}\sigma_{T,T_\infty} = \frac{3h}{\rho c r_0}\sigma_{T_\infty}^2 - \left(\frac{3h}{\rho c r_0} + a_1\right)\sigma_{T,T_\infty} \tag{58}$$

$$\frac{d}{dt}\sigma_{T_\infty}^2 = -2a_1\sigma_{T_\infty}^2 + \sigma_W^2 \tag{59}$$

As T_∞ is stationary, $\sigma^2_{T_\infty}$ is not a function of time so that Eq. (59) reduces to

$$\sigma^2_{T_\infty} = \frac{\sigma^2_W}{2a_1} \tag{60}$$

The solution of Eq. (58) can readily be found through direct integration:

$$\sigma_{T,T_\infty} = \frac{3h/\rho c r_0}{3h/\rho c r_0 + a_1}\,\sigma^2_{T_\infty}\left\{1 - \exp\left[-\left(\frac{3h}{\rho c r_0} + a_1\right)t\right]\right\} \tag{61}$$

After substitution of Eq. (61) in Eq. (57) and subsequent integration, we can derive the following expression for σ^2_T:

$$\begin{aligned}\sigma^2_T = {} & \frac{3h/\rho c r_0}{3h/\rho c r_0 + a_1}\sigma^2_{T_\infty} + \frac{3h/\rho c r_0}{3h/\rho c r_0 - a_1}\,\sigma^2_{T_\infty}\exp\left(-\frac{6h}{\rho c r_0}t\right) \\ & - \frac{18h^2/\rho^2 c^2 r_0^2}{(3h/\rho c r_0 + a_1)(3h/\rho c r_0 - a_1)}\,\sigma^2_{T_\infty}\exp\left[-\left(\frac{3h}{\rho c r_0} + a_1\right)t\right]\end{aligned} \tag{62}$$

In the special case of a random variable, we simplify the above expression by putting $a_1 = 0$, so that we have

$$\sigma^2_T = a^2_{T_\infty} + \sigma^2_{T_\infty}\exp\left(-\frac{6h}{\rho c r_0}t\right) - 2\sigma^2_{T_\infty}\exp\left[-\left(\frac{3h}{\rho c r_0}\right)t\right] \tag{63}$$

A sample of the random process ambient temperature and the corresponding temperature course in the sphere are shown in Figure 4. The parameter values were as follows: $\rho = 1000$ kg/m^3, $c = 4189$ J/kg °C, $r_0 = 0.01$ m, $T_0 = 20$°C, $h = 10$ W/m^2 °C, $\bar{T}_\infty = 80$°C, $\sigma_{T_\infty} = 5$°C, and $\theta = 600$ sec.

A Crank–Nicolson finite–difference scheme in the time domain was used to solve Eq. (41). The high frequency fluctuations are smoothed by because of the thermal inertia of the sphere. There was a very good agreement between the mean temperature of the sphere calculated by means of the Monte Carlo and the variance propagation algorithm (not shown). In Figure 5, the time course of the variance of the temperature of the sphere is shown. The results obtained by means of the variance propagation algorithm and the Monte Carlo method with 1000 or 5000 runs were comparable. However, the variances obtained by means of the Monte Carlo method with 100 runs are scattered.

The mean value and 95% confidence interval for the temperature prediction in the sphere are shown in Figure 6.

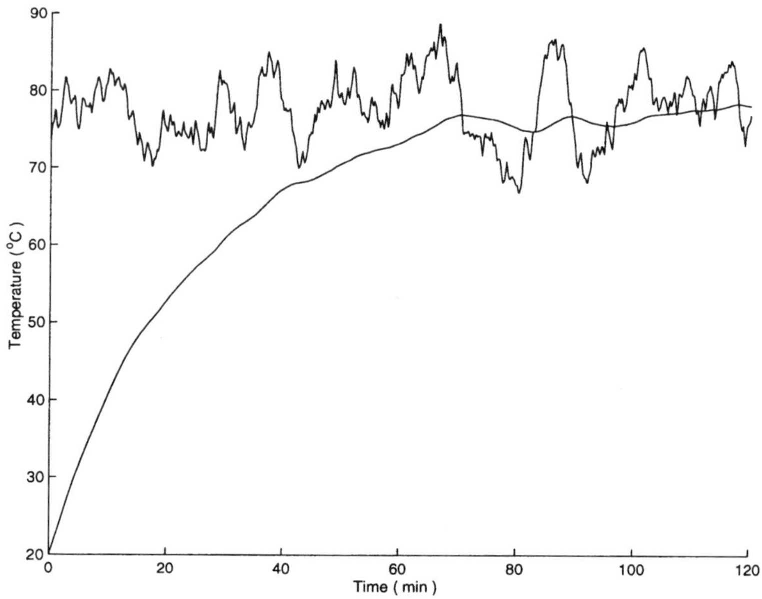

FIGURE 4 Realization of AR(1) ambient temperature and corresponding temperature in a sphere.

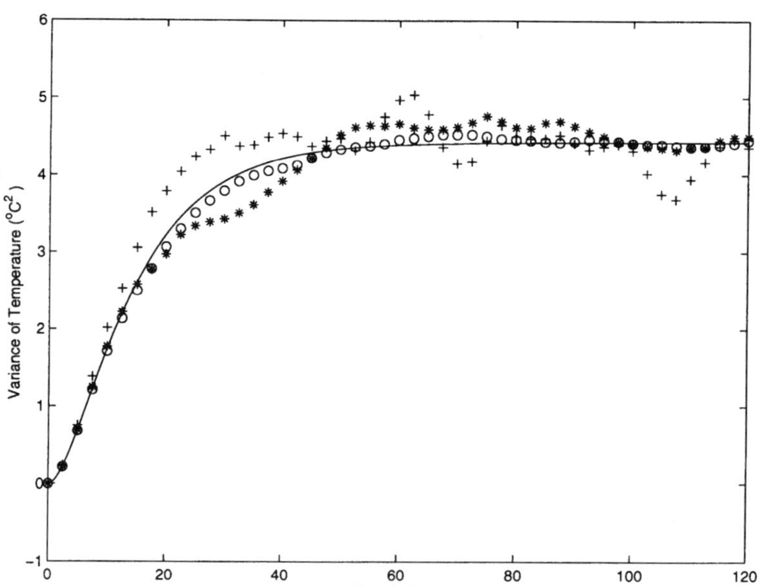

FIGURE 5 Temperature variance in sphere subjected to random process ambient temperature: (——) variance propagation; (*) Monte Carlo ($n_{MC} = 1000$); (+) Monte Carlo ($n_{MC} = 100$); (○) Monte Carlo ($n_{MC} = 5000$).

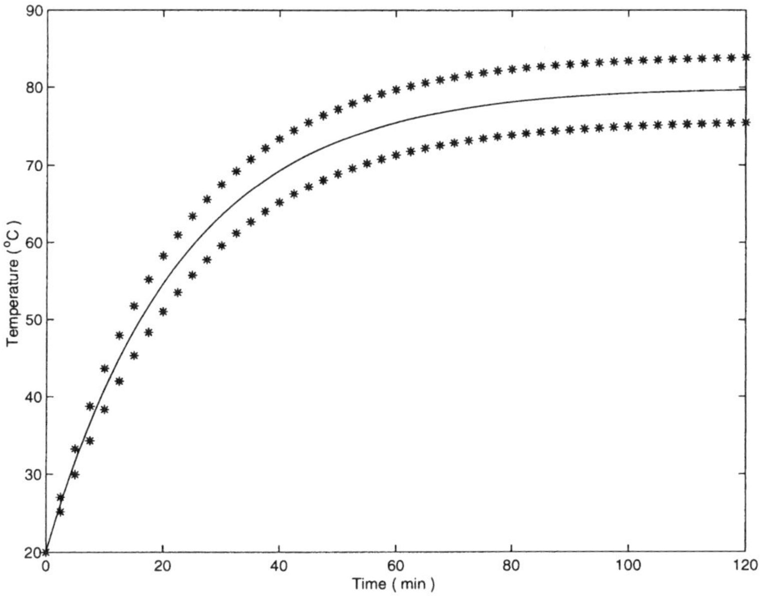

FIGURE 6 Mean value (—) and 95% confidence interval (∗) for temperature prediction in a sphere subjected to random process ambient temperature.

5.2 Heat-Conduction Problems

For the extension of the variance propagation to conduction-limited problems, we will start from the spatially discretized system (5). We will further assume that T_∞, h, and Q are autoregressive processes of order m_{T_∞}, m_h, and m_Q, respectively, as defined by the following state-space equations:

$$\frac{d}{dt}\mathbf{x}_{T_\infty}(t) = \mathbf{A}_{T_\infty}\mathbf{x}_{T_\infty}(t) + \mathbf{B}_{T_\infty}W_{T_\infty}(t) \tag{64}$$

$$\frac{d}{dt}\mathbf{x}_h(t) = \mathbf{A}_h\mathbf{x}_h(t) + \mathbf{B}_h W_h(t) \tag{65}$$

$$\frac{d}{dt}\mathbf{x}_Q(t) = \mathbf{A}_Q\mathbf{x}_Q(t) + \mathbf{B}_Q W_Q(t) \tag{66}$$

with W_{T_∞}, W_h, W_Q white-noise processes of, in general, different covariances. As the thermophysical properties k and ρc usually do not change as a function of time, they are modeled as random variables by means of the following trivial differential equation:

$$\frac{d}{dt}k = \frac{d}{dt}\rho c = 0 \tag{67}$$

with appropriate initial conditions. Obviously, T_0 is modeled as a random variable as well.

As with the lumped-capacitance problem, a first step is to write the stochastic heat conduction in the form of Eq. (44). It is easy to see that this can be accomplished through the following choice of **x**, **g**, and **h**:

$$\mathbf{x} = \begin{bmatrix} \mathbf{u} \\ \mathbf{x}_{T_\infty} \\ \mathbf{x}_h \\ \mathbf{x}_Q \\ k \\ \rho c \end{bmatrix} \tag{68}$$

$$\mathbf{g} = \begin{bmatrix} \mathbf{C}^{-1}(-ku + \mathbf{f}) \\ \mathbf{A}_{T_\infty}\mathbf{x}_{T_\infty} + \mathbf{B}_{T_\infty}\bar{W}_{T_\infty} \\ \mathbf{A}_h\mathbf{x}_h + \mathbf{B}_h\bar{W}_h \\ \mathbf{A}_Q\mathbf{x}_Q + \mathbf{B}_Q\bar{W}_Q \\ 0 \\ 0 \end{bmatrix} \tag{69}$$

$$\mathbf{h} = \begin{bmatrix} \mathbf{0} & \mathbf{0} & \mathbf{0} \\ \mathbf{B}_{T_\infty} & \mathbf{0} & \mathbf{0} \\ \mathbf{0} & \mathbf{B}_h & \mathbf{0} \\ \mathbf{0} & \mathbf{0} & \mathbf{B}_Q \\ 0 & 0 & 0 \\ 0 & 0 & 0 \end{bmatrix} \tag{70}$$

$$\mathbf{w} = \begin{bmatrix} W_{T_\infty} - \bar{W}_{T_\infty} \\ W_h - \bar{W}_h \\ W_Q - \bar{W}_Q \end{bmatrix} \tag{71}$$

with **0** null vectors of appropriate dimension, and

$$\mathbf{V}_{\mathbf{w},\mathbf{w}}(\tau) \triangleq \mathrm{E}[\mathbf{w}(t)\mathbf{w}^T(t+\tau)] = \delta(\tau) \begin{bmatrix} \sigma^2_{W,T_\infty} & 0 & 0 \\ 0 & \sigma^2_{W,h} & 0 \\ 0 & 0 & \sigma^2_{W,Q} \end{bmatrix} \tag{72}$$

After substitution of Eqs. (68)–(71) in Eqs. (48) and (49) and subsequent rearrangement, the following system is obtained:

$$\frac{d}{dt}\,\bar{\mathbf{u}} = \bar{\mathbf{C}}^{-1}(-\bar{\mathbf{K}}\bar{\mathbf{u}} + \bar{\mathbf{f}}) \tag{73}$$

$$\begin{aligned}\frac{d}{dt}\,\mathbf{V}_{\mathbf{u},\mathbf{u}} = {} & \bar{\mathbf{C}}^{-1}\Bigg[-\bar{\mathbf{K}}\mathbf{V}_{\mathbf{u},\mathbf{u}} - \frac{\partial\mathbf{K}}{\partial k}\,\bar{\mathbf{u}}\mathbf{V}^T_{\mathbf{u},k} - \frac{\partial\mathbf{C}}{\partial\rho c}\,\frac{d\bar{\mathbf{u}}}{dt}\,\mathbf{V}^T_{\mathbf{u},\rho c} \\ & + \frac{\partial\mathbf{f}}{\partial T_\infty}\,\mathbf{V}^T_{\mathbf{u},T_\infty} + \left(\frac{\partial\mathbf{f}}{\partial h} - \frac{\partial\mathbf{K}}{\partial h}\,\bar{\mathbf{u}}\right)\mathbf{V}^T_{\mathbf{u},h} + \frac{\partial\mathbf{f}}{\partial Q}\,\mathbf{V}^T_{\mathbf{u},Q}\Bigg] \\ & + \Bigg[-\bar{\mathbf{K}}\mathbf{V}_{\mathbf{u},\mathbf{u}} - \frac{\partial\mathbf{K}}{\partial k}\,\bar{\mathbf{u}}\mathbf{V}^T_{\mathbf{u},k} - \frac{\partial\mathbf{C}}{\partial c}\,\frac{d\bar{\mathbf{u}}}{dt}\,\mathbf{V}^T_{\mathbf{u}\rho c} \\ & + \frac{\partial\mathbf{f}}{\partial T_\infty}\,\mathbf{V}^T_{\mathbf{u},T_\infty} + \left(\frac{\partial\mathbf{f}}{\partial h} - \frac{\partial\mathbf{K}}{\partial h}\,\bar{\mathbf{u}}\right)\mathbf{V}^T_{\mathbf{u},h} + \frac{\partial\mathbf{f}}{\partial Q}\,\mathbf{V}^T_{\mathbf{u},Q}\Bigg]^T \bar{\mathbf{C}}^{-T}\end{aligned} \tag{74}$$

$$\frac{d}{dt}\,\mathbf{V}_{\mathbf{u},\mathbf{x}_{T_\infty}} = \bar{\mathbf{C}}^{-1}\left(-\bar{\mathbf{K}}\mathbf{V}_{\mathbf{u},\mathbf{x}_{T_\infty}} + \frac{\partial\mathbf{f}}{\partial T_\infty}\,\mathbf{V}_{T_\infty,\mathbf{x}_{T_\infty}}\right) + \mathbf{V}_{\mathbf{u},\mathbf{x}_{T_\infty}}\mathbf{A}^T_{T_\infty} \tag{75}$$

$$\frac{d}{dt}\,\mathbf{V}_{\mathbf{u},\mathbf{x}_h} = \bar{\mathbf{C}}^{-1}\left[-\bar{\mathbf{K}}\mathbf{V}_{\mathbf{u},\mathbf{x}_h} + \left(\frac{\partial\mathbf{f}}{\partial h} - \frac{\partial\mathbf{K}}{\partial h}\,\bar{\mathbf{u}}\right)\mathbf{V}_{h,\mathbf{x}_h}\right] + \mathbf{V}_{\mathbf{u},\mathbf{x}_h}\mathbf{A}^T_h \tag{76}$$

$$\frac{d}{dt}\,\mathbf{V}_{\mathbf{u},\mathbf{x}_Q} = \bar{\mathbf{C}}^{-1}\left(-\bar{\mathbf{K}}\mathbf{V}_{\mathbf{u},\mathbf{x}_Q} + \frac{\partial\mathbf{f}}{\partial Q}\,\mathbf{V}_{Q,\mathbf{x}_Q}\right) + \mathbf{V}_{\bar{\mathbf{u}},\mathbf{x}_Q}\mathbf{A}^T_Q \tag{77}$$

$$\frac{d}{dt}\,\mathbf{V}_{\mathbf{u},k} = \bar{\mathbf{C}}^{-1}\left(-\bar{\mathbf{K}}\mathbf{V}_{\mathbf{u},k} - \sigma_k^2\,\frac{\partial\mathbf{K}}{\partial k}\,\bar{\mathbf{u}}\right) \tag{78}$$

$$\frac{d}{dt}\,\mathbf{V}_{\mathbf{u},\rho c} = \bar{\mathbf{C}}^{-1}\left(-\bar{\mathbf{K}}\mathbf{V}_{\mathbf{u},\rho c} - \sigma_{\rho c}^2\,\frac{\partial\mathbf{C}}{\partial\rho c}\,\frac{d}{dt}\,\bar{\mathbf{u}}\right) \tag{79}$$

where the notation $\bar{\mathbf{C}}^{-T}$ denotes the transpose of the inverse of $\bar{\mathbf{C}}$. $\bar{\mathbf{C}}$, $\bar{\mathbf{K}}$, and $\bar{\mathbf{f}}$ are assembled using the mean values of ρc, k, T_∞, h, and Q. The initial condition for Eq. (73) is given by

$$\bar{\mathbf{u}}(t=0) = \bar{\mathbf{u}}_0 \tag{80}$$

$$\mathbf{V}_{\mathbf{u},\mathbf{u}} = \sigma^2_{T_0}\mathbf{I} \tag{81}$$

where $\mathbf{I}$ is an $n_{\text{nod}} \times n_{\text{nod}}$ unity matrix. Further, because the initial temperature is uncorrelated with k, ρc, h, T_∞, and Q, the other initial conditions are equal to null matrices of appropriate dimension. Equations (73)–(81) constitute the variance propagation algorithm for stochastic heat-conduction problems.

Observe that the above algorithm can be extended to take into account nonlinear heat conduction with temperature-dependent thermal properties because Eqs. (48) and (49) are applicable to general nonlinear systems. The corresponding algorithm has a similar overall structure as the above algorithm and has been described in detail in Ref. 37. As it is essentially based on a linearization of the finite-element formulation of the (nonlinear) heat-conduction equation, it can, however, be expected to be sufficiently accurate for smooth nonlinear heat-conduction problems only. The applicability of this algorithm for heat-conduction problems with phase changes is currently being investigated by the authors. A more extended variance propagation algorithm for heat-conduction problems with random process and random wave parameters was described recently [25].

Equation (74) is of the general form

$$\frac{d}{dt}\mathbf{V}(t) = \mathbf{A}\mathbf{V}(t) + \mathbf{V}(t)\mathbf{A}^T + \mathbf{B}(t)$$

with **V**, **A**, and **B** square matrices of equal dimension, and is called a *Lyapunov* matrix differential equation. Equations (75)–(77) are of the general form

$$\frac{d}{dt}\mathbf{V}(t) = \mathbf{A}\mathbf{V}(t) + \mathbf{V}(t)\mathbf{B} + \mathbf{C}(t) \tag{82}$$

with **A** and **B** square matrices and **V** and **C** matrices which are, in general, not square. Equation (82) is called a *Sylvester* matrix differential equation. Equations (78) and (79) are of the form

$$\bar{\mathbf{C}}\frac{d}{dt}\mathbf{V}(t) + \bar{\mathbf{K}}V(t) = \mathbf{h}(t)$$

with $\bar{\mathbf{C}}$ and $\bar{\mathbf{K}}$ the finite-element matrices and **V** and **h** vectors of dimension n_{nod}. This structure is similar to that of Eq. (73), and later it will be outlined that this fact can be exploited advantageously.

The matrices $\mathbf{V}_{\mathbf{x}_{T_\infty},\mathbf{x}_{T_\infty}}$, $\mathbf{V}_{\mathbf{x}_h,\mathbf{x}_h}$, and $\mathbf{V}_{\mathbf{x}_Q,\mathbf{x}_Q}$ in Eqs. (75)–(77) can be computed by straightforward application of the variance propagation algorithm to the Eqs. (64)–(66), respectively, which yields, for example, for T_∞,

$$\frac{d}{dt}\mathbf{V}_{\mathbf{x}_{T_\infty},\mathbf{x}_{T_\infty}} = \mathbf{A}_{T_\infty}\mathbf{V}_{\mathbf{x}_{T_\infty},\mathbf{x}_{T_\infty}} + \mathbf{V}_{\mathbf{x}_{T_\infty},\mathbf{x}_{T_\infty}}\mathbf{A}_{T_\infty}^T + \mathbf{B}_{T_\infty}\sigma_{T_\infty}^2\mathbf{B}_{T_\infty}^T \tag{83}$$

It can be proven that AR(m) processes driven by *stationary* white noise are stationary [36]. This implies that the mean and the covariance of the AR(m) process does not change in time. Consequently, the time derivatives in the left-hand sides of Eq. (83) vanish and the following algebraic matrix Lyapunov equation is obtained.

$$\mathbf{A}_{T_\infty}\mathbf{V}_{\mathbf{x}_{T_\infty},\mathbf{x}_{T_\infty}} + \mathbf{V}_{\mathbf{x}_{T_\infty},\mathbf{x}_{T_\infty}}\mathbf{A}_{T_\infty}^T + \mathbf{B}_{T_\infty}\sigma_{T_\infty}^2\mathbf{B}_{T_\infty}^T = \mathbf{0} \tag{84}$$

TABLE 2 Variance-Propagation Algorithm

Step 1	Compute $\bar{\mathbf{u}}$ from Eq. (73) with initial condition (80)
Step 2	Solve the Lyapunov matrix equations (84)–(86)
Step 3	Compute $\mathbf{V}_{\mathbf{u},\mathbf{x}_{T\infty}}$, $\mathbf{V}_{\mathbf{u},\mathbf{x}_h}$, $\mathbf{V}_{\mathbf{u},\mathbf{x}_Q}$, $\mathbf{V}_{\mathbf{u},k}$ and $\mathbf{V}_{\mathbf{u},\rho c}$ from Eqs. (75)–(79)
Step 4	Compute $\mathbf{V}_{\mathbf{u},\mathbf{u}}(t)$ by solving the Lyapunov matrix differential equation (74)

Similarly,

$$\mathbf{A}_h\mathbf{V}_{\mathbf{x}_h,\mathbf{x}_h} + \mathbf{V}_{\mathbf{x}_h,\mathbf{x}_{T\infty}}\mathbf{A}_h^T + \mathbf{B}_h\sigma_{T_\infty}^2\mathbf{B}_h^T = \mathbf{0} \tag{85}$$

$$\mathbf{A}_Q\mathbf{V}_{\mathbf{x}_q,\mathbf{x}_Q} + \mathbf{V}_{\mathbf{x}_Q,\mathbf{x}_{T\infty}}\mathbf{A}_Q^T + \mathbf{B}_Q\sigma_{T_\infty}^2\mathbf{B}_Q^T = \mathbf{0} \tag{86}$$

The above equations can now be combined conveniently in the algorithm outlined in Table 2.

Step 2 is calculated in advance, as well as the partial derivatives of **K**, **C**, and **f** with respect to the random parameters (see section 5.4). The other steps are merged into a time-stepping scheme in which the mean temperature vector and all covariance matrices are updated at each time step. The linear differential systems (73), (77)–(79) can be solved using a similar time-stepping algorithm as in the case of a deterministic problem. For an implicit Euler finite-difference algorithm, the following recursive relationship can be used:

$$\left(\frac{\bar{\mathbf{C}}}{\Delta t} + \bar{\mathbf{K}}\right)\bar{\mathbf{u}}_{t+\Delta t} - \frac{\bar{\mathbf{C}}}{\Delta t}\bar{\mathbf{u}}_t = \bar{\mathbf{f}}_{t+\Delta t} \tag{87}$$

Observe that the matrix $\bar{\mathbf{C}}/\Delta t + \bar{\mathbf{K}}$ is to be triangularized only once.

5.3 Algorithm for Random Variable Parameters

If all parameters are random variables, the mean value and the covariance matrix can be calculated by means of the first-order perturbation algorithm which was derived in Ref. 22. The variance propagation algorithm is then equivalent with the perturbation algorithm. Without loss of generality, this equivalence will be proven below for the simple case of a random variable initial condition and thermal conductivity. This fact can be exploited advantageously, as the perturbation algorithm involves only the solution of vector differential equations.

Proof. The first-order perturbation algorithm for heat-conduction problems with random variable parameters starts with a system of differential equations which describe the sensitivity of the nodal temperature vector with respect to the random parameters. For the case of a random variable initial condition and thermal conductivity, the following system is obtained [22]:

$$\bar{\mathbf{C}} \frac{d}{dt} \bar{\mathbf{u}} + \bar{\mathbf{K}}\bar{\mathbf{u}} = \bar{\mathbf{f}} \tag{88}$$

$$\bar{\mathbf{C}} \frac{d}{dt} \left(\frac{\partial \mathbf{u}}{\partial k} \right) + \bar{\mathbf{K}} \frac{\partial \mathbf{u}}{\partial k} = - \frac{\partial \mathbf{K}}{\partial k} \bar{\mathbf{u}} \tag{89}$$

$$\bar{\mathbf{C}} \frac{d}{dt} \left(\frac{\partial \mathbf{u}}{\partial T_0} \right) + \bar{\mathbf{K}} \frac{\partial \mathbf{u}}{\partial T_0} = \mathbf{0} \tag{90}$$

subject to the initial conditions

$$\frac{\partial \mathbf{u}}{\partial k} = \mathbf{0}$$

$$\frac{\partial \mathbf{u}}{\partial T_0} = [1 \ \ 1 \cdots 1]^T$$

The covariance matrix is then calculated from

$$\mathbf{V}_{\mathbf{u},\mathbf{u}} = \frac{\partial \mathbf{u}}{\partial k} \left(\frac{\partial \mathbf{u}}{\partial k} \right)^T \sigma_k^2 + \frac{\partial \mathbf{u}}{\partial T_0} \left(\frac{\partial \mathbf{u}}{\partial T_0} \right)^T \sigma_{T_0}^2 \tag{91}$$

Differentiaition of Eq. (91) with respect to time yields

$$\begin{aligned} \frac{d}{dt} \mathbf{V}_{\mathbf{u},\mathbf{u}} = {} & \frac{d}{dt} \left(\frac{\partial \mathbf{u}}{\partial k} \right) \left(\frac{\partial \mathbf{u}}{\partial k} \right)^T \sigma_k^2 + \left(\frac{\partial \mathbf{u}}{\partial k} \right) \frac{d}{dt} \left(\frac{\partial \mathbf{u}}{\partial k} \right)^T \sigma_k^2 \\ & + \frac{d}{dt} \left(\frac{\partial \mathbf{u}}{\partial T_0} \right) \left(\frac{\partial \mathbf{u}}{\partial T_0} \right)^T \sigma_{T_0}^2 + \left(\frac{\partial \mathbf{u}}{\partial T_0} \right) \frac{d}{dt} \left(\frac{\partial \mathbf{u}}{\partial T_0} \right)^T \sigma_{T_0}^2 \end{aligned} \tag{92}$$

Further, Eq. (89) and (90) can be rearranged as

$$\frac{d}{dt} \left(\frac{\partial \mathbf{u}}{\partial k} \right) = -\mathbf{C}^{-1} \bar{\mathbf{K}} \frac{\partial \mathbf{u}}{\partial k} - \mathbf{C}^{-1} \frac{\partial \mathbf{K}}{\partial k} \bar{\mathbf{u}} \tag{93}$$

$$\frac{d}{dt} \left(\frac{\partial \mathbf{u}}{\partial T_0} \right) = -\mathbf{C}^{-1} \bar{\mathbf{K}} \frac{\partial \mathbf{u}}{\partial T_0} \tag{94}$$

Substitution of Eqs. (93) and (94) in Eq. (92) results in

$$\frac{d}{dt}\mathbf{V}_{\mathbf{u},\mathbf{u}} = -\mathbf{C}^{-1}\left(\bar{\mathbf{K}}\frac{\partial \mathbf{u}}{\partial k} + \frac{\partial \mathbf{K}}{\partial k}\right)\left(\frac{\partial \mathbf{u}}{\partial k}\right)^T \sigma_k^2$$
$$- \frac{\partial \mathbf{u}}{\partial k}\left(\bar{\mathbf{K}}\frac{\partial \mathbf{u}}{\partial k} + \frac{\partial \mathbf{K}}{\partial k}\bar{\mathbf{u}}\right)^T \mathbf{C}^{-T}\sigma_k^2$$
$$- \mathbf{C}^{-1}\bar{\mathbf{K}}\frac{\partial \mathbf{u}}{\partial T_0}\left(\frac{\partial \mathbf{u}}{\partial T_0}\right)^T \sigma_{T_0}^2 - \frac{\partial \mathbf{u}}{\partial T_0}\left(\frac{\partial \mathbf{u}}{\partial T_0}\right)^T \bar{\mathbf{K}}^T\mathbf{C}^{-T}\sigma_{T_0}^2 \tag{95}$$

The perturbation algorithm is based on the following first-order Taylor expansion of $\mathbf{u}(k,\ T_0)$ around $\bar{\mathbf{u}}$:

$$\mathbf{u} \cong \bar{\mathbf{u}} + \frac{\partial \mathbf{u}}{\partial k}\Delta k + \frac{\partial \mathbf{u}}{\partial T_0}\Delta T_0 \tag{96}$$

where

$$\Delta k = k - \bar{k}$$
$$\Delta T_0 = T_0 - \bar{T}_0$$

From Eq. (96) it follows that

$$\mathbf{V}_{\mathbf{u},k} = \mathrm{E}[(\mathbf{u} - \bar{\mathbf{u}})(k - \bar{k})]$$
$$= \frac{\partial \mathbf{u}}{\partial k}\sigma_k^2 \tag{97}$$
$$\mathbf{V}_{\mathbf{u},T_0} = \mathrm{E}[(\mathbf{u} - \bar{\mathbf{u}})(T_0 - \bar{T}_0)]$$
$$= \frac{\partial \mathbf{u}}{\partial T_0}\sigma_{T_0}^2 \tag{98}$$

After substitution of Eqs. (91), (97), and (98) in Eq. (95), the following equation is obtained:

$$\frac{d}{dt}\mathbf{V}_{\mathbf{u},\mathbf{u}} = -\mathbf{C}^{-1}\left(\bar{\mathbf{K}}\mathbf{V}_{\mathbf{u},\mathbf{u}} + \frac{\partial \mathbf{K}}{\partial k}\bar{\mathbf{u}}\mathbf{V}_{\mathbf{u},k}^T\right)$$
$$- \left(\bar{\mathbf{K}}\mathbf{V}_{\mathbf{u},\mathbf{u}} + \frac{\partial \mathbf{K}}{\partial k}\bar{\mathbf{u}}\mathbf{V}_{\mathbf{u},k}^T\right)^T \mathbf{C}^{-T} \tag{99}$$

Further, right multiplication of Eq. (93) by σ_k^2 and using Eq. (97) gives

$$\frac{d}{dt}\mathbf{V}_{\mathbf{u},k} = -\mathbf{C}^{-1}\bar{\mathbf{K}}\mathbf{V}_{\mathbf{u},k} - \mathbf{C}^{-1}\frac{\partial \mathbf{K}}{\partial k}\bar{\mathbf{u}}\sigma_k^2 \tag{100}$$

Equations (99) and (100) are equivalent to Eqs. (74) and (78) for the given stochastic specifications. This concludes the proof. ■

5.4 Derivatives of C, K, and f with Respect to Random Parameters

The derivatives of **C**, **K**, and **f** with respect to the random parameters can be computed by differentiation of the element matrices and subsequent incorporation in the global derivative matrices. The following expressions are easily derived from Eqs. (7)–(9):

$$\frac{\partial \mathbf{C}^j}{\partial \rho c} = \int_{V^j} \boldsymbol{\phi}^j \boldsymbol{\phi}^{j^T} \, dV$$

$$\frac{\partial \mathbf{K}^j}{\partial k} = \int_{V^j} \mathbf{B}^j \mathbf{B}^{j^T} \, dV$$

$$\frac{\partial \mathbf{K}^j}{\partial h} = \int_{S^j} \boldsymbol{\phi}^j \boldsymbol{\phi}^{j^T} \, dS$$

$$\frac{\partial \mathbf{f}^j}{\partial h} = \int_{S^j} T_\infty \boldsymbol{\phi}^j \, dS$$

$$\frac{\partial \mathbf{f}^j}{\partial T_\infty} = \int_{S^j} h \boldsymbol{\phi}^j \, dS$$

$$\frac{\partial \mathbf{f}^j}{\partial Q} = \int_{V^j} \boldsymbol{\phi}^j \, dV$$

6 NUMERICAL SOLUTION OF LYAPUNOV AND SYLVESTER DIFFERENTIAL EQUATIONS

6.1 Algebraic Lyapunov and Sylvester Equations

The variance-propagation algorithm requires the numerical solution of Sylvester matrix differential equations of the form

$$\frac{d}{dt}\mathbf{V}(t) = \mathbf{A}(t)\mathbf{V}(t) + \mathbf{V}(t)\mathbf{B}(t) + \mathbf{C}(t) \tag{101}$$

where **A**, **B**, and **C** are real matrices of dimensions $r \times r$, $s \times s$, and $r \times s$, respectively, so that **V** is of dimension $r \times s$. If $\mathbf{B} = \mathbf{A}^T$ and $r = s$, the Sylvester differential equation (101) reduces to a Lyapunov equation.

As with *vector* differential equations, implicit as well as explicit methods can be applied for the numerical solution of *matrix* differential equations. A first-order explicit algorithm is obtained by substitution of the differential operator in Eq. (101) by a first order forward difference operator:

$$\mathbf{V}(t+\Delta t) = [\mathbf{I} + \Delta t\mathbf{A}(t)[\mathbf{V}(t) + \Delta t\mathbf{V}(t)\mathbf{B}(t) + \Delta t\mathbf{C}(t) \tag{102}$$

The algorithm involves matrix multiplication and addition and is particularly simple to implement. However, as in the case of ordinary differential equations, it will be shown later that the algorithm is only conditionally stable provided that a suitable time step has been chosen.

An unconditionally stable implicit algorithm is obtained by substitution of the differential operator in Eq. (101) by a first-order backward difference operator:

$$\begin{aligned}[\tfrac{1}{2}\mathbf{I} - \Delta t\mathbf{A}(t+\Delta t)\mathbf{V}(t+\Delta t) + \mathbf{V}(t+\Delta t)[\tfrac{1}{2}\mathbf{I} - \Delta t\mathbf{B}(t+\Delta t)]\\ = \mathbf{V}(t) + \Delta t\mathbf{C}(t+\Delta t)\end{aligned} \tag{103}$$

where the equality $\mathbf{V} = \frac{1}{2}\mathbf{V} + \frac{1}{2}\mathbf{V}$ is used. Equation (103) can be written as

$$\mathbf{DX} + \mathbf{XE} = \mathbf{F} \tag{104}$$

with

$$\mathbf{D} = [\tfrac{1}{2}\mathbf{I} - \Delta t\mathbf{A}(t+\Delta t)]$$

$$\mathbf{E} = [\tfrac{1}{2}\mathbf{I} - \Delta t\mathbf{B}(t+\Delta t)]$$

$$\mathbf{F} = \mathbf{V}(t) + \Delta t\mathbf{C}(t+\Delta t)]$$

$$\mathbf{X} = \mathbf{V}(t+\Delta t)$$

An equation of the form (104) is called an *algebraic Sylvester equation* and is solved as follows [39].

First $\mathbf{D}$ is reduced to lower real Schur form $\mathbf{D}'$ by an orthogonal similarity transformation $\mathbf{U}$:

$$\mathbf{D}' = \mathbf{U}^T\mathbf{D}\mathbf{U} = \begin{bmatrix} \mathbf{D}'_{1,1} & \mathbf{0} & \cdots & \mathbf{0} \\ \mathbf{D}'_{2,1} & \mathbf{D}'_{2,2} & \cdots & \mathbf{0} \\ \ddots & & \vdots & \ddots \\ \mathbf{D}'_{r,1} & \mathbf{D}'_{r,2} & \cdots \mathbf{D}'_{r,r} & \end{bmatrix}$$

where the diagonal submatrices $\mathbf{D}'_{i,i}$ are of order at most 2 and $\mathbf{U}\mathbf{U}^T = \mathbf{I}$. Similarly, $\mathbf{E}$ is reduced to upper real Schur form $\mathbf{E}'$ by an orthogonal similarity transformation $\mathbf{U}'$:

$$\mathbf{E}' = \mathbf{U}'^T\mathbf{E}\mathbf{U}' = \begin{bmatrix} \mathbf{E}'_{1,1} & \mathbf{E}'_{1,1} & \cdots & \mathbf{E}'_{1,s} \\ \mathbf{0} & \mathbf{E}'_{2,2} & \cdots & \mathbf{E}'_{2,s} \\ \ddots & & \ddots & \vdots \\ \mathbf{0} & \mathbf{0} & \cdots & \mathbf{E}'_{s,s} \end{bmatrix}$$

where, again, $\mathbf{E}'_{i,i}$ is of order at most 2. Substitution of $\mathbf{D}$ and $\mathbf{E}$ in Eq. (104) by $\mathbf{U}\mathbf{D}'\mathbf{U}^T$ and $\mathbf{U}'\mathbf{E}'\mathbf{U}'^T$, respectively, premultiplication by $\mathbf{U}^T$, and postmultiplication by $\mathbf{U}'$ yields the following system:

$$\mathbf{D}'\mathbf{X}' + \mathbf{X}'\mathbf{E}' = \mathbf{F}' \tag{105}$$

with

$$\mathbf{F}' = \mathbf{U}^T\mathbf{F}\mathbf{U}' \tag{106}$$

$$\mathbf{X}' = \mathbf{U}^T\mathbf{X}\mathbf{U}' \tag{107}$$

The advantage of the transformation of Eq. (104) to Eq. (105) is that the latter equation can be written as a system of mutually uncoupled algebraic matrix equations of order at most 2. These are equivalent to an ordinary algebraic system of at most four equations which can be solved using appropriate techniques (e.g., variants of the method of Gauss). The solution must then, of course, be backtransformed using Eq. (107).

Other algorithms of the Runge–Kutta and BDF (backward difference formula) type of order 1–6 for the solution of Lyapunov equations are described in Ref. 40.

6.2 Convergence and Stability Analysis

A convergence and stability analysis of the explicit and implicit Euler methods for the solution of Lyapunov matrix differential equations is now presented for a stochastic heat-conduction problem with random variable initial temperature. Although this is a very simple case, it allows one to investigate some interesting features of the algorithms. A more comprehensive stability and convergence analysis for stochastic heat-conduction problems with random process ambient and initial temperature is described elsewhere [41].

From Eq. (74), follows that the variance-propagation algorithm for linear heat-conduction problems with random field initial temperature is given by the following Lyapunov equation:

$$\frac{d}{dt}\mathbf{V}(t) = \mathbf{A}\mathbf{V}(t) + \mathbf{V}(t)\mathbf{A}^T \tag{108}$$

$$\mathbf{V}(t) = \mathbf{V}_0 \quad \text{at } t = 0 \tag{109}$$

with

$$\mathbf{A} = -\bar{\mathbf{C}}^{-1}\bar{\mathbf{K}}$$

$\mathbf{A}$ and $\mathbf{V}$ are square matrices of dimension $n_{\text{nod}} \times n_{\text{nod}}$. Note that $\mathbf{A}$ is constant if the surface heat transfer coefficient does not change overtime. Equation (108), subject to the initial condition (109), can be solved numerically using the explicit or implicit Euler method as outlined. These methods are subcases of the more general class of linear multi step methods which are among the most popular methods for the numerical solution of differential equations. The convergence and stability theory of linear multistep methods is well established for scalar and vector differential equations [42,43] and can be readily extended to matrix differential equations.

For future use, a general expression for the exact solution of Eq. (108) is now derived. Assume that $\mathbf{A}$ has n_{nod} distinct eigenvalues. $\mathbf{A}$ can than be written as

$$\mathbf{A} = \mathbf{H}\mathbf{\Lambda}\mathbf{H}^{-1} \tag{110}$$

where $\mathbf{H}$ is the matrix of eigenvectors and $\mathbf{\Lambda}$ is the diagonal matrix of eigenvalues. Note that if there are eigenvalues with multiplicity larger than 1, $\mathbf{A}$ can no longer be diagonalized. The analysis can then be based on the Jordan canonical form, but this is not elaborated further here. Substitution of Eq. (110) in Eq. (108), premultiplication with $\mathbf{H}^{-1}$, and postmultiplication with $\mathbf{H}^{-T}$ yields the following matrix differential equation:

$$\frac{d}{dt}\mathbf{W} = \mathbf{\Lambda}\mathbf{W} + \mathbf{W}\mathbf{\Lambda} \tag{111}$$

where

$$\mathbf{W} \triangleq \mathbf{H}^{-1}\mathbf{V}\mathbf{H}^{-T} \tag{112}$$

Because Eq. (111) is completely uncoupled, it can be derived that its solution is given by

$$W_{i,j} = c_{i,j}\ \exp[(\lambda_i + \lambda_j)t] \tag{113}$$

where the $c_{i,j}$ are the integration constants which can be determined by imposing the initial condition

$$\mathbf{W}_0 = \mathbf{H}^{-1}\mathbf{V}_0\mathbf{H}^{-T},$$

with λ_i and λ_j the eigenvalues of $\mathbf{A}$ for $i, j = 1, \ldots, n_{\text{nod}}$.

The general multistep matrix method is now introduced. For this purpose, assume the following matrix differential equation:

$$\frac{d}{dt}\mathbf{V}(t) = \mathbf{F}(t, \mathbf{V})$$

with an appropriate initial value. A general linear multi step matrix method of order k is defined as

$$\sum_{l=0}^{k} \alpha_l \mathbf{V}'(r_{n+l}) = \Delta t \sum_{l=0}^{k} \beta_l \mathbf{F}'(t_{n+l}) \tag{114}$$

where

$t_{n+l} = (n + l)\Delta t$

$\mathbf{V}'(t_{n+l})$ is the approximation of $\mathbf{V}(t_{n+l})$

$\mathbf{F}'(t_{n+1}) = \mathbf{F}(t_{n+l},\ \mathbf{V}'t_{n+l})$

and α_l and β_l are the coefficients of the method. For both the explicit and the implicit Euler methods, $k = 1$. The values of the coefficients α_l and β_l are given in Table 3. An initial sequence $\mathbf{V}'(t_j)$, $j = 0,\ k - 1$ must be provided for the algorithm to start.

Application of Eq. (114) to the Lyapunov equation (108) gives

$$\sum_{l=0}^{k} \alpha_l \mathbf{V}'(t_{n+l}) = \Delta t \sum_{l=0}^{k} \beta_l [\mathbf{A}\mathbf{V}'(t_{n+l}) + \mathbf{V}'(t_{n+l})\mathbf{A}^T] \tag{115}$$

An obvious property to be met by the general linear matrix method is that, in the limit $\Delta t \to 0$, the approximate solution $\mathbf{V}'(t_n), n = 1, \ldots, N = t_f/\Delta t$, converges to the exact solution $\mathbf{V}(t)$, $t \in [0,\ t_f]$. The time final t_f is hereby kept constant, so that at the same time $n \to \infty$. This can be stated more precisely as follows:

Table 3 Coefficients of the Linear Multistep Method for the Explicit and Implicit Euler Methods

	Explicit	Implicit
α_0	−1	−1
α_1	1	1
β_0	1	0
β_1	0	1

Definition 1 The linear multistep method defined by Eq. (115) is said to be *convergent* if

$$\lim_{\substack{\Delta t \to 0 \\ t_n \text{ fixed}}} \mathbf{V}'(t_n) = \mathbf{V}(t_n) \tag{116}$$

holds for all n and for all starting values $\mathbf{V}'(r_1)$ for which

$$\lim_{\Delta t \to 0} \mathbf{V}'(t_l) = \mathbf{V}(t_l), \quad l = 0, k-1 \tag{117}$$

The conditions for convergence of the linear matrix multistep method are summarized in the following theorem.

Theorem 1 *The method defined by Eq.* (115) *is convergent if and only if it is consistent*

$$\sum_{l=0}^{k} \alpha_l = 0; \qquad \sum_{l=0}^{k} l\alpha_l = \sum_{l=0}^{k} \beta_1 \tag{118}$$

and zero-stable, which means that no root ζ_i *of the characteristic polynomial*

$$\sum_{l=0}^{k} \alpha_l \xi_i^l = 0 \tag{119}$$

is larger than 1 *in modulus, and every root with modulus* 1 *is simple.*

Proof: Let $\mathbf{V}_i'$ be the ith column of $\mathbf{V}'$. Equation (115) can be rearranged as

$$\sum_{l=0}^{k} \alpha_l \mathbf{V}'^*(t_{n+l}) = \Delta t \sum_{l=0}^{k} \beta_l \mathbf{A}'^*(t_{n+l}) \tag{120}$$

where

$$\mathbf{V}'^* = \begin{bmatrix} \mathbf{V}_1' \\ \vdots \\ \mathbf{V}_{n_{\text{nod}}}' \end{bmatrix} \tag{121}$$

$$\mathbf{A}'^* = \begin{bmatrix} \mathbf{A}V_1' + \sum_{i=1}^{n_{\text{nod}}} \mathbf{V}_i' \mathbf{A}_{1,i} \\ \vdots \\ \mathbf{A}V_{N_{\text{nod}}}' + \sum_{i=1}^{n_{\text{nod}}} \mathbf{V}_i' \mathbf{A}_{n_{\text{nod}},i} \end{bmatrix} \tag{122}$$

Equation (120) is a linear multistep *vector* algorithm. From Ref. 43, it follows that under the conditions (118) and (119),

$$\lim_{\substack{\Delta t \to 0 \\ t_n \text{ fixed}}} \mathbf{V}'^*(t_n) = \mathbf{V}^*(t_n) \tag{123}$$

for all $t_n, n = 1, 2, \ldots$, provided that the initial sequence is chosen such that

$$\lim_{\Delta t \to 0} \mathbf{V}^*(t_l) = \mathbf{V}'^*(t_l), \quad l = 0, \ldots, k-1 \tag{124}$$

Equation (123) and (124) are obviously equivalent to Eqs. (116) and (117), so that the theorem is proven. ■

Where Theorem 1 deals with the behavior of the approximate solution $\mathbf{V}'$ if Δt tends to zero, it is also interesting to investigate whether for a *fixed* time step, the local errors are accumulating in an adverse fashion. This is the subject of the linear stability theory [43]. Before proceeding further, the following lemma is proven.

Lemma 1 *The eigenvalues of* $\mathbf{A} = -\bar{\mathbf{C}}^{-1}\bar{\mathbf{K}}$ *are real and negative.*

Proof: Let λ and $\mathbf{x}$ be an eigenvalue–eigenvector pair of $\mathbf{A}$. Then, by definition,

$$\mathbf{A}\mathbf{x} = -\bar{\mathbf{C}}^{-1}\bar{\mathbf{K}}\mathbf{x} = \lambda\mathbf{x}$$

Left multiplication of both sides by $\mathbf{x}^T\bar{\mathbf{C}}$ yields

$$-\mathbf{x}^T\bar{\mathbf{K}}\mathbf{x} = \mathbf{x}^T\bar{\mathbf{C}}\mathbf{x}\lambda$$

Because $\bar{\mathbf{K}}$ and $\bar{\mathbf{C}}$ are both positive definite, the following relations holds for any vector $\mathbf{x}$ (real or imaginary) and thus also if $\mathbf{x}$ is an eigenvector

$$\mathbf{x}^T\bar{\mathbf{K}}\mathbf{x} > 0 \tag{125}$$

$$\mathbf{x}^T\bar{\mathbf{C}}\mathbf{x} > 0 \tag{126}$$

and both expressions are real scalars. As a consequence, λ must be real and negative. By repeating the above derivation for each eigenvalue–eigenvector pair of $\mathbf{A}$, the proof is completed. ■

Using Lemma 1, it follows from Eq. (112) and (113) that for $t \to \infty$, all solutions $\mathbf{V}(t)$ of Eq. (108) satisfy

$$\|\mathbf{V}(t)\| \to 0$$

The following stability definition is now stated.

Definition 2 The linear multistep matrix method (115) is said to be absolutely stable if, for a given Δt, the approximate solution $\mathbf{V}'(t_n)$ of Eq. (108) satisfies

$$||\mathbf{V}'(t_n)|| \to 0 \tag{127}$$

if $n \to \infty$.

The conditions for absolute stability of the linear multistep matrix method (115) are given by the following theorem:

Theorem 2 *Let* $\mathbf{A}$ *have* n_{nod} *distinct eigenvalues. The linear multistep method* (115) *is absolutely stable if and only if the roots of the polynomial*

$$\sum_{l=0}^{k} [\alpha_l - \Delta t \beta_l (\lambda_i + \lambda_j)] p^l$$

are less than 1 in modulus for all i and j.

Proof: Because $\mathbf{A}$ has, by assumption, n_{nod} distinct eigenvalues λ_i, $i = 1, \ldots, n_{\text{nod}}$, the eigendecomposition (110) exists. After premultiplication of by $\mathbf{H}^{-1}$ and postmultiplication by $\mathbf{H}^{-T}$, the following equation is obtained from Eq. (115):

$$\sum_{l=0}^{k} \left[\left(\frac{\alpha_l}{2} - \Delta t \beta_l \boldsymbol{\Lambda} \right) \mathbf{W}'(t_{n+l}) + \mathbf{W}'(t_{n+l}) \left(\frac{\alpha_l}{2} - \Delta t \beta_l \boldsymbol{\Lambda} \right) \right] = 0 \tag{128}$$

with $\mathbf{W}'$ defined by

$$\mathbf{W}' \triangleq \mathbf{H}^{-1} \mathbf{V}' \mathbf{H}^{-T} \tag{129}$$

Equation (128) represents an *uncoupled* set of $n_{\text{nod}} \times n_{\text{nod}}$ equations:

$$\sum_{l=0}^{k} \left[\left(\frac{\alpha_l}{2} - \Delta t \beta_l \lambda_i \right) W'_{i,j}(t_{n+l}) + \left(\frac{\alpha_l}{2} - \Delta t \beta_l \lambda_j \right) W'_{i,j}(t_{n+l}) \right] = 0$$

or

$$\sum_{l=0}^{k} [\alpha_l - \Delta t \beta_l (\lambda_i + \lambda_j) W'_{i,j}(t_{n+l})] = 0, \quad i, j = 1, \ldots, n_{\text{nod}} \tag{130}$$

By Eq. (129), $||\mathbf{V}'|| \to 0$ as $n \to \infty$, if and only if $||\mathbf{W}'|| \to 0$ as $n \to \infty$, and, hence, Eq. (127) is satisfied if and only if all solutions $W'_{i,j}(t_n)$ of Eq. 9130) satisfy

$$|W'_{i,j}(t_n)| \to 0 \quad \text{if } n \to \infty, \quad i, j = 1, n_{\text{nod}} \tag{131}$$

The solutions of each of the difference equations in Eq. (130) are given by

$$W'_{i,j}(t_n) = \sum_{l=1}^{k} c_{i,j,l}\, p_{i,j,l}^{n} \tag{132}$$

where $c_{i,j,l}$ are arbitrary coefficients and $p_{i,j,l}$, $l = 1, \ldots, k$, are the roots of the polynomial

$$\sum_{l=0}^{k} [\alpha_l - \Delta t \beta_l(\lambda_i + \lambda_j)]\, p^l \tag{133}$$

Clearly, Eq. (131) and, consequently, Eq. (127) are satisfied if the roots $p_{i,j,l}$ satisfy

$$|p_{i,j,l}| < 1$$

for all i, j, and l. This concludes the proof. ∎

Both the implicit and explicit Euler methods are of order $k = 1$, so that the single root of Eq. (133) is equal to

$$p = -\frac{\alpha_0 - \Delta t \beta_0(\lambda_i + \lambda_j)}{\alpha_1 - \Delta t \beta_1(\lambda_i + \lambda_j)}$$

Substitution of α_l and β_l with the values given in Table 3 results in the following conditions:

Explicit:

$$|1 + \Delta t(\lambda_i + \lambda_j)| < 1 \tag{134}$$

Implicit:

$$\left| \frac{1}{1 - \Delta t(\lambda_i + \lambda_j)} \right| < 1 \tag{135}$$

for all i and j. Using Lemma 1, the following conclusions regarding the stability of the implicit and explicit Euler methods can be drawn. From Eq. (135), it follows that the stability conditions for the implicit Euler method will always be satisfied, as the eigenvalues are negative and real. On the other hand, the explicit Euler method will only be stable if and only if the condition

$$-2 < \Delta t(\lambda_i + \lambda_j) < 0$$

is met because the eigenvalues are negative and real. In this case, Δt cannot be chosen freely.

7 APPLICATION TO THERMAL STERILIZATION PROCESSES

In order to illustrate the above algorithms, we will now analyze a typical thermal food process with a random variable ambient temperature. The problem consists of a cylindrical container (radius $r_0 = 3.41$ cm, height $L = 10.02$ cm) filled with a 30% solids content tomato concentrate with $k = 0.542$ W/m °C and $\rho c = 3.89 \times 10^6$ J/m^3 °C. The following process conditions were applied: $T_0 = 65$°C and $h = 100$ W/m^2 °C. The ambient temperature is now described by means of an AR(1) process with $\bar{T}_\infty = 125$°C, $\sigma_{T_\infty} = 1$°C, and $a_1 = 0.00277$ sec^{-1}. An implicit Euler finite-difference method in the time domain was used to integrate the differential systems. For the finite-element analysis, the region $[0, r_0] \times [0, L/2]$ is subdivided in 100 axisymmetric linear quadrilateral elements. The time step is set equal to 36 sec.

The Monte Carlo and variance-propagation algorithms were programmed on top of the existing finite-element code DOT [44].

In Figure 7 the mean temperature at three different positions in the centerplane of the can are shown as calculated by means of the Monte Carlo method with 100 and 1000 runs and the variance-propagation algorithm. It is clear that there is excellent agreement between the different methods.

In Figure 8 the temperature variance at three different positions in the centerplane of the can are shown. The agreement between the Monte Carlo

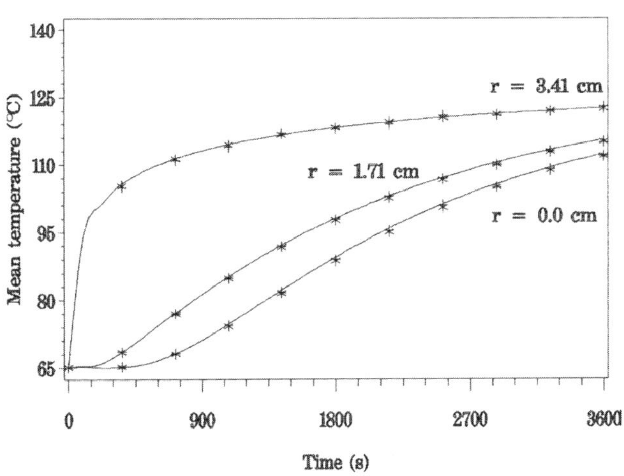

FIGURE 7 Mean temperature as a function of time in a heated Al can with random process ambient temperature at three different positions: (+) Monte Carlo ($n_{MC} = 1000$); (*) Monte Carlo ($n_{MC} = 100$); (—) variance-propagation algorithm.

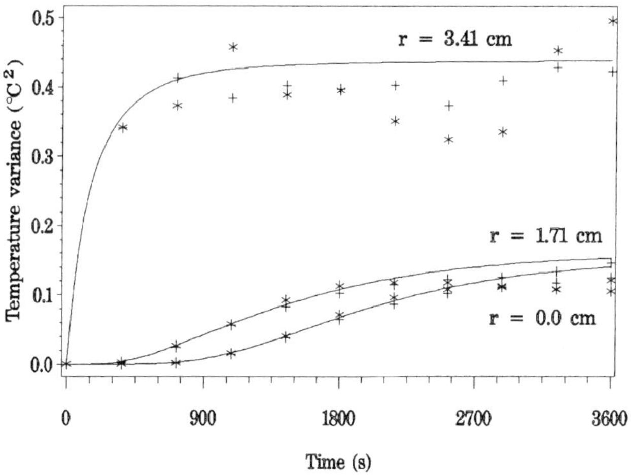

Figure 8 Temperature variance as a function of time in a heated Al can with random process ambient temperature at three different positions: (+) Monte Carlo ($n_{MC} = 1000$); (*) Monte Carlo ($n_{MC} = 100$); (—) variance-propagation algorithm.

method with 1000 runs and the variance-propagation algorithm is good, but the Monte Carlo method with 100 runs is not very accurate. The relative CPU time (total CPU time divided by CPU time for deterministic simulation) was equal to 74, 242, and 2426 for the variance propagation, Monte Carlo with 100 runs, and Monte Carlo with 1000 runs, respectively.

8 CONCLUSIONS

In this chapter, some algorithms for stochastic heat transfer analysis are outlined. In the *Monte Carlo method*, a large number of process samples is obtained by solving the heat transfer model for artificially generated random parameter samples. Straightforward statistical analysis of the simulation results yields the mean values and variances of the temperature. The *variance-propagation algorithm* is based on stochastic systems theory and was originally developed for systems of ordinary differential equations. The formalism here is applied to the spatially discretized heat-conduction equation to yield a system of matrix differential equations which can be solved numerically.

The Monte Carlo method in general requires a large amount of computer time to obtain results with an acceptable accuracy. Also, it requires a complete stochastic specification of the random parameters, whereas for the variance-propagation algorithm, only the mean values of the parameters and their covariance matrices must be known. However, the latter algorithm can provide only limited statistical information such as the mean value and the variance, whereas the Monte Carlo method can also be applied to derive other statistical characteristics such as the probability density function. Also, as the variance-propagation algorithm is essentially based on a linearization of the governing equations around their mean solution, it is only applicable if the variability is relatively small (coefficient of variation smaller than 0.2).

NOMENCLATURE

a_i	Coefficient of autoregressive process or wave
c	Heat capacity (J/kg°C)
$\mathbf{C}$	Finite-element capacity matrix
E	Mean value operator
f	Probability density function
$\mathbf{f}$	Finite-element thermal load vector
$\mathbf{g}$	Vector-valued function
h	Surface heat transfer coefficient (W/m^2 °C)
$\mathbf{h}$	Vector-valued function
k	Thermal conductivity (W/m °C)
$\mathbf{K}$	Finite-element stiffness matrix
L	Half-height of can (m)
n	Outward normal
n_{nod}	Number of nodes
n_{MC}	Number of Monte Carlo runs
$\mathbf{0}$	Null matrix
r_0	Radius (m)
S	Surface
t	Time (sec)
T	Temperature (°C)
T_0	Initial temperature (°C)
T_∞	Retort temperature (°C)
$\mathbf{u}$	Nodal temperature vector
V	Covariance function, volume
$\mathbf{V}$	Covariance matrix
W	White-noise process
$\mathbf{X}$	Random vector
$\mathbf{Y}$	Auxiliary random process
$\mathbf{z}$	Position vector
$\mathbf{Z}$	Discrete-time white-noise process

Δt	Time step
Γ	Convection surface
ρ	Density (kg/m^3)
σ	Standard deviation
τ	Separation time

ACKNOWLEDGMENTS

The European Communities (FAIR project FAIR-CT96-1192) and the Flemish Minister of Science and Technology are gratefully acknowledged for financial support. Author Bart Nicolaï is a Research Associate with the Flanders Fund for Scientific Research.

REFERENCES

1. AA Teixeira, JR Dixon, JW Zahradnik, GE Zinsmeister. Computer optimization of nutrient retention in thermal processing of conduction-heated foods. Food Technol 23(6):137–140, 1969.
2. D Naveh, IJ Kopelman, IJ Pflug. The finite element method in thermal processing of foods. J Food Sci 48:1086–1093, 1983.
3. AK Datta, AA Teixeira. Numerical modeling of natural convection heating in canned liquid foods. Trans ASAE 30(5):1542–1551 1987.
4. AK Datta, AA Teixeira. Numerically predicted transient temperature and velocity profiles during natural convection heating of canned liquid foods. J Food Sci 53(1):191–195, 1988.
5. D Naveh. Analysis of transient conduction heat transfer in the thermal processing of foods using the finite element method. PhD thesis, University of Minnesota, St. Paul, 1982.
6. AK Datta, AA Teixeira, JE Manson. Computer-based control logic of on-line correction of process deviations J Food Sci. 51:480–483, 1986.
7. JR Banga, JM Perez-Martin, JM Gallardo, JJ Casares. Optimization of the thermal processing of conduction-heated canned foods: Study of several objective functions. J Food Eng 14:25–51, 1991.
8. C Silva, M Hendrickx, F. Oliveira, P Tobback. Critical evaluation of commonly used objective functions to optimize overall quality and nutrient retention of heat-preserved foods. Food Eng 17:241–258, 1992.
9. BM Nicolaï, N Scheerlinck, P Verboven, J De Baerdemaeker. Stochastic perturbation analysis of thermal food processes with random field parameters. Trans ASAE, 43:131–138, 2000.
10. H Patino, JR Heil. A statistical approach to error analysis in thermal process calculations. Food Sci. 50:1110–1114, 1985.
11. HF Meffert. Story, aims, results and future of thermophysical properties work within COST-90. In: R. Jowitt, F Escher, B Hallstrom, H. Meffert, W Spiess,

G Vos, eds. *Physical Properties of Foods* London: Applied Science Publishers, pp. 229–267.
12. M Sheard, C Rodger. Optimum heat treatments for "sous vide" cook–chill products. In: T Martens, M Schellekens, eds. Proceedings of the First European "Sous Vide" Cooking Symposium, Leuven, Belgium, 1993, pp 118–126. ALMA Universiteit-srestaurants v.z.w.
13. T Martens. Mathematical model of heat processing in flat containers. PhD thesis, Katholieke Universiteit Leuven, Belgium, 1980.
14. T Ohlsson. Progress in pasteurization and sterilization. In: T Yano, R Matsuno, K Nakamura, eds. Developments in Food Engineering. London: Blackie Academic & Professional, 1994, pp. 18–23.
15. H Ramaswamy, S Campbell, C Passey. Temperature distribution in a standard 1-basket water–cascade retort. Can Inst Food Sci. Technol 24:19–26, 1991.
16. A Van Loey. Enzymic time temperature integrators for the quantification of thermal processes in terms of food safety. PhD thesis, Katholieke Universiteit Leuven, Leuven, Belgium, 1996.
17. MK Lenz, DB Lund. The lethality–Fourier number method: Confidence intervals for calculated lethality and mass-average retention of conduction-heating, canned foods. J Food Sci 42(4):1002–1007, 1977.
18. MK Lenz, DB Lund. The lethality–Fourier number method. Heating rate variations and lethality confidence intervals for forced-convection heated foods in containers. J Food Process Eng 2:227–271, 1978.
19. DB Lund. Statistical analysis of thermal process calculations. Food Technol 76–78, March 1978.
20. K Hayakawa, P De Massaguer, R Trout. Statistical variability of thermal process lethality in conduction heating food—computerized simulation. J Food Sci 53(6):1887–1893, 1988.
21. J Wang, RR Wolfe, K Hayakawa. Thermal process lethality variability in conduction-heated foods. J Food Sci 56:1424–1428, 1991.
22. BM Nicolaï, J De Baerdemaeker. Finite element perturbation analysis of non-linear heat conduction problems with random field parameters. Int J Numer Methods Heat Fluid Flow 7(5):525–544, 1997.
23. BM Nicolaï, J De Baerdemaeker. Simulation of heat transfer in foods with stochastic initial and boundary conditions. Trans IChemE Part C 70:78–82, 1992.
24. BM Nicolaï, J De Baerdemaeker. A variance propagation algorithm for the computation of heat conduction under stochastic conditions. Int J Heat Mass Transfer 42:1513–1520, 1999.
25. BM Nicolaï, P Verboven, N. Scheerlinck, J De Baerdemaeker. Numerical analysis of the propagation of random parameter fluctuations in time and space during thermal food processes. J Food Eng 38:259–278, 1999.
26. L Segerlind. Applied Finite Element analysis. 2nd ed. New York: John Wiley & Sons, 1984.
27. H Jiang, DR Thompson, RV Morey. Finite element model of temperature distribution in broccoli stalks during forced-air precooling. Trans ASAE 30(5):1473–1477, 1987.

28. JC Pan, SR Bhowmik. The finite element analysis of transient heat transfer in fresh tomatoes during cooling. Trans ASAE 34(3):972–976, 1991.
29. E Vanmarcke. Random Fields—Analysis and Synthesis. Cambridge, MA: The MIT Press, 1983.
30. RM Gray, LD Davisson. Random Processes—A Mathematical Approach for Engineers. Englewood Cliffs, NJ: Prentice-Hall, 1986.
31. WL Brogan. Modern Control Theory. 2nd ed. Englewood Cliffs, NJ: Prentice-Hall, 1985.
32. RY Rubinstein. Simulation and the Monte Carlo Method. New York: John Wiley & Sons, 1981.
33. MR Spiegel. Theory and Problems of Probability and Statistics. New York: McGraw-Hill, 1980.
34. WH Press, SA Teukolsky, WT Vetterling, BP Flannery. Numerical Recipes in C. 2nd ed. Cambridge: Cambridge University Press, 1995.
35. GE Box, ME Muller. A note on the generation of random numbers. Anns Math Statist 29:610–611, 1958.
36. JL Melsa, AP Sage. An Introduction to Probability and Stochastic Processes. Englewood Cliffs, NJ: Prentice-Hall, 1973.
37. BM Nicolaï. Modeling and Uncertainty Propagation Analysis of Thermal Food Processes. PhD thesis, Katholieke Universiteit Leuven, Leuven, Belgium, 1994.
38. FP Incropera, D De Witt. Fundamentals of Heat and Mass Transfer. 3rd ed. New York: John Wiley Sons, 1990.
39. RH Bartels, GW Steward. Solution of the matrix equation $AX + XB = C$. Commun ACM 15(9):820–826, 1972.
40. N Scheerlinck, P Verboven, J De Baerdemaeker, BM Nicolaï. A variance propagation algorithm for stochastic heat and mass transfer problems in food processes. 1998 ASAE Annual International Meeting, Orlando, FL, 1998, paper No. 983176.
41. N Scheerlinck, P Verboven, J De Baerdemaeker, BM Nicolaï. Numerical solution of Lyapunov differential equations which appear in stochastic finite element conductive heat transfer problems. 1998 ASAE Annual International Meeting, Orlando, FL, 1998, paper No. 983163.
42. KE Atkinson. An Introduction to Numerical Analysis. 2nd ed. New York: John Wiley & Sons, 1989.
43. JD Lambert. Numerical Methods for Ordinary Differential Systems. New York: John Wiley & Sons, 1991.
44. RM Polivka, EL Wilson. Finite element analysis of nonlinear heat transfer problems—DOT user's manual. Technical Report UC SESM 76-2, University of California, Berkeley, 1976.

10

Neural Networks Approach to Modeling Food Processing Operations

Vinod K. Jindal and Vikrant Chauhan
Asian Institute of Technology, Bangkok, Thailand

1 INTRODUCTION

In recent years, neural networks have turned out as a powerful method for numerous practical applications in a wide variety of disciplines. Neural networks are increasingly used as effective general-purpose, nonlinear regression tools and for developing models governed by complex relationships.

Food processing operations may face problems due to uncertainties and complexities of chemical and biological changes taking place in a particular process. Accurate modeling and control of food processing operations could be beneficial in increasing the process efficiency and maintaining the uniform quality of the final product.

Neural network models are constructed by interconnecting many nonlinear computational elements, known as neurons or nodes, operating in parallel and arranged in patterns similar to biological networks. Because neural networks learn from known patterns or examples without requiring explicit functional relationships, they offer several advantages over conventional techniques in process modeling and control.

Neural networks can be applied to datasets obtained from many food processing operations without going through the rigor of extensive training and understanding the mathematical background of a problem. This chapter reviews the background and basic principles related to neural networks and their possible applications in food processing through selected case-study examples.

2 BRIEF HISTORICAL PERSPECTIVE

The human brain is unique in its ability to think, remember, and recognize things. Scientists have been attempting in the past to understand the functioning of the nervous system and possibly develop computer models for its operation. The study of biological neurons has led to the development of artificial neural networks representing the simplified models of the human brain. Artificial neural networks are increasingly used to solve problems which may be difficult to solve by conventional methods.

It is believed that the basic mechanism for information processing in the human brain is based on the transfer of activation patterns in highly interconnected groups of neurons working in parallel through synapses [1,2]. Artificial neural networks refer to computing systems utilizing a large number of interconnected neurons or nerve cells similar to those found in biological systems. It is common knowledge that a large amount of visual information is processed in the human brain with very little effort. Therefore, an artificial neural network is often viewed as a parallel, distributed information processing structure patterned after the brain and consisting of processing elements or nodes with unidirectional interconnections. The processing speed of biological neurons is much slower than that of digital computers. However, the massive parallel-processing power of neurons offers a distinct advantage over digital computers in their ability to learn and generalize and to organize data into clusters having similar characteristics.

A neural network consists of interconnected neurons which perform simple computations and convey a signal from one neuron to another through the connections. The signal may be amplified or diminished depending on the "connection strength" or "weight." A learning algorithm is then used to determine the values of the weights to match the output of network with the desired result Artificial neural network models combine a highly interconnected network architecture with a simple neuron model. The knowledge specific to a problem is stored in the weights of connections between neurons through the use of a learning algorithm.

Artificial neural networks are also referred to as "neural nets," "artificial neural systems," "parallel distributed processing systems," and "connectionist systems" [3]. The prefix "artificial," which is used to distinguish them from the biological or real systems, is often dropped. Also, the terms "neurons," "nodes," "processing units," and "computing elements" are used interchangeably.

Following the introduction of a first mathematical model of a single biological neuron using simple binary thresholding functions by McCullock and Pitts [4], some of the significant developments included the learning rule by Hebb [5], "perceptron" neural model and associated learning rule by Rosenblatt [6], and introduction of "Adaline" by Widrow and Hoff [7] trained by a gradient descent rule to minimize the mean squared error. The research activities in neural networks were adversely affected following the criticism of Minsky and Papert [8] directed at the single-layer perceptron. However, the limitations of the early neural networks were overcome with the introduction of back-propagation learning algorithms for multilayer feedforward networks [9–11]. In addition, their popularization by Rumelhart et al. [9] had a major impact on the practical applications of neural networks in the 1980s in many disciplines. Improved supervised learning algorithms for feedforward networks using radial basis node function followed next for a variety of applications. In recent years, major contributions have been made in the field of neural networks, especially in Kohonen self-organizing maps and in associative memories networks [3]. More information on the history of neural networks and related developments can be found in Refs. 12–14.

3 BIOLOGICAL NEURON

The human brain contains over 100 billion computing elements or neurons [15]. These neurons are thought to convey information to other neurons by transmitting and receiving electrical signals. A highly simplified general structure of neurons is hypothesized as shown schematically in Figure 1. A more detailed description of the functions of biological neurons has been presented by Churchland and Sejnowski [16].

A typical biological neuron is made up of soma or cell body, dendrites, and axon. The incoming signals generated by other neurons are received by dendrites surrounding the soma where they are added up over time. Each neuron may be connected to hundreds of surrounding neurons through a network of dendrites. The axon, which is a long, thin tube starting at the base of soma, splits into multiple endbulbs called boutons that almost touch the dendrites of other neurons. The small gaps between the end bulbs and dendrites of other neurons are called synapses across which information is passed. Thus, a neuron sends an electrical impulse down its axon to the dendrites of other neurons via synaptic gaps when firing. The arrival of an

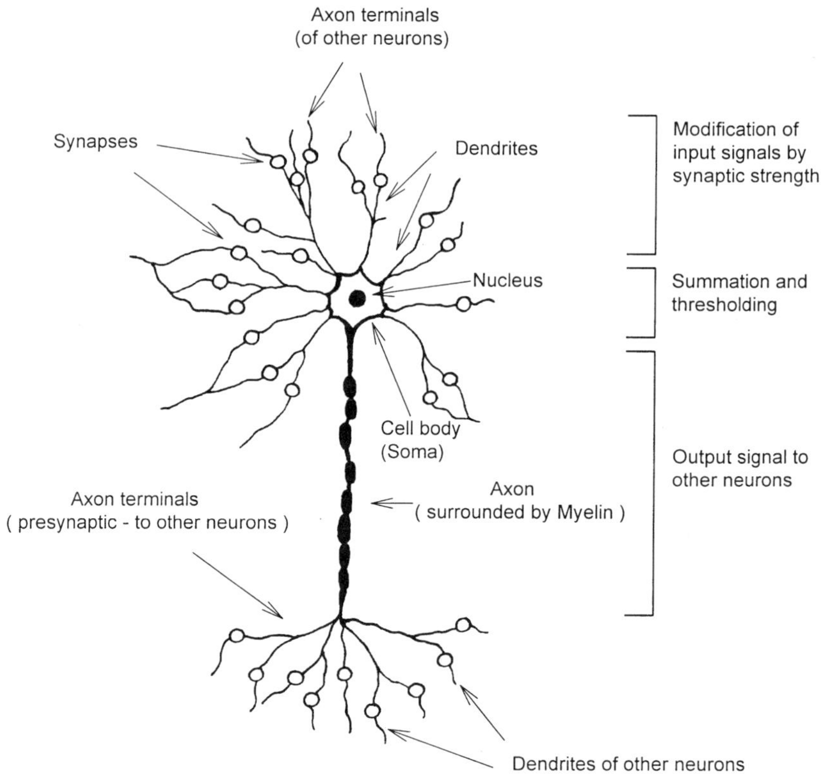

FIGURE 1 Basic structure of a biological neuron.

impulse (action potential) initiates chemical activity in the synaptic gap causing the electrical signal to be transmitted to the receiving dendrites.

The presynaptic side of the synapse receives the signals from a neuron, whereas the postsynaptic side sends the signals to a neuron. The magnitude of the signal received by a neuron depends on the efficiency of the synaptic transmission, which may be either inhibitory or excitatory, and is often thought of as the strength of the connection between the neurons. The activation of excitatory synapses increases the voltage in the postsynaptic neurons, whereas the activation of inhibitory synapses decreases the action potential. The increase in the intracellular voltage of the neuron will depend on the net contribution of excitatory and inhibitory synaptic actions. If the summation of incoming signals in the soma exceeds a certain activation level, called the threshold, the neuron fires and a new action potential is propagated along its axon. If the summation is lower than the threshold, the neuron remains inactive. The neurons cannot be excited for a short time, known as refractory period, following the generation of action potential. The activity of a neuron is often measured in terms of its firing frequency of electrical impulses, which, in turn, affect other neurons.

Artificial neural networks represent extremely simplified formal models of biological neurons and their interconnections without making any attempt to model the biological system itself. Their importance lies in the fact that artificial neural networks are brain-inspired computational tools for solving complex problems.

4 OPERATION OF A SINGLE ARTIFICIAL NEURON

The first mathematical model of an artificial neuron was originally proposed by McCullock and Pitts [4]. An artificial neuron is an elementary processing unit with several inputs and a single output as illustrated in Figure 2. The inputs to a neuron can be outputs from other neurons or simple external inputs. The single output from a neuron could be input to several other neurons.

The input signals to the neurons are modified by the weights representing the strengths of synapses and associated with each input. For example, a number of inputs $x_1, x_2, \ldots, x_n$ associated with respective weights $w_{j,1}, w_{j,2}, \ldots, w_{j,n}$ form a combined input net_j to the jth neuron, which is expressed as the weighted sum of the inputs.

$$\text{net}_j = \sum w_{j,i} x_i \tag{1}$$

Sometimes a bias is added to the net input

$$\text{net}_j = \sum w_{j,i} x_i + \text{bias}_j \tag{2}$$

This sum of weighted signals, net_j, is transformed into an activation level using a transfer or activation function to produce an output signal y_j only

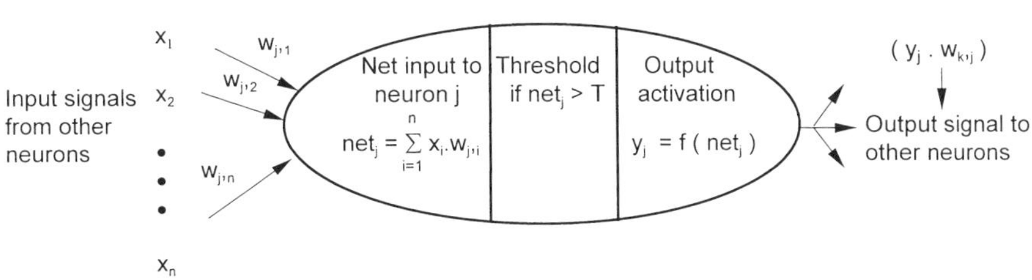

FIGURE 2 Schematic representation of the operation of a single artificial neuron.

when exceeding a certain threshold as follows:

$$y_j = f(\text{net}_j) = f\left(\sum w_{j,i} x_i\right) \tag{3}$$

The output of a neuron is determined by the nature of its activation function. If the weighted sum or action potential does not exceed the threshold, nothing is transmitted.

The most frequently used activation functions are the identity, the linear threshold, the sigmoid, and the hyperbolic tangent shown in Figure 3. The threshold function is defined only for the positive output values. Thus,

$$y_j = \begin{cases} 1 & \text{if net}_j > 0 \text{ (threshold value)} \\ 0 & \text{otherwise} \end{cases} \tag{4}$$

The step function is more suitable as a class identifier because its fixed output is independent of the magnitude of net input. The threshold value may be considered a bias and the application of activation function results in positive binary outputs 0 and 1 only. The bias or threshold often represents the weight which is associated with a dummy input constant equal to unity. Thus, it is convenient to include the threshold along with the weight vector of a neuron.

The ramp function identifies the minimum and maximum values of net output similar to the step function. However, the neuron, when activated by a minimum threshold, results in an output in the form of a continuous ramp until it reaches its maximum limit.

The sigmoid represents a convenient form and is most frequently used:

$$y_j = \frac{1}{1 + \exp[(-\text{net}_j + \beta)/\theta]} \tag{5}$$

where β is the bias or weight of an imaginary neuron that is always firing. A positive value of β shifts the activation function to the left along the horizontal axis and θ modifies the shape of the activation function. The large values of θ make the curve flatter, whereas small values result in a steeper curve. The sigmoid and hyperbolic tangent function approach the limiting minimum and maximum values asymptotically. They are frequently used in forms that transform the net inputs ranging between $[-\infty, +\infty]$ to real numbers between [0, 1] and $[-1, +1]$, respectively. Also, their continuously differentiable nature meets the requirement of learning algorithms that follow the gradient descent method. There are other activation functions such as piecewise linear functions and Gaussian function [3].

In summary, a single artificial neuron functions as an extremely simple processing element. It essentially produces a weighted sum of inputs and

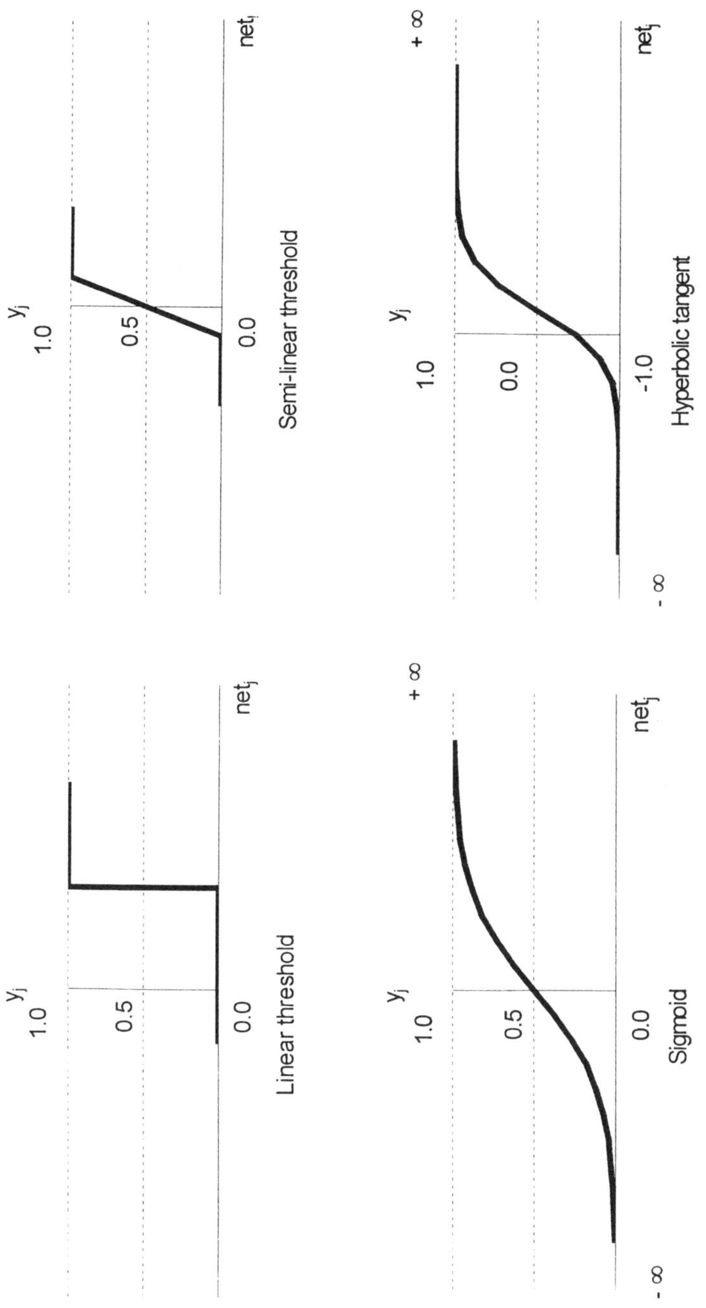

FIGURE 3 Some activation functions used in neural networks.

transforms it into binary or continuous output values depending on the nature of the activation function.

5 LEARNING IN NEURAL NETWORKS

There is not much known on how the brain learns and trains itself to produce information. The learning by the brain is associated with the synaptic efficiency, or modification of the strength, or weight of the interconnections [17]. In artificial neural networks, the processing element or neuron computes the output by applying the transfer function to the weighted sum of inputs. Learning implies the adjustment of weights or strength of the connections to match network output with the actual result for a given training set of data. A training set is often called a pattern or example. Thus, the neural network learns by changing the weighted input through modification of the weights according to a so-called learning law without being programmed. The learning by a neural network may take place in a supervised or unsupervised mode. In supervised learning, neural networks are presented with known input and output patterns. The error between the desired result and the computed output is minimized continuously by modifying the weights. Thus, a neural network attempts to come up with an internal representation of data which may be generalized. In unsupervised learning, the neural networks are presented only with a series of input patterns and no information is provided about their expected performance. Thus, there is no defined criterion to adjust the weights based on the specific or target outputs. Instead, the network attempts to group input patterns that are similar to each other and adapts according to a particular organization scheme.

5.1 Associative and Nonassociative Learning

The learning in the human brain is considered to be in associative or nonassociative form. Associative or correlation learning, which is also known as Hebbian learning [5], is based on the relationships between pairs of input and output patterns. This forms the basis for the supervised learning in the associative memories' neural networks by the weight modifications of connections between neurons. In nonassociative learning, also known as competitive learning, the repetition of various input patterns makes it possible to learn about their properties in the form of selected neural responses. When an input pattern is presented to a network, different nodes may compete to be winners until a single winner emerges. This allows the modifications of the connections between the input nodes and the winner node, thus improving the chances of the same winner when similar input patterns are presented. Thus, each winning node may be associated with a set of similar patterns. This forms the basis for unsupervised learning in Kohonen networks [18].

5.2 Error Feedback-Based Learning

This is based on the principle that a measure of performance evaluation or feedback error may lead to the modification of weights for generalizing the relationship between given input and output patterns. Usually, the changes made in weights are very small to make the network learn effectively.

Multilayer feedforward networks with nonlinear node function, such as a sigmoid using back-propagation of error in their learning mechanism, are most commonly used. These are also referred to as back-propagation networks. Back-propagation is similar to the least mean squared error (MSE) learning algorithm, also known as the Delta rule [7]. A gradient descent is used to diminish the MSE at every iteration because its derivative exists everywhere. In back-propagation, weights are modified in the direction corresponding to the negative gradient of an error measure such as a MSE based on the generalization of Delta rule. The errors at a higher (outer) layer of a multilayer network are propagated backward to nodes at lower (inner) layers to allow the gradients to modify the weights associated with the connections of hidden nodes.

5.2.1 Delta Rule for Supervised Learning in Single-Layer Networks

Single-layer neural networks, commonly known as perceptrons, consist of one layer of one or more neurons or nodes for computation. The simplest perceptron is represented by a single node used to classify each input pattern into two classes corresponding to node output of 0 and 1. Training or learning at a node in a supervised mode consists of adjusting the weights of interconnections and threshold (bias) to obtain a desired output.

The Delta rule was developed by Widrow and Hoff [7] for the classification of linearly independent training patterns. The artificial neural network structure in such cases consists of input nodes connected to output nodes with modifiable connecting weights. The Delta rule, which is based on the gradient descent method, minimizes the squares of the differences between the actual and the desired outputs by adjusting the values of the connecting weights. The net error for an input pattern consisting of N inputs $(1 < i < N)$ is represented by the MSE given by

$$E(p) = \frac{1}{2} \sum_{j=1}^{K} (y_j - o_j)^2 \tag{6}$$

Where $E(p)$ is the MSE for pattern p, y_j is the computed output for node j, o_j is the target or desired output of node j, and j is the number of output nodes $(1 < j < K)$. The weight $w_{j,i}$ is associated with the connection from a node i in the source layer to node j in the target layer. The Delta rule allows the modification of the weights so that the MSE diminishes at every iteration. The gradient descent can only be performed if the error is a continuous function of the weights. Because the MSE is a quadratic function whose derivative exists everywhere, it can be shown that

$$\Delta w_{j,i} = -\eta\left(\frac{\partial E(p)}{\partial w_{j,i}}\right) = \eta\ \delta_j(p)x_i(p) \tag{7}$$

where η is a proportionality constant called the learning rate with values greater than 0 but less than 1. The $\delta_j(p)$ is the difference between the computed and target output vector for pattern p:

$$\delta_j(p) = y_j - o_j \tag{8}$$

The following equation is then used for calculating new weights:

$$w_{j,i}(\text{new}) = w_{j,i}(\text{old}) + \eta\delta_j(p)x_i(p) \tag{9}$$

After adjusting all weights and thresholds, the whole calculation procedure is repeated for all input training patterns. This eventually leads to a set of connection weights that minimizes the error between the computed and target outputs.

The Delta rule works fine if the training patterns are linearly independent. If nonlinear relationships exist between the inputs and outputs, a multilayer perceptron model is used for mapping the nonlinear functions. It is possible to generalize the Delta rule to simultaneously train more than a single layer by performing a gradient descent on the errors at the intermediate layers. This procedure is known as the error back-propagation algorithm based on the generalization of the Delta rule.

5.2.2 Generalized Delta Rule for Multilayer Feedforward Networks

A multilayer feedforward neural network consisting of N nodes in the input layer, M nodes in an intermediate or hidden layer, and K nodes in the output layer is shown in Figure 4. The input layer simply directs the inputs to the hidden layer. The processing only takes place in the hidden and output layers. The position of a node in input, hidden, and output layers is denoted by the subscripts i, j, and k, respectively. Thus, x_i is the value of the ith node in the input layer $(1 < i < N)$, y_j is the value of the jth node in the hidden layer $(1 < j < M)$, and y_k is the value of the node in the output layer $(1 < k < K)$. Similarly, $w_{j,i}$ represents the weight associated with the connection from the ith input node to the jth node in the next layer. The sets of

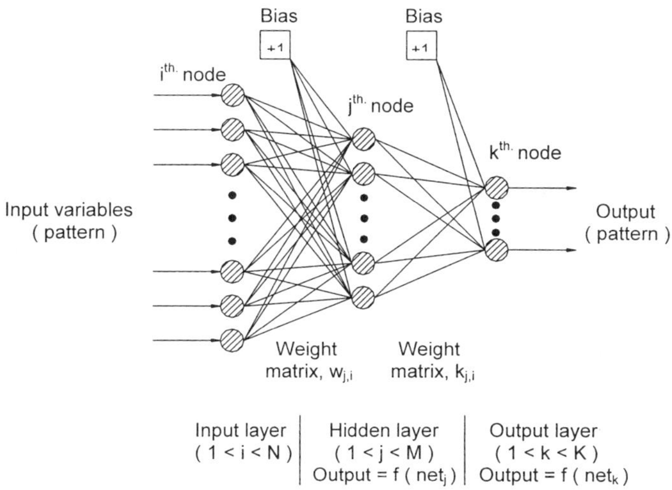

FIGURE 4 A multilayer feedforward neural network architecture showing one hidden layer.

inputs, their associated weights, and outputs in a given layer are designated in vector form. The training patterns consist of an array of input vectors $(1 < p < P)$. The length of input vector equals the number of nodes in the input layer. The input vector often includes a dummy input node with constant input equal to 1 so that the bias or threshold term can be treated just like other weights in the network. The number of weight vectors is equal to the number of nodes to be processed. Thus, any input vector is modified by the weight vector to produce a weighted sum which is transformed by an activation function into an output vector. The choice of a sigmoid activation function, which is continuous and differentiable everywhere, is often made for application to all nodes in the hidden and output layers:

$$f(\mathrm{net}_j) = \frac{1}{1 + \exp(-\mathrm{net}_j)]} \tag{10}$$

For each input vector, there is a desired or target output vector. The goal of the training is to modify the weights so that the difference between the computed and target vectors is within acceptable limits. The following steps are involved in the error backpropagation learning algorithm:

1. The net input to the jth node in the hidden layer including the threshold, net_j, is transformed by the sigmoid function to produce the output

$$y_j(1 < j < M) = f(\mathrm{net}_j) = f\left(\sum_{i=1}^{N} w_{j,i} x_i\right) \tag{11}$$

2. The net input to the kth node of the output layer including the threshold, net_k, after transformation becomes

$$y_k(1 < k < K) = f(\mathrm{net}_k) = f\left(\sum_{i=1}^{M} w_{k,j} y_j\right) \tag{12}$$

3. The learning involves the adjustment of both weight vectors $w_{j,i}$ and $w_{k,j}$ corresponding to the hidden and output layer, respectively. The MSE between the computed and target outputs is minimized by the generalization of the Delta rule. The MSE for the kth node in the output layer for each input pattern $p(1 < p < P)$ is

$$E(p) = \frac{1}{2} \sum_{k=1}^{K} (y_k - o_k)^2 \tag{13}$$

As in the case of single-layer network, the weight change at the outer layer is given by

$$\Delta w_{k,j} = -\eta \left(\frac{\partial E(p)}{\partial w_{k,j}} \right) = \eta \delta_k y_j \tag{14}$$

where

$$\delta_k = (y_k - o_k) f'(\mathrm{net}_j) = (y_k - o_k) f(\mathrm{net}_k)[1 - f(\mathrm{net}_k)] \tag{15}$$

4. The weight change for the inner layer is given by

$$\Delta w_{j,i} = -\eta \left(\frac{\partial E(p)}{\partial w_{j,i}} \right) = \eta \mu_j y_i \tag{16}$$

where

$$\mu_j = \left(\sum \delta_k w_{k,j}\right) f'(\mathrm{net}_j) = \left(\sum \delta_k w_{k,j}\right) f(\mathrm{net}_j)[1 - f(\mathrm{net}_j)] \tag{17}$$

5. The new weights can then be calculated by

$$w_{k,j}(\mathrm{new}) = w_{k,j}(\mathrm{old}) + \Delta w_{k,j} \quad \text{(for outer layer)} \tag{18}$$

$$w_{j,i}(\mathrm{new}) = w_{j,i}(\mathrm{old}) + \Delta w_{j,i} \quad \text{(for inner layer)} \tag{19}$$

In the forward pass, the output of nodes in the hidden layer is found for each input pattern. It is then followed by the determination of the outputs of nodes in the output layer. The estimation of MSE between the target and computed patterns leads to the determination of δ_k and $w_{k,j}$ for each output node. In the backward pass, the δ_k and $w_{k,j}$ obtained from the output nodes are used for adjusting the weights of the connections between the hidden and input layer nodes. In this way, the errors are propagated backward, making the training of hidden nodes possible for which no desired outputs are available. The internal thresholds for nodes are also adjusted along with the weights. After adjusting all weights and thresholds, the whole calculation procedure is repeated for other input training patterns. This procedure eventually leads to a set of connection weights that minimize the error between the actual and desired outputs.

For speeding up the learning during training, the following modification is often adopted:

$$\Delta w_{k,j}(\text{new}) = \eta \delta_k y_j + \lambda_j \Delta w_{k,j}(\text{old}) \tag{20}$$

where λ is known as momentum.

6 NEURAL NETWORK ARCHITECTURES

The various architectures of artificial neural networks are patterned after our limited understanding of biological neural networks. The nature of nodes and the structure of the interconnections define the artificial neural network architecture. They may have a feedforward or feedback (recurrent) architecture, or a combination of both. A feedforward network has a layered structure where nodes in each layer receive their inputs from the previous layer (lower) and send their outputs to succeeding (higher) layer. Networks with feedback architecture have feedback loops in their interunit connection path. Fully connected symmetric networks are used in associative memory applications in which intralayer connections may exist except in the input layer. Acyclic networks are layered networks with no intralayer connections. They are further simplified into commonly used feedforward neural networks when connections are allowed from a lower layer to an adjacent or next higher layer only. The number of nodes in each layer are designated sequentially. For example, a 2-3-1 feedforward neural network contains two nodes in the input layer, three nodes in the hidden layer, and one node in the output layer. Sometimes, neural networks may be combined in small modules and organized subsequently with sparse connections between modules to provide a modular architecture for specific applications.

The network receives the training data at the nodes in the input layer. The number of nodes in the input layer depends on the number of variables or categories in the input data. The hidden layers are between the input and output layers. The number of hidden layers and the number of neurons in a layer can only be determined through experimentation. The learning takes place within the hidden layers and the output layer by internal mapping of the input data to detect possible relationships with the output. The output layer presents the response of the network to a given training pattern. A frequently used three-layered neural network architecture is shown in Figure 4.

7 TYPES OF NEURAL NETWORKS

A neural network is characterized by its architecture, activation function for the node output, and learning algorithm for determining the weights of the connections. In principle, neural networks can be classified according to their supervised or unsupervised mode of learning. Both McCullock and Pitts's threshold logic unit [4] and Rosenblatt's [6] simple perceptron used only binary inputs and outputs, and the threshold activation function. However, the Widrow and Hoff scheme [7] introduced the Delta rule and an iterative learning procedure using a linear activation function forcing the output to be the linear weighted sum of all inputs. The Adaline was similar to perceptron and was used for pattern-recognition tasks. An extension of the Adaline algorithm for multilayer networks led to the development of the Madaline algorithm with a slightly different threshold function and weight update rule [3]. Finally, the introduction of supervised learning procedures based on the back-propagation of error opened the way for their applications to a variety of practical problems. There are many different variations of the back-propagation algorithm which continue to be the most widely used supervised learning method for classification, pattern recognition, interpolation, predicting and forecasting, and process modeling. Two effective approaches for accelerating the back-propagation convergence by reducing the number of iterations are Quickprop [19] and the conjugate gradient method [20]. A brief description of other types of neural networks follows next.

7.1 Radial Basis Function Networks

They are multilayer feedforward networks, developed from the Delta rule with only one hidden layer and they are easier to train than back-propagation networks. The primary difference between back-propagation networks and RBF networks is in the nature of hidden layer. In radial basis function (RBF) networks, the nodes in the hidden layer have a radial "basis function" or a statistical transformation based on the Gaussian activation function with a center corresponding to a maximal output for any given input value. The output of such a unit is equal to 1 when the input is "centered" or at zero distance, and it falls to smaller values as the distance from the center increases. They are suitable for applications such as pattern recognition, pattern classification, forecasting, and so forth [21,22].

7.2 Learning Vector Quantization Networks

These networks provide discrete outputs, and are used mainly for classification problems based on competitive unsupervised learning. Their performance is generally better than the multilayer feedforward or radial basis function networks for certain classification problems [23,24]. The learning vector quantization (LVQ) networks contain code book vectors which are associated with different output classes. The number of processing elements in code book vectors is equal to the number of inputs to the neural network. These networks aim at finding the code book vector closest to the input vector based on the minimization of classification errors. The code book vector then provides the designated output of the algorithm.

7.3 Kohonen Networks

The Kohonen network, also known as self-organizing feature mapping, uses competitive unsupervised learning to detect existing similarities in the input patterns and thus finding clusters and structure in the data [18,24,25].

There is a variety of Kohonen networks that follow a vector-quantization technique to represent multidimensional data in a much lower-dimensional space. In a Kohonen network, nodes are arranged in a two-dimensional grid referred to as Kohonen layer or feature map with the possibility of using one or more dimensions Each node in the input layer is connected to every node in the Kohonen layer. Feedback is limited to lateral interconnections to nearest-neighboring nodes in the Kohonen layer.

When an input pattern is presented to the Kohonen layer, competitive learning takes place. This simply implies that a node is selected as the winner from all nodes in the Kohonen layer to give the best response based on the smallest minimum distance between the presented input vector and its weight vector. This winner node and its immediate neighbors are the only nodes allowed to learn a particular input pattern. Learning for the nodes in the neighborhood of the winning node is carried out by adjusting their weights closer to the input vector. Thus, the learning algorithm allows the organization of the nodes in the Kohonen layer into local neighborhoods or clusters, which indicate the classes present in the input data.

7.4 Hopefield Networks

Hopefield networks have been used as associative memory tools for solving optimization problems. They have recurrent architecture in which the output of a processing node is fed back through the connecting weights to all the other nodes except itself. This results in the current input to the processing node equal to the sum of the weighted external input and weighted output from every node and enables them to store information with time. This makes them especially suitable for forecasting applications and representing the past history of a time series for making predictions [26,27].

The autoassociative memory networks may be viewed as two multilayer networks connected back-to-back. The number of nodes in the middle layer must be less than those in the input layer. They use input and output datasets which are identical in their training. When presented with incomplete information or invalid data, they are able to recall the correct output based on the closeness of the incorrect data to correct values. The autoassociative networks are also used in data-compression applications. They show a remarkable degree of fault tolerance in recalling the correct output and are frequently used in the storage and recognition of image files. A generalization of the Hopefield network also allows the implementation of heteroassociative tasks such as the matching of names and corresponding telephone numbers with a limited degree of fault tolerance.

7.5 Adaptive Resonance Theory Networks

Adding new training patterns to already trained multilayer feedforward neural networks may change the weight values drastically, making them unstable and requiring retraining of the network. The neural network topology based on adaptive resonance theory (ART) addresses the problem of learning new information or adapting to new input patterns without disturbing the previously learned knowledge as it occurs in biological systems [28,29]. This trade-off between continued learning and buffering of old memories is achieved by identifying stable clusters in the input patterns using unsupervised learning and self-organization. There is a family of ART neural networks designated by ART1, ART2, ART3, and so forth [30,31]. The intrinsic stability of an ART system permits rapid learning of new information while preserving the essential components previously learned patterns. The application of ART networks allows the number of clusters to vary with the problem size along with some control over the degree of similarity between members of the same class.

Although back-propagation networks are most commonly used, numerous other types of neural network based on several variations in network architecture and learning algorithms are presented in Ref. 3. For example, counterpropagation neural networks use supervised and competitive unsupervised learning for training in a significantly different way from back-propagation [32]. They are designed to function as a self-programming lookup table with the additional ability to interpolate between inputs. The LVQ and counterpropagation networks may be useful for classification when identifiable clusters are present in relatively less noisy training data. Back-propagation may generalize better in the case of noisy training data without the presence of any obvious clusters.

8 GENERAL ASPECTS OF NEURAL NETWORK APPLICATIONS

There are no universal guidelines for the application of neural networks to a given problem. The development of a neural network model often requires experimentation depending on the type and nature of the problem and the complexities involved. There are several articles which offer guidelines and so-called rules of thumb for neural network applications in general [33–37] and in the context of specific problems [38–40]. A brief discussion of some main points related to neural network applications is presented next.

8.1 Training Method

A training pass consisting of the entire set of training samples is known as an epoch. The training of a neural network may be carried out based on the change of weights for an individual input–output pattern or after presenting the complete set of input–output patterns. The training is continued until a reasonable low error is achieved. The results from these two training methods may be different. The training of each input–output pattern obviously involves the change of weights far more frequently. In addition, the chances of generalization may be higher when training the entire set in a single pass [3].

8.2 Number of Samples and Data Preproeessing

The starting point is to obtain data for the variables to be modeled (outputs and inputs). In general, it is better to have as much data as possible. A rule of thumb is to have at least 5–10 times as many training samples as the number of weights to be trained. In another estimation, at least 10–40 training samples should be provided for each input variable. If there are, for example, 10 variables (both input and output variables) to be used in building a model, a minimum of 100–400 data points for each variable is needed. The range of values of each variable must cover the operating conditions for which the model will be used. Also, the data must be cleaned up in order to remove bad values which may have adverse effects on training.

8.3 Normalization of Training Patterns

The use of a sigmoid transfer function in the back-propagation algorithm necessitates the scaling of the output values of the training patterns between 0 and 1. Therefore, the output values of training pattern should fall within these limits. Also, infinitely large weights will be required for the outputs to reach values close to 0 or 1 because of the asymptotic nature of the transfer function. In addition, high numerical values of input patterns will tend to push the output of the sigmoid function toward 0 and 1. Such a situation may be avoided by initializing the weights to small random values. However, it is considered a better practice to normalize both the input and output pattern data between 0.1 and 0.9 using the minimum and maximum values.

8.4 Number of Hidden Layers and Nodes

There are no fixed rules for determining the required number of hidden layers and nodes. The three-layer back-propagation network with sigmoidal activation function has been most widely used to solve problems in many areas. In general, one hidden layer has been found to be adequate, and only in some cases, slight advantage may be gained by using two hidden layers [41]. In addition, there is no direct way of knowing the most appropriate number of nodes required in each hidden layer. Sometimes, it is possible to determine their suitable number by observing the changes in the training and test set error with a different number of nodes in a hidden layer and arbitrarily fixing the learning runs. According to one estimation, the number of hidden nodes should be about two-thirds of the total number of inputs and outputs. A trial-and-error approach is often handy depending on a specific application.

8.5 Weight Initialization and Choice of Learning Parameters

Initially, small values of weights are randomly chosen ranging between $[-1, +1]$ or $[-0.5, +0.5]$ to prevent network outputs close to 0 and 1 and, thus, large training times. Sometimes, it is possible to initialize the weights, allowing the inputs to hidden layers to be approximately of the same magnitude [4].

The magnitudes of the desired weight changes depend on the value of the learning rate in the training of neural network based on gradient descent. The values of the learning rate (η) generally range between 0.1 and 0.9. A large value of η may imply faster learning but with the possibility of oscillations in weight determinations. A low value of η makes the learning very slow. Some possible approaches for selecting appropriate values of η in supervised learning of multilayer networks have been discussed in Ref. 3.

In back-propagation neural networks, there is always a possibility that a local minimum of the mean squared error may prevent the determination of optimum weights based on a global minimum. This may be prevented by considering the weight change over a small region and comparing the average gradient of mean squared error rather than a point value. Such a modification allows smooth learning of networks by overcoming the influence of some local minima. A weight update rule which makes use of weight changes in the preceding iteration to estimate the weight changes in a current iteration, is often used. Introduction of the momentum term necessitates the determination of its optimum value which can only be done adaptively, depending on the application. An appropriately selected value of momentum coefficient may often accelerate the convergence.

8.6 Performance Assessment

A trained neural network should accurately respond to previously unseen input–output data as well. This capability of a neural network is referred to as generalization. On the other hand, memorization, which refers to the reproduction of results due to overtraining of the neural network, should be avoided. It is very important to assess the performance of a neural network in estimating the results for a new input dataset not used in training. Thus, emphasis should be on learning and a network's ability to generalize rather than memorize.

The data available for the development of a neural network are divided into training and test set data. The training set is used in the model development, and the network performance is evaluated both for training and test set data. A properly trained neural network should respond with comparable error measures to both training and testing data. A common measure of error may be root mean square deviations of actual and estimated outputs. Initially, the error decreases with increasing number of iterations and the network performance improves both for training and test set data. However, the testing set error starts to increase with the increasing number of iterations while the training set error continues to decrease. The training should be stopped at this point to avoid the tendency of a network to memorize due to overtraining. Often it is assumed that optimum learning and generalization takes place close to the global minimum of the testing data set error.

8.7 Advantages and Limitations of Neural Networks

Training a neural network may sometimes take a very long time depending on the network size and the complexity of the problem. However, they take very little time in their execution after the completion of training and can be easily applied to solve a variety of problems. It is a good practice to keep a neural network model as simple as possible in terms of its size, consisting of the number of nodes, connections, and layers to maintain generalizability and shorter training time. The main limitations of neural networks include the requirement of large amount of data to build a model, the inability to extrapolate outside the region of data used to build the model, and their dependency on historical data which may be unstable or even chaotic. In addition, the choice of neural network architecture and selection of input and output variables may significantly affect the outcome.

Neural network models do not eliminate the limitations and dangers of conventional modeling techniques. It is generally accepted that the performance of a well-designed multilayer neural network is generally comparable with but no better than classical statistical techniques. However, there are cases where neural network modeling has a distinct edge over statistical analysis. Neural networks score over classical techniques in their much reduced development time, their ability to adapt to changing situations, and their ability to make use of related information. However, neural networks may not offer a dramatic increase in performance for the problems that have been extensively analyzed by statistics.

9 NEURAL NETWORK APPLICATIONS IN FOOD PROCESSING

Over the past decade, the number of reported publications on the applications of artificial neural networks in food processing related areas has been steadily growing. Most studies have used the feedforward multi-layer neural networks trained with backpropagation algorithm. The food related neural network applications may be broadly classified into four categories such as product grading and classification, food quality assessment, food process modeling, and process control. These studies clearly reflect that the impact of neural network technologies on food processing research and development is going to be enormous in the near future [42,43]. A brief description of selected applications is presented below.

9.1 Product Grading and Classification

Neural networks have been successfully and frequently applied to the grading and classification of various agricultural and food products using machine vision. These applications include the detection of fertility in hatching eggs [44], corn kernel breakage classification [45], fat estimation in beef ultrasound images [46], inspection of potatoes and apples [47,48], identification of seeds by color imaging [49], grading of pistachio nuts [50], fruit firmness assessment [51], and detection of blood spots and cracks in eggs [52,53]. The accuracy of classification has been reported well above 90% in many studies [44,45,50]. Also, neural-network-based predictions appear to outperform the statistical predictions [49,54]. Interestingly, a multistructure neural network classifier consisting of four parallel multilayer neural networks, each corresponding to a particular variety of pistachio nuts, was shown to perform considerably better than a conventional multilayer feedforward neural network [50].

9.2 Food Quality Assessment

There are inherent difficulties in statistical modeling due to the complex nature of interrelationships between sensory and instrumental measurements. Therefore, neural networks have been widely applied in assessing the quality attributes of a variety of foods. It has been asserted that neural networks will increasingly play a significant role in the sensory evaluation of foods and product development [55,56]. In general, artificial neural networks have been reported to perform better than statistical methods when solving nonlinear problems.

Human sensory judgments have been related to physical measurements of external color in terms of chroma, lightness, and hue angle for tomatoes and peaches using both neural networks and statistical regression. It was shown that both neural network predictions and statistically estimated values were similar due to the linear nature of relationships [57]. In another approach, the apple quality was estimated, using near-infrared spectra in terms of its sugar content by subjecting the data first to principal component analysis and then applying both multiple regression and multilayer feedforward neural network modeling [58]. The sensory evaluation of beef quality has been reported using ultrasonic spectral features [54,59]. The results from a neural network model were better than those from statistical analysis.

A back-propagation neural network was used for predicting sensory scores of olive oils by chromatographic analysis of the volatile fractions extracted from head-space gas [60]. Other applications of neural networks include prediction of foaming and emulsifying properties of protein [61] and prediction of rheological properties of Swiss-type cheeses from their composition [62]. In an interesting study, the snack eating quality attributes were predicted from the image features used as input in a back-propagation neural network [63]. The complex nonlinear relationship between the image features and the sensory attributes of the snack food was modeled by neural network with relative ease.

Neural networks have been applied to the automated assessment of the baking quality of biscuits, resulting in significant improvement. The changes in color during actual baking of biscuits were analyzed using a Kohonen self-organizing map for their subsequent classification as underbaked, correctly baked, and overbaked samples [64,65]. The characteristics bake curves were then used for classifying the biscuit color data obtained in developing of an automated assessment by a feedforward neural network trained by back-propagation [66,67]. The overall performance of the automated on-line inspection system for the grading of biscuits was shown to be considerably superior to that of a trained human inspector.

The applications of electronic nose technology in the food industry are also being reported for detecting spoilage and product-quality control [68,69]. Electronic noses, also called sniffers, combine conductive polymer sensors with neural-network-based pattern-recognition techniques for the classification of aroma quality of foods.

9.3 Modeling of Complex Processes

The neural network approach has been successful in situations where conventional modeling of food processing operation could not be carried out due to the complex relationships. Such applications include the prediction of the performance of an industrial dryer [70], optimum processing parameters during the drying of cooked rice [71], and bioprocess variable estimation and state prediction [72]. In another study, it was demonstrated that a neural network approach could be successfully applied for modeling the thermal processing of canned foods [73]. In training of a 3-8-8-3 back-propagation neural network, the computed values of output parameters (optimum sterilization temperature, process time, and F value) were used for arbitrarily selected inputs (characteristic can dimension, thermal diffusivity of product, and z value) to the model. The performance of the trained neural network was judged satisfactory. Also, the feasibility of neural network modeling of heat transfer to liquid particle mixtures in cans subjected to end-over-end processing was demonstrated in Ref. 74.

Back-propagation networks have been used for predicting the farinograph peak, extensibility, and maximum resistance of dough from the mixer torque curve consisting of 400 points [75]. In a later study [76], the raw spectra of mixer power-consumption curves were transformed into frequency domain data using fast Fourier transform and power spectral density techniques. When used in the training of neural networks, the prediction accuracy of rheological properties of dough was improved. Also, neural networks have been applied to the analysis of ultrasonic spectra for determining palatability attributes of beef [77].

Other recent studies are related to the characterization of non-Newtonian flow of liquid foods in tubes [78], modeling of a continuous snack food frying process [79], estimation of psychrometric parameters [80], and prediction of the loaf volume of breads made from different wheat cultivars [81]. In addition, the prediction of wheat loaf volume based on neural network modeling was found to be far more accurate than principal component regression analysis [81].

9.4 Process Control Applications

Process modeling and control are important to the automation of the food processing industry. A good general description of the model-based process control using neural networks is given in Ref. 82. The applications of neural networks to the control of food processing operations have been emerging rather slowly [42,43,82].

In internal model control, the feedforward neural network model is placed in parallel with the actual process to receive similar inputs. The difference between the actual process output and neural network model output is used for feedback control through an inverse model [82]. The challenge is to predict the best set of input conditions for the desired outputs or target functions for optimizing the process. This requires inverting the network. However, a neural network may not be easy to invert, especially when many input parameters and a few target functions are used in the training of the original network. Also, a neural network model is like a blackbox without any knowledge of causal relationships. This may lead to unexpected difficulties when attempting to invert a neural network.

Neural network modeling of real-time variable estimation and multi-step-ahead prediction of enzyme activity and biomass dry matter has been shown to be feasible [83]. In another study, the development and implementation of a prototype neural-network-based supervisory control system has been reported to achieve maximum microorganism growth [84]. Initially, a back-propagation neural network was trained to predict the optical density of the growth medium. The control objective was to achieve highest optical density within the shortest fermentation time. In operation, the neural network controller searched for desired control set points for medium temperature and pH value through a multivariable search process based on predicted optical density. The setpoints of the main proportional-integral-derivative (PID) controller were then adjusted according to process conditions for optimal performance. Other reported studies include the use of a machine vision sensor with a neural net supervised controller for microbial cultivations [85] and the modeling of the kinetics of *Bacillus subtilis*–amylase inactivation [86].

In the context of food processing, neural networks have been applied to extrusion process identification and control [87] and the development a prototype internal model control system for a continuous-food-frying operation [88]. They also represent a set of very powerful techniques for modeling, control, and optimization because of their ability to learn process dynamics directly from historical data. However, neural network models do not eliminate the limitations of conventional modeling techniques. Also, they are not appropriate for predicting and controlling unstable processes [89].

10 SELECTED APPLICATION EXAMPLES

Selected applications of feedforward neural networks mainly based on the back-propagation learning algorithm are described in this section. A commercially available software NeuroShell2 (Ward Systems Group, Inc., Frederick, MD) was used in these applications.

10.1 Grading of Tomatoes Based on RGB Color Components

Ninety-five tomatoes at different ripening stages were selected randomly and graded subsequently based on the sensory judgment of color in four separate classes as shown in Table 1. The tomatoes were digitally imaged using a machine vision system after placing them upon a white background with proper illumination to minimize the shadow effects. Also, the image background was painted black using Paint Shop software to eliminate the effect of illumination variations caused by the ac power. Finally, the three primary color components red (R), green (G), and blue (B) were obtained from the images of individual tomatoes in each color class. All color measurements were replicated thrice, and the average values of R, G, and B components were determined. Two schemes were later developed for color grading of tomatoes using neural network and statistical regression analysis [90].

A back-propagation 3-6-1 neural network with three input neurons corresponding to R, G, and B color components, six neurons in the hidden layer, and a single neuron in the output layer indicating the numerical value of color grade was found to be most appropriate. The experimental data were separated into training set (75%), test set (15%), and production set (10%). The learning threshold was fixed at 20,000 epochs with no further improvement in error for the test set, which was saved for further use. In the case of statistical analysis, the following relationship was developed for estimating the color grade of tomatoes from R, G, and B color components:

$$Y = \exp(a + b\mathrm{R} + c\mathrm{G} + d\mathrm{B}) \quad (R^2 = 0.889) \tag{21}$$

where Y represents the numerical value of various color grades (red for $Y \leq 1$, light red for $1 < Y \leq 2$, pinkish yellow for $2 < Y \leq 3$, and yellowish green for $Y > 3$). The values of regression coefficients were determined to be $a = -9.1575 \times 10^{-1}$, $b = -4.5288 \times 10^{-3}$, $c = 1.2713 \times 10^{-2}$, and $d = 7.3748 \times 10^{-4}$.

The performance of the neural network trained on the pattern file consisting of both known and unknown samples presented in random order as well as the statistical regression analysis is presented in Table 1 In both cases, the performance was defined as the proportion of correctly classified samples in each grade of tomatoes. It was apparent that the performance of the neural network for grading tomatoes was much closer to the human sensory response than that of statistical-regression-based approach. There were only 6 misclassifications out of 95 tomatoes when using a neural network as compared to 17 misclassifications from statistical estimation. It was apparent that the misclassification of samples was primarily due to the difficulty encountered in differentiating the samples close to the boundary regions of various color grades.

10.2 Estimation of Sensory Stickiness Scores of Cooked Rice

In a study on textural characterization of cooked rice, Limphanudom [91] recently developed several statistical correlations for sensory evaluation of

TABLE 1 Number and Percentages of Correctly Classified Tomatoes

		Predicted success		Correctly classified (%)	
Color grade	Total	Neural network	Statistical regression	Neural network	Statistical regression
1 (Red)	24	24	21	100	87.5
2 (Pinkish red)	26	23	17	88.5	65.4
3 (Pinkish yellow)	28	25	26	89.3	92.9
4 (Yellowish green)	17	17	14	100	82.3
All grades	95	89	78	93.7	82.1

eating quality of cooked rice from the physicochemical properties of raw rice (amylose content, gel consistency, and alkali spreading value). Because the water-to-rice ratio had a marked influence on the sensory properties of cooked rice, it was also included as an independent variable in the study. The eating quality of cooked rice was described in terms of hardness, stickiness, and acceptability scores determined by the taste panelists for 10 Thai rice varieties and 7 different water-to-rice ratios. In this example, the results of stepwise multiple-regression analysis are compared with the neural-network-based approach only for estimating the sensory stickiness scores.

A three-layer back-propagation neural network was considered for this case [90]. The input layer consisted of four neurons corresponding to the amylose content, gel consistency, alkali spreading value, and water-to-rice ratio. The output layer had three neurons corresponding to the sensory hardness, stickiness, and acceptability of cooked rice. The complete dataset consisted of 70 patterns of individual input and output data. About 20% of the data was extracted randomly and was used as a test set to check the error level during training of the remaining data. A sigmoid logistic activation function was used. The inputs were linearly scaled between [−1, 1] and the initial weights were kept at 0.3 for all links.

In the training, the number of neurons in the hidden layer was varied from 2 to 14 by keeping the number of learning runs constant. The root mean square error (RMSE) associated with the prediction of sensory stickiness scores is shown in Figure 5. The calculated error converged

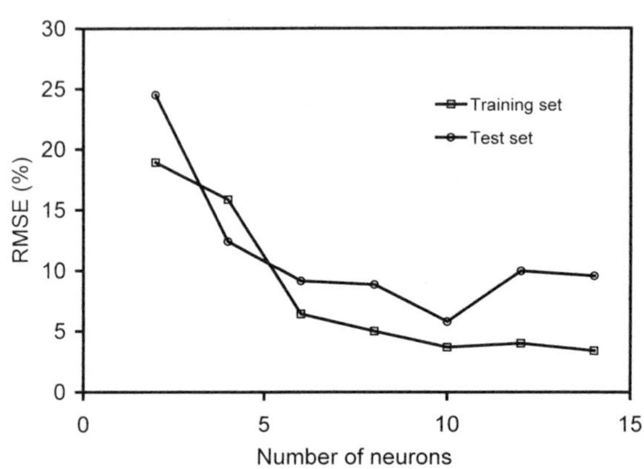

FIGURE 5 Change in root mean square error with the number of neurons in the hidden layer in the neural network development for estimating sensory stickiness scores of cooked rice.

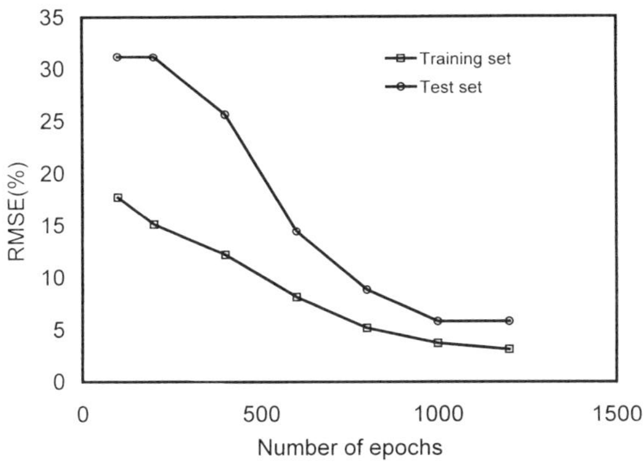

FIGURE 6 Change in root mean square error with the number of epochs in neural network development for estimating sensory stickiness scores of cooked rice.

to a minimum value corresponding to 10 neurons both for the training and test sets. Thereafter, an increase in the number of neurons resulted in larger training time and greater error for the test set. Similar trends were observed for sensory hardness and acceptability scores of cooked rice [90].

In the next step, the number of neurons in the hidden layer was fixed, and the optimum values of the learning rate and momentum were determined to be 0.3 and 0.5, respectively, by trial and error, indicating improved network performance and faster learning. Finally, the number of learning epochs was varied from 100 to 1200, and the respective values of RMSE were determined both for training and test set data as shown in Figure 6. It appeared that approximately 1000 learning epochs were required for the test set error to reach a minimum level.

Thus, the neural network configuration was fixed with 10 neurons in the hidden layer, 0.3 and 0.5 as the learning rate and momentum, respectively, and 1000 as the number of learning epochs. The performance of the selected network on the training and test datasets is presented in Figure 7 for sensory stickiness scores of cooked rice. These results corresponded to 5.14% RMSE for the training set data.

A comparison of experimental sensory stickiness scores of cooked rice estimated by statistical analysis [91] and the neural network approach is presented in Figure 8. A distinct improvement in the estimated values of sensory stickiness scores indicated a significant reduction in the RMSE from 13.7% to 6.02% obtained from statistical analysis and neural network approach, respectively.

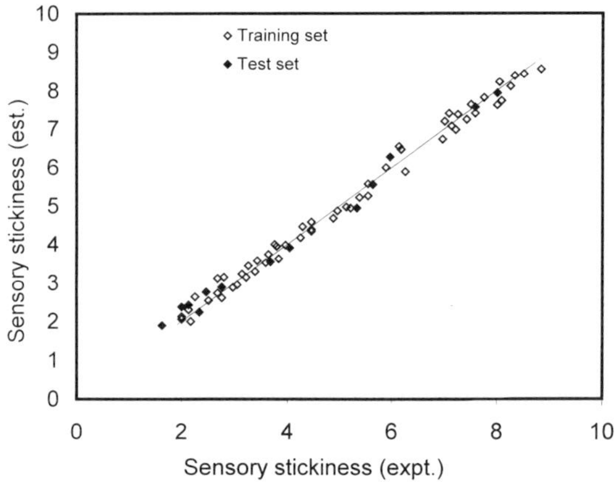

FIGURE 7 Comparison of experimental and neural network estimated sensory stickiness scores of cooked rice both for training and test data sets corresponding to 5.14% root mean square error.

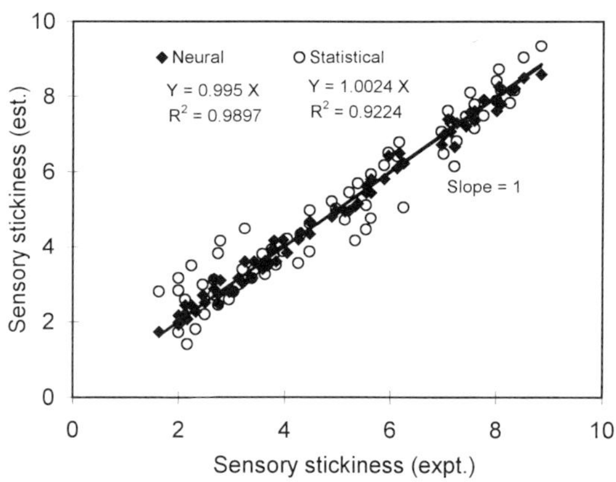

FIGURE 8 Comparison of experimental and estimated values of sensory stickiness scores of cooked rice based on neural network and statistical analysis.

10.3 Moisture Content Changes in Deep-Bed Drying of Rough Rice

In this example, the ability of neural networks to model a complex process such as the drying of rough rice in a deep bed due to forced aeration is demonstrated. A limitation of a feedforward neural network is that it has no "memory" of previous inputs in recall mode. Hence, it is necessary to include the time history in the inputs to a neural network.

The dataset used in the training of a neural network was generated from the Drying Simulation software package for deep-bed drying of grains developed at the Asian Institute of Technology [92]. The input dataset comprised of drying air temperature (30°C, 40°C, and 50°C), air relative humidity (30% and 50%), airflow rate (0.2 and 0.3 $m^3/s{\cdot}m^2$), depth of grain bed (0.5, 1.0, 1.5, and 2.0 m), initial moisture content (22% and 25% wet basis, wb), and drying time. A drying time step of 3 h was selected for computing moisture contents at various depths in the grain bed. Drying was allowed to continue until the final moisture content in the grain bed reached approximately 9% wb. Thus, a total of 1115 input datasets were used to obtain the predicted moisture content profiles in the grain bed. However, the output dataset in the training of the neural network consisted of the final moisture contents at the bottom and top layers only by dividing the grain bed into 10-cm-thick layers.

A 4-layer back-propagation neural network with 6 neurons in the input layer, 16 neurons in each of the two hidden layers, and 2 neurons in the output layer was found to be satisfactory. The learning rate, momentum, and the initial weights were fixed at 0.3, 0.5, and 0.3, respectively, for each link. All inputs were normalized linearly in the [−1, 1] range and a sigmoid logistic activation function was used. The testing and production datasets comprised approximately 20% and 5% respectively, of complete dataset. The training was stopped when the root mean square error of the estimated and actual moisture contents in the bottom and top layers approached 3.0% and 4.0%, respectively, for the test set data containing 117 values.

A comparison of the changes in moisture content in the bottom and top layers of a 1-m-thick bed of rough rice with drying time estimated by neural network model and simulation program is presented in Figure 9. The drying conditions included inlet air temperature, relative humidity, and flow rate of 40°C, 30%, and 0.2 $m^3/s{\cdot}m^2$, respectively. These results clearly showed that the developed neural network model was able to map the complex relationships in deep-bed drying of rough rice satisfactorily.

10.4 Modeling of Transient Conduction Heating and Cooling in a Sphere

This example illustrates the development of a neural network model for transient conduction heating and cooling in a sphere similar to a commonly used Heisler chart. The study was limited to temperature ratio values above

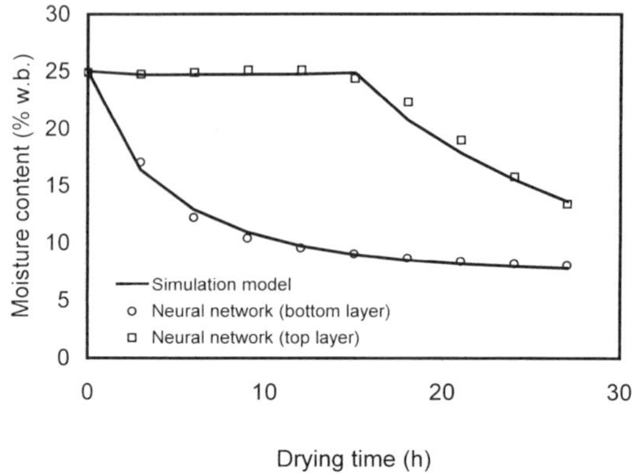

FIGURE 9 Comparison of simulated and neural network estimated changes in the moisture content with time for the drying of rough rice in the top and bottom layers of a 1-m-thick bed (drying air temperature = 40°C; relative humidity = 30%; air velocity = 0.2 m/sec).

0.5 due to the changing nature of the relationships between the temperature ratio and Fourier number from concave to convex shapes for different Biot numbers. The training set data were obtained with a computer program written for the analytical solution of the Fourier transient heat-conduction equation [93]. The temperature ratio at the center of a sphere was computed as a function of Fourier and Biot numbers. The values of the Fourier number and Biot number were selected in the range 0.1–3.5 and 0.01–50, respectively. Two hundred sixty-six sets of data were generated and used in the training and testing of the neural network.

The neural network algorithm found suitable for this problem was the polynomial net having one input, one output, and a hidden layer. There were 213 neurons in the hidden layer and their number was equal to the number of datasets in the training file. The inputs and outputs were linearly scaled between [−1, 1]. The training of the network was stopped when the root mean square error reached about 2.0%. A comparison of temperature ratios based on the analytical solution and the neural network model is presented as a function of Fourier number for selected values of Biot number in Figure 10. It is obvious that a neural network representation of transient conduction heating and cooling can be carried out with reasonable accuracy using a polynomial network structure. In this particular case, the back-propagation neural network did not succeed in properly accounting for the changes in the temperature ratio with Fourier number during transient conduction heat transfer.

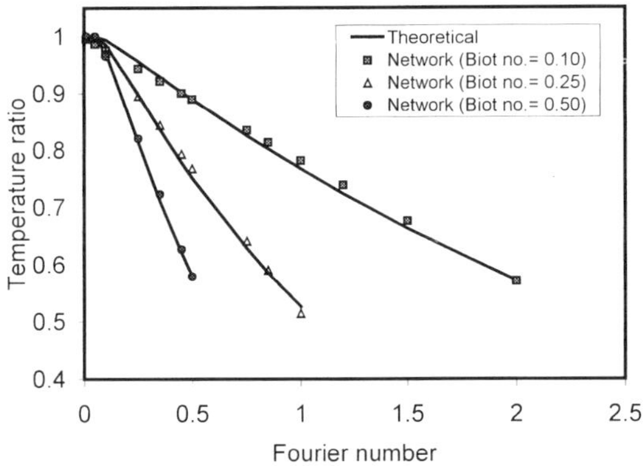

FIGURE 10 Comparison of analytical and neural network estimated changes in temperature ratio at the center of a sphere with time for selected values of the Biot number.

11 CONCLUDING REMARKS

Artificial neural networks are capable of learning complex relationships and generalizing solutions from known patterns of input/output data. Thus, they are good for the modeling of complex systems for which exact models or expected performance are not known. Some neural networks have the capacity to self-organize and adapt to the changes taking place. They offer several advantages over conventional computational techniques and may sometimes have a considerable edge for solving problems with a history of imprecise or incomplete data. Their main applications include pattern recognition, classification, clustering, function approximation, forecasting, process control, and optimization. Once trained, they can quickly solve the complex problems. Neural networks are particularly suitable for interpreting the imprecise response of an array of sensors such as those used in electronic noses to detect the underlying patterns for aroma scanning.

Neural networks are not suitable for precise numerical computations or following the logical sequence of operations. In general, their successful application depends on the size, quality, and preprocessing of the training data, type, and structure of the neural network and the learning algorithm for a particular problem. Because neural networks can develop solutions to most problems using historical data without conceptual understanding like a "black box," they may be used indiscreetly. As they do not provide a formal representation of the relationship between input and output data, it is essential to check their performance by statistical tests or symbolic reasoning. In this respect, the potential of neural networks for hypothesis generation needs to be exploited.

Significant advances have been made in the application of artificial neural networks in a variety of fields [94–97], especially over the past two decades. There is immense literature on neural networks comprised of numerous books and hundreds of papers written both in professional journals and popular magazines. It is a formidable task to review such material on neural networks and their applications in a large number of academic disciplines. In food-processing-related areas, their applications go back to early 1990s, covering a range of diverse topics such as product grading and classification, quality evaluation, and process modeling and control. The material presented here provides only a basic introduction to the most widely used neural networks and their underlying principles through selected application examples.

There are several good websites on the Internet that provide a great deal of information and useful links related to various aspects of neural networks [98–102]. A number of inexpensive microcomputer-based user-friendly software packages are available in the market for neural network applications. In addition, many shareware versions of neural network programs can be downloaded for free from the Internet. These developments have enabled nonspecialists to utilize the enormous potential of neural networks without requiring extensive training and understanding of the underlying mathematical principles. Despite some limitations, the applications of neural networks in the modeling and control of food processing operations and food quality assessment are likely to increase rapidly in near future.

NOMENCLATURE

x_i	Input vector component
$w_{j,i}$	Weight of link from node i to node j
net_j	Weighted sum of inputs in node j
y_j	Output of node j
o_j	Target or desired output of node j
$E(p)$	Mean square error for pattern p
β	Bias in a sigmoid function
θ	Parameter associated with the slope of sigmoid function
$\delta_j(p)$	Delta error for actual and desired output for node j
$\Delta w_{j,i}$	Change in weight
η	Learning rate
λ	Momentum
i	Index of a node in the input layer
j	Index of a node in the hidden or output layer

k	Index of a node in the output layer
p	Index of the pattern number
N	Number of nodes in the input layer
M	Number of nodes in the hidden layer
K	Number of nodes in the output layer
P	Number of training patterns

REFERENCES

1. I Aleksander, H Morton. An Introduction to Neural Computing. London: Chapman & Hall, 1990.
2. R Hecht-Nielsen. Neurocomputing. Reading, MA: The MIT Press, 1990.
3. KG Mehrotra, CK Mohan, S Ranka. Elements of Artificial Neural Networks. Cambridge, MA: The MIT Press, 1996.
4. WW McCullock, W Pitts. A logical calculus of ideas imminent in nervous activity. Bull Math Biophy 5:115–133, 1943.
5. DO Hebb. The Organization of Behavior. New York: John Wiley & Sons, 1949.
6. F Rosenblatt. The perceptrons: a probabilistic model for information storage and organization of the brain. Psychol Rev 65:386–408, 1958.
7. B Widrow, M. Hoff. Adaptive switching circuits. Western Electronic Show and Convention, Convention Record, Institute of Radio Engineers (now IEEE), 1960, Vol. 4, pp 96–104.
8. M Minisky, S Papert. Perceptrons: An Introduction to Computational Geometry. Cambridge, MA: The MIT Press, 1969.
9. DE Rumelhart, GE Hinton, RJ Williams. Learning internal representation by error propagation. In: DE Rumelhart, JL McClellant, eds. Parallel Distributed Processing: Exploration in the Microstructure of Cognition. Vol. 1: Foundations. Cambridge, MA: The MIT Press, 1986.
10. PJ Werbos. Beyond regression: New tools for prediction and analysis in the behavioral sciences. PhD dissertation, Harvard University, Cambridge, MA, 1974.
11. DB Parker. A comparison of algorithms for neuron-like cells. In: JS Denker, ed. Neural Networks for Computing. New York: American Institute of Physics, 1986, pp 327–332.
12. JA Anderson, E Rosenfled, eds. Neurocomputing: Foundations of Research. Cambridge, MA: The MIT Press, 1988.
13. JA Anderson, A Pellionisz, E Rosenfled, eds. Neurocomputing 2: Directions for Research. Cambridge, MA: The MIT Press, 1990.
14. S. Haykin. Neural Networks, a Comprehensive Foundation. New York: Macmillan, 1994.
15. PD Wassermann. Neural Computing: Theory and Practice. New York: Van Nostrand Reinhold, 1989.

16. PS Churchland, TJ Sejnowski. The Computational Brain. Cambridge, MA: The MIT Press, 1992.
17. GE Hinton. Connectionist learning procedures. Carnegie-Mellon University Technical Report CMU-CS-87-115, Pittsburgh, PA, 1987.
18. T Kohonen. Self-organized formation of topologically correct feature maps. Biol Cybern 43:59–69, 1982.
19. SE Fahlman. An empirical study of learning speed in backpropagation networks. Carnegie-Mellon University Technical Report CMU-CS-88-162, Pittsburgh, PA, 1988.
20. JR Shewchuk. An introduction to the conjugate gradient method without agonizing pain. Carnegie-Mellon University Technical Report CMU-CS-94-125, Pittsburgh, PA, 1994.
21. JE Moody, C Darken. Fast learning in networks of locally-tuned processing units. Neural Computat 1:281–294, 1989.
22. C Bishop. Improving the generalisation properties of radial basis function neural networks. Neural Computat 3:579–588, 1991.
23. T Kohonen. Learning vector quantization for pattern recognition. Helsinki University of Technology Technical Report TKK-F-A601, 1986.
24. T Kohonen. The self-organizing map. Proc IEEE 78:1464–1480, 1990.
25. T Kohonen. Self-Organization and Associative Memory. 3rd ed. New York: Springer-Verlag, 1988.
26. JJ Hopfield. Neural networks and physical systems with emergent collective computational abilities Proc Natl Acad Sci 79:2554–2558, 1982.
27. J Hopefleld, D Tank. Neural computation of decisions in optimization problems. Biol Cybern 52:141–152, 1985.
28. S Grossberg. Competitive learning: From interactive to adaptive resonance. Cogn Sci 1:23–63, 1987.
29. GA Carpenter, S Grossberg. The ART of adaptive pattern recognition by a self-organizing neural network. IEEE Computer 21:77–88, 1988.
30. GA Carpenter, S Grossberg. ART 2: Stable self-organization of pattern recognition codes for analog input patterns. Appl Opt 26:4919–4930, 1987.
31. GA Carpenter, S Grossberg. ART 3: Hierarchical search using chemical transmitters in self-organizing pattern recognition architectures. Neural Networks 3:129-152, 1990.
32. R Hecht-Nielsen. Application of counterpropagation networks. Neural Networks 1:131–139, 1987.
33. M Caudill. Neural network training tips and techniques. AI Expert 6:56–61, 1991.
34. SP Chitra. Use neural networks for problem solving. Chem Eng Prog 89:44–52, 1993.
35. JF Barthelemy, CW Carpenter. Common misconception about neural networks as approximators. J Computing Civil Eng 8:345–358, 1994.
36. D Bullock, D Whittaker, J Brown, D Cook. Neural networks for your toolbox. Agric Eng 73:10–12, 31, 1992.

37. D Mukesh. Hate statistics? Try neural networks. Chem Eng 104(3):96–99, 101–102, 104, 1997.
38. YR Chen, M Nguyen. Improving performance of neural network classifiers by input data pretreatment with principle component analysis. ASAE Paper No. 973050, 1997.
39. F Salehi, R Lacroix, KM Wade. Effects of learning parameters and data presentation on the performance of backpropagation networks for milk yield prediction. Trans ASAE 41:253–259, 1998.
40. R Lacroix, F Salehi, XZ Yang, KM Wade. Effects of data preprocessing on the performance of artificial neural networks for dairy yield prediction and cow culling operation. Trans ASAE 40:839–846, 1997.
41. R Hecht-Nielsen. Theory of the backpropagation neural network. Proceedings of International Joint Conference on Neural Networks, Washington, DC, 1989, Vol. I, pp 593–605.
42. T Eerikäinen, S Linko, P Linko, T Siimes, YH Zhu. Fuzzy logic and neural network applications in food science and technology. Trends Food Sci Technol 4:237–242, 1993.
43. G Montague, J Morris. Neural-network contribution in biotechnology. Trends Biotechnol 12:312–323, 1994.
44. K Das, MD Evans. Detecting fertility of hatching eggs using machine vision II: Neural network classifiers. Trans ASAE 35:2035–2041, 1992.
45. K Liao, MP Paulsen, JF Reid, BC Ni, EP Bonifacio-Maghirang. Corn kernel breakage classification using image analysis and neural networks. Trans ASAE 36:1949–1953, 1993.
46. JD McCauley, BR Thane, AD Whittaker Fat estimation in beef ultrasound images using texture and adaptive logic networks. Trans ASAE 37:997–1002, 1994.
47. S Deck, CT Morrow, PH Heinemann, HJ Sommer III. Comparison of a neural network and traditional classifier for machine vision inspection. Appl Eng. Agric 11:319–326, 1995.
48. Y Tao, PH Heinemann, Z Varghese, CT Morrow, HJ Sommer III Machine vision for color inspection of potatoes and apples. Trans ASAE 38:1555–1561, 1995.
49. Y Chitioui, D Bertrand, Y Dattee, M Devaux. Identification of seeds by colour imaging: Comparison of discriminant analysis and artificial neural network. J Sci Food Agric 71:431–433, 1996.
50. A Ghazanfari, J Irudayaraj, A Kusalik. Grading pistachio nuts using a neural network approach. Trans ASAE 39:2319–2324, 1996.
51. V Steinmetz, M Crochon, VB Maurel, JLG Fernandez, PB Elorza, L Verstreken. Sensors for fruit firmness assessment: Comparison and fusion. J Agric Eng Res 64:15–28, 1996.
52. VC Patel, RW McClendon, JW Goodrum. Detection of blood spots and dirt stains in eggs using computer vision and neural networks. Appl Eng Agric 12:253–258, 1996.

53. VC Patel, RW McClendon, JW Goodrum. Detection of cracks in eggs using color computer vision and artificial neural networks. AI Applic 10(3):19–28, 1996.
54. B Park, YR Chen, AD Whittaker, RK Miller, DS Hale. Neural network modeling for beef sensory evaluation. Trans ASAE 37:1547–1553, 1994.
55. PM Smith, LG Walter. Neural networks in sensory perception. In: HT Lawless, BP Klein, eds. Sensory Science Theory and Applications in Foods. New York: Marcel Dekker, 1991, pp 207–222.
56. M Bomio. Neural networks and the future of sensory evaluation. Food Technol 52:62–63, 1998.
57. CN Thai, RL Shewfelt. Modeling sensor color quality of tomato and peach: Neural networks and statistical regression. Trans ASAE 34:50–955, 1991.
58. L Bochereau, P Bourgine, B Palagos. A method for prediction by combining data analysis and neural networks: Application to prediction of apple quality using near infra red spectra. J Agric Eng Res 51:207–216, 1992.
59. Y Huang, RE Lacey, AD Whittaker. Neural network prediction modeling based on ultrasonic elastograms for meat quality evaluation. Trans ASAE 41:1173–1179, 1998.
60. F Angerosa, LD Giacinto, R Vito, S Cumitini. Sensory evaluation of virgin olive oils by artificial neural network processing of dynamic head-space gas chromatographic data. J Sci Food Agric 72:323–328, 1996.
61. GE Arteaga, S Nakai. Predicting protein function with artificial neural networks: Foaming and emulsifying properties. J Food Sci 58:1152–1156, 1993.
62. H Ni, S Gunasekaran. Food quality prediction with neural networks. Food Technol 52:60–65, 1998.
63. MS Sayeed, AD Whittaker, ND Kehtarnavaz. Snack quality evaluation method based on image features and neural network prediction. Trans ASAE 38:1239–1245, 1995.
64. LGC Hamey, JCH Yeh, T Westcott, SKY Sung. Pre-processing colour images with a self-organising map: Baking curve identification and bake image segmentation. Proceedings of the 14th International Conference on Pattern Recognition, Brisbane, 1998, pp 1771–1775.
65. T RayChaudhuri, JCH Yeh, LGC Hamey, CT Westcott. Baked product classification with the use of a self-organising map. Proceedings of the Sixth Australian Conference on Neural Networks. M Charles, C Latimer, eds, Sydney, 1995, pp 152–155.
66. T RayChaudhuri, JCH Yeh, LGC Hamey, SKY Sung, T Westcott. A connectionist approach to quality assessment of food products. Proceedings of the Eighth Australian Joint Conference on Artificial Intelligence, X Yao, ed., Canberra, Australia, 1995, pp 435–441.
67. JCH Yeh, LGC Hamey, CT Westcott and SKY Sung. Colour bake inspection system using hybrid artificial neural networks. Proceedings of the IEEE International Conference on Neural Networks, Vol. 1, Perth, Australia, 1995, pp 37–42.

68. A Jonsson, F Winquist, J Schnurer, H Sundgren, I Lundstrom. Electronic nose for microbial quality classification of grains. Int J Food Microbiol 35:187–193, 1997.
69. J Haugen, K Kvaal. Electronic nose and artificial neural network. Meat Sci 49:S273–S286, 1998.
70. B Huang, AS Mujumdar. Prediction of industrial dryer performance using neural network. In: AS Mujumdar, ed. Drying '92. Amsterdam: Elsevier Science Publishers, 1992, pp 1124–1132.
71. MN Ramesh, MA Kumar, PN Srinivasa Rao. Application of artificial neural networks to investigate the drying of cooked rice. J Food Process Eng 19:321–329, 1996.
72. P Linko, YH Zhu. Neural network programming in bioprocess variable estimation and state prediction. J Biotechnol 21:253–270, 1991.
73. SS Sablani, HS Ramaswamy, SO Prasher. A neural network approach for thermal processing applications. J Food Process Preserv 19:283–301, 1995.
74. SS Sablani, HS Ramaswamy, S Sreekanth, SO Prasher. Neural network modeling of heat transfer to liquid particle mixtures in cans subjected to end-over-end processing. Food Res Int 30:105–116, 1997.
75. R Ruan, S Almaer, J Zhang. Prediction of dough rheological properties using neural networks. Cereal Chem 72(3):308–311, 1995.
76. R Ruan, S Almaer, C Zou, PL Chen. Spectrum analysis of mixing power curves for neural network prediction of dough rheological properties. Trans ASAE 40:677–681, 1997.
77. B Park, AD Whittaker, RK Miller, DS Hale. Ultrasonic spectral analysis for beef palatability attributes. J. Food Sci 59:697–701, 724, 1994.
78. PP Singh, VK Jindal. Pressure drop estimation in tube flow of non-Newtonian fluids by neural networks. ASAE Paper No. 976009, 1997.
79. Y Huang, AD Whittaker, RE Lacey. Neural network prediction modeling for a continuous snack food frying process. Trans ASAE 41:1511–1517, 1998.
80. S Sreekanth, HS Ramaswamy, S Sablani. Prediction of psychrometric parameters using neural networks. Drying Technol 16:825–837, 1998.
81. Y Horimoto, T Durance, S Nakai, OM Lukow. Neural networks vs. principal component regression for prediction of wheat flour loaf volume in baking tests. J Food Sci 60:429–433, 1995.
82. AJ Morris, GA Montague, MJ Willis. Artificial neural networks: Studies in process modelling and control. Trans IChemE, Part A 72:3–19, 1994.
83. P Linko, YH Zhu. Neural network modelling for real time variable estimation and prediction in the control of glucoamylase fermentation. Process Biochem 27:275–283, 1992.
84. Q Zhang, JF Reid, JB Litchfield, J Ren, SW Chang. A prototype neural network supervised control system for *Bacillus thuringiensis* fermentation process. Biotechnol Bioeng 43:483–489, 1994.
85. Q Zhang, JB Litchfield, JF Reid, J Ren, SW Chang. Coupling a machine vision sensor and neural net supervised controller: Controlling microbial cultivations. J Biotechnol 38:219–228, 1995.

86. AH Geeraerd, CH Herremans, LR Ludikhuyze, ME Hendrickx, JF Van Impe. Modeling the kinetics of isobaric–isothermal inactivation of *Bacillus subtilis*–amylase with artificial neural networks. J Food Eng 36:263–279, 1998.
87. T Eerikäinen, YH Zhu, P Linko. Neural networks in extrusion process identification and control. Food Control 5:111–119, 1994.
88. Y Huang, AD Whittaker, RE Lacey. Internal model control for a continuous snack food flying process using neural networks. Trans ASAE 41:1519–1525, 1998.
89. WT Miller, RS Sutton, PJ Werbos. Neural Networks for Control. Cambridge, MA: The MIT Press, 1990.
90. V Chauhan. Modeling of selected food engineering operations by neural networks. MEng thesis, Asian Institute of Technology, Bangkok, 1998.
91. P Limphanudom. Textural characterization of cooked rice. DTechSc dissertation, Asian Institute of Technology, Bangkok, 1997.
92. VK Jindal, RC Martinez, LV Diep. A PC-based, user oriented decision support system for in-store drying and aeration of grains. In: BR Champ, E Highley, GI Johnson, eds. Grain Drying in Asia. Canberra, Australia: ACIAR, 1996, pp 282–283.
93. JP Holman. Heat Transfer. 7th ed. New York: McGraw-Hill, 1997.
94. DR Baughman, YA Liu. Neural Networks in Bioprocessing and Chemical Engineering. San Diego, CA: Academic Press, 1996.
95. BHV Topping, A. Bahreininejad. Neural Computing for Structural Mechanics. Edinburgh: Saxe-Coburg Publications, 1997.
96. RJ Van Eyden. The Application of Neural Network in the Forecasting of Share Prices. Haymarket, VA: Finance & Technology Publishing, 1996.
97. TS Dillon, D Niebur, eds. Neural Networks Applications in Power Systems. London: CRL Publishing, 1996.
98. WS Sarle, ed. Neural Network FAQ, Part 1 of 7: Introduction, multipart periodic posting to the Usenet newsgroup comp.ai.neural-nets, URL: ftp://ftp.sas.com/pub/neural/FAQ.html, last-modified: 1999-04-01.
99. K Gurney. Neural Nets, http://www.shef.ac.uk/psychology/gurney/notes/index.html, modified by JP Frisby@sheffield.ac.uk on 29-Sep-97.
100. DTI NeuroComputing Web, http://www.clients.globalweb.co.uk/nctt/internet/, updated Mon 11 Mar 1996.
101. PE Keller. Neural Networks at Pacific Northwest National Laboratory (PNNL), URL: http://www.emsl.pnl.gov:2080/proj/neuron/web/neural/homepage.html, last revised on Monday 1 March 1999.
102. DD Olmsted. History and Principles of Neural Networks, URL: http://www.neurocomputing.org/history.htm, last revised 11-14-98.

Index